Flood Issues in Contemporary Water Management

NATO Science Series

A Series presenting the results of activities sponsored by the NATO Science Committee. The Series is published by IOS Press and Kluwer Academic Publishers, in conjunction with the NATO Scientific Affairs Division.

A. Life Sciences	IOS Press
B. Physics	Kluwer Academic Publishers
C. Mathematical and Physical Sciences	Kluwer Academic Publishers
D. Behavioural and Social Sciences	Kluwer Academic Publishers
E. Applied Sciences	Kluwer Academic Publishers
F. Computer and Systems Sciences	IOS Press

1. Disarmament Technologies	Kluwer Academic Publishers
2. Environmental Security	Kluwer Academic Publishers
3. High Technology	Kluwer Academic Publishers
4. Science and Technology Policy	IOS Press
5. Computer Networking	IOS Press

NATO-PCO-DATA BASE

The NATO Science Series continues the series of books published formerly in the NATO ASI Series. An electronic index to the NATO ASI Series provides full bibliographical references (with keywords and/or abstracts) to more than 50000 contributions from international scientists published in all sections of the NATO ASI Series.
Access to the NATO-PCO-DATA BASE is possible via CD-ROM "NATO-PCO-DATA BASE" with user-friendly retrieval software in English, French and German (WTV GmbH and DATAWARE Technologies Inc. 1989).

The CD-ROM of the NATO ASI Series can be ordered from: PCO, Overijse, Belgium

Series 2. Environment Security – Vol. 71

Flood Issues in Contemporary Water Management

edited by

Jiri Marsalek

National Water Research Institute,
Burlington, Ontario, Canada

W. Ed Watt

Queen's University,
Kingston, Ontario, Canada

Evzen Zeman

Hydroinform a.s. Prague,
Prague, Czech Republic

and

Friedhelm Sieker

University of Hannover,
Hannover, Germany

Kluwer Academic Publishers

Dordrecht / Boston / London

Published in cooperation with NATO Scientific Affairs Division

Proceedings of the NATO Advanced Research Workshop on
Coping with Flesh Floods: Lessons Learned from Recent Experience
Malenovice, Czech Republic
May 16-21, 1999

A C.I.P. Catalogue record for this book is available from the Library of Congress.

ISBN 0-7923-6451-1 (HB)
ISBN 0-7923-6452-X (PB)

Published by Kluwer Academic Publishers,
P.O. Box 17, 3300 AA Dordrecht, The Netherlands.

Sold and distributed in North, Central and South America
by Kluwer Academic Publishers,
101 Philip Drive, Norwell, MA 02061, U.S.A.

In all other countries, sold and distributed
by Kluwer Academic Publishers,
P.O. Box 322, 3300 AH Dordrecht, The Netherlands.

Printed on acid-free paper

Printed in the Netherlands.

TABLE OF CONTENTS

PREFACE

The year of 1997 was particularly devastating with respect to floods in Central and Eastern Europe. Between July 4 and 7, 1997, a number of precipitation records were set in the northeastern Czech Republic and southern Poland, with the highest precipitation total observed at Lysa Mountain in the Czech Republic, 595 mm. The resulting flood, with three distinct waves, was devastating as it moved from the Czech Republic through Poland and Germany. A large number of human lives were lost and property damage was estimated in billions of dollars. Other floods were reported in the same year in many other countries. The year of 1998 was even worse, with the flood on the Yangtze River resulting in the death of more than 3000 people, dislocation of 230 million people, and direct damages of more than $45 billion. In the aftermath of these catastrophic floods, both the public and the experts ask what can we learn from these recent events, recognizing that even though floods are random events and hence inevitable, in most cases, we can prevent or at least reduce the associated loss of life and flood damages. To answer this question, we have proposed to, and received support from, NATO to hold a workshop on floods.

The main objectives of the NATO Advanced Research Workshop on 'Coping with Floods: Lessons Learned from Recent Experience' were to review the state of the art in managing flood risks, and to learn from experts which measures worked well during the 1997 flood and which shortcomings should be avoided in the future. With respect to the 1997 flood, the workshop fully succeeded, with leading experts from the Czech Republic, Poland and Germany attending the workshop and presenting timely reports on flood fighting and both successes and shortcomings. Their experiences were further extended for those of experts from 13 other countries, and a broad overview of flood risk management was developed. This overview covered the ecosystem approach to flood management, including socio-economic issues, flood impacts on water quality, human health, and natural ecosystems. This broad view is reflected in the workshop proceedings, which include 43 selected papers. Whencver trade, product or firm names are used in the proceedings, it is for descriptive purposes only and does not imply endorsement by the Editors, Authors or NATO.

After establishing the ARW organizing committee, applying for and receiving the NATO support grant, the organizers recruited keynote speakers and workshop participants, finalized the workshop program, and held the workshop at Malenovice, Czech Republic, with 53 experts from 16 countries in attendance. The venue for the workshop was selected quite appropriately; Hotel P. Bezruc is located at the foot of Lysa (Bare) Mountain, where the record precipitation of 595 mm, causing the 1997 flood, was recorded.

Only the formal workshop presentations are reflected in the proceedings that follow. Besides these papers and extensive discussions, there were many other ways of sharing and exchanging information among the participants, in the form of new or renewed collaborative links, professional networks and personal friendships. The success of the workshop was acknowledged by workshop participants in the evaluation questionnaire and by the NATO EST Program Director, Dr. A.H. Jubier, who opened the workshop. For this success, we would like thank all who helped stage the workshop and produce its proceedings, and particularly those who are listed in the Acknowledgment.

Jiri Marsalek
Burlington, Ontario, Canada

Ed Watt
Kingston, Ontario, Canada

Evzen Zeman
Prague, Czech Republic

Friedhelm Sieker
Hannover, Germany

ACKNOWLEDGEMENT

This Advanced Research Workshop (ARW) was directed by Dr. Jiri Marsalek, National Water Research Institute (NWRI), Environment Canada, Burlington, Canada, and Dr. Evzen Zeman, Hydroinform a.s., Prague, Czech Republic. They were assisted by two other members of the workshop Organizing Committee, Prof. W.E. Watt, Queen's University, Kingston, Ontario, Canada, and Prof. F. Sieker, University of Hannover, Hannover, Germany. The ARW was sponsored by NATO, in the form of a grant, and by employers of the members of the Organizing Committee, who provided additional resources required to prepare the workshop and its proceedings. Special thanks are due to Dr. A.H. Jubier, Programme Director, Environmental and Earth Science & Technology, NATO, who provided liaison between the workshop organizers and NATO, and personally assisted with many tasks.

The compilation of the proceedings typescript was done by Karen MacIntyre (Queen's University, Kingston, Canada) and Quintin Rochfort (National Water Research Institute, Burlington, Canada). All local arrangements were done by the local organizing committee, headed by Evzen Zeman, with assistance from Pavlina Nesvadbova and other staff of Hydroinform, a.s. Special thanks are due to all these contributors, and above all, to all the participants, who made this workshop a memorable and interactive learning experience for all involved.

LIST OF PARTICIPANTS

Directors

Marsalek, J. — National Water Research Institute
867 Lakeshore Road, Burlington, Ontario, L7R 4A6
CANADA

Zeman, E. — Hydroinform Ltd.
Na Vrsick 5, 100 00 Prague 10
CZECH REPUBLIC

Key Speakers

Arsov, R. — University of Architecture, Civil Engineering and Geodesy, Sofia
1 Chr. Smirnenski Blvd., 1421 Sofia
BULGARIA

Watt, E. — Dept. of Civil Engineering
Queen's Unvirsity, Kingston, Ontario K7L 3N6
CANADA

Biza, P. — Povodi Moravy, a.s.
Drevarska 11, 601 75 Brno
CZECH REPUBLIC

Jirasek, V — Povodi Labe, a.s.
Vita Nejedleho 951, 500 03 Hradec Kralove 3
CZECH REPUBLIC

Kubat, J. — Czech Hydrometeorological Institute
Na Sabatce 17, 143 06 Prague 4
CZECH REPUBLIC

Narwa, P. — District Office Rychnov n/Kn
Havlickova 136, 516 01 Rychnov nad Kneznou
CZECH REPUBLIC

Babovic, V. — Danish Hydraulic Institute
Agern Alle 5, DK-2970 Horsholm
DENMARK

Host-Madsen, J. — Danish Hydraulic Institute
Agern Alle 5, DK-2970 Horsholm
DENMARK

Bronstert, A.	Potsdam Institute for Climate Impact Research P.O. Box 610203, 14412 Potsdam GERMANY
Ostrowski, M.W.	FG Ingenierhydrologie und Wasserbewirtschaftung Technical University of Darmstadt, Petersenstr. 13, 64287 Darmstadt GERMANY
Sieker, F.	Institut fur Wasserwirtschaft, Universitat Hannover Appelstr. 9A, D 30167 Hannover GERMANY
Todini, E.	Universita di Bologna, Dipartmento di Scienze della Terra e Geologico Ambientali, Via Zamboni 67, I-40127, Bologna ITALY
Kowalczak, P.	Institute of Meteorology and Water Management Poznan Branch Ul Dabrowskiego 174/176, 60-594 Poznan POLAND
Kulig, A.	Institute of Environmental Engineering Systems Warsaw University of Technology Nowowiejska 20, 00-653 Warsaw POLAND
Szamalek, K.	National Security Bureau 10 Wiejska St., 00-902 Warszawa POLAND
Price, R.K.	International Institue for Infrasturctuaral, Hydraulic and Environmental Engineering Westvest 7, PO Box 3015, 2601 DA Delft THE NETHERLANDS
Moore, R.J.	Institute of Hydrology Wallingford, Oxon, OX10 8BB UNITED KINGDOM
Pasterick, E.T.	Federal Insurance Administration, Federal Emergency Management Agency 500 C St. SW, Washington, D.C. 20472 USA

Riebau, M.A.	Short Elliott Hendrickson 421 Frenette Drive, Chippewa Falls, WI 54703 USA

Other Participants

Dodov, B.	University of Architecture, Civil Engineering and Geodesy, Sofia 1 Chr. Smirnenski Blvd. 1421 Sofia BULGARIA
Bowering, R.	Manitoba Water Resources 200 Saulteaux Crescent, Winnipeg, Manitoba, R3J 3W3 CANADA
Kosacky, V.	Povodi Moravy, a.s. Drevarska 11, 601 75 Brno CZECH REPUBLIC
Kutalek, P.	Aquatis Ltd. Botanicka 56, 656 32 Brno CZECH REPUBLIC
Metelka, T.	Hydroprojekt a.s. Taborska 31, 140 00 Praha 4 CZECH REPUBLIC
Puncochar, P.	Ministry of Agriculture Department of Water Management Tesnov 17, 117 05 Praha 1 CZECH REPUBLIC
Stary, M.	HYSOFT Brno, Rolnicka 9/46, 625 00 Brno CZECH REPUBLIC
Turecek, B.	Povodi Odry Varenska 49, 701 26 Ostrava CZECH REPUBLIC
Voznica, P.	Army of the Czech Republic Territorial Defense Committee PO Box 61/70, 390 61 Tabor CZECH REPUBLIC

Zezulak, J.	Czech Agriculture University, Forestry Faculty Kamycka, 165 21 Praha 6 CZECH REPUBLIC
Engel, H.	Bundesanstalt fur Gewasserkunde Kaiserin-Augusta-Anlagen 15-17, 56068 Koblenz GERMANY
Loster, T.R.	Munich Reinsurance Company Koeniginstr. 107, D-80 791 Munich GERMANY
Sieker, H.	Ingenieurgesellschaft Prof. Sieker mbH Berliner Str. 71, D 15366 Dahlwitz-Hoppengarten GERMANY
Van der Ploeg, R	Institute of Soil Science, University of Hannover Herrenhaeuser Str.2, 30419 Hannover GERMANY
Kranicz, I.	National Water Authority Marvany u. 1/c, H-1012 Budapest HUNGARY
Szlavik, L.	VITUKI PO Box 27, H-1453 Budapest HUNGARY
Trombetta, M.J.	Institute for International Political Studies, Milan Via Duca d'Aosta 16, 34074 Manfalcone GO ITALY
Lewandowski, A.	GEOMOR, Geoscience and Marine Research Koscierska 5, 80-328 Gdansk POLAND
Sowinski, M.	Institute of Environmental Engineering. Poznan University of Technology ul. Piotrowo 5, 60 965 Poznan POLAND
David, L.	Sanitary Engineering, Hydraulics Department, National Laboratory of Civil Engineering Av. Do Brasil 101, P-1799 Lisboa Codex PORTUGAL

Klenov, V.	State Institute for Applied Ecology, 98-5-80 Profsoyuznaya, 117485 Moscow RUSSIA
Cunderlik, J.	Department of Land and Water Resources Management, Slovak University of Technology, Faculty of Civil Engineering Radlinskeho 11, 813 68 Bratislava SLOVAK REPUBLIC
Kohnnova, S.	Department of Land and Water Resources Management, Slovak University of Technology, Faculty of Civil Engineering Radlinskeho 11, 813 68 Bratislava SLOVAK REPUBLIC
Andriansee, M.	RIZA Maerlant 16, PO Box 17, 8200 AA Leystad THE NETHERLANDS
Jak, M.	Ministry of Transport, Public Works and Water Management PO Box 5044, NL-2600 GA Delft THE NETHERLANDS
Kok, M.	HKV LIJN IN WATER / HKV Consultants PO Box 2120, 8203 AC Leylstad THE NETHERLANDS
Ozdemir, O.	Civil Engineering Department, Faculty of Engineering and Architecture, Gazi University 06570 Maltepe, Ankara TURKEY
Lees, M.	Department of Civil and Environmental Engineering, Imperial College of Science, Technology and Medicine London, SW7 2BU UNITED KINGDOM
Shulyarenko, A.	Dept. of Hydrology & Hydrochemistry Taras Shevchenk Kyiv University Vasilkivska st. 90, Kiev 252022 UKRAINE

Udovyk, O.	Ukrainian Research Institute of Environment and Resources
Chokolivski Blvd. 13, Kiev 252180
UKRAINE

Pinter, N.	Department of Geology, Southern Illinois University
Carbondale, IL 62901-4324
USA

Urbonas, B.	Urban Drainage & Flood Control District
2480 W. 26th Avenue, 156 B, Denver, CO, 80211
USA

OVERVIEW OF FLOOD ISSUES IN CONTEMPORARY WATER MANAGEMENT

J. MARSALEK

National Water Research Institute
Burlington, Ontario L7R 4A6, Canada

1. Introduction

One of the most important goals of water management is the protection of the society and material property against the harmful effects of water, including flooding and disruption of economic activities by flood waters. Such a goal is particularly important in densely developed urban areas with high concentrations of population, material properties, and economic activities, and potentially high flood damages [1].

While flood protection has been practiced as a single-purpose water management task for thousands of years, increasing competition for water resources and conflicts among their uses call for a more holistic approach to this problem. Such an approach can be defined as the ecosystem approach to water management, in which the social, economic and environmental interests are equally important. The fundamental qualities of the ecosystem approach are its holistic nature, which recognizes the system's complexity and inter-connectivity of its elements, demonstrated by an exchange of information, energy and matter, and the style of planning actions [2]. Other characteristics of the ecosystem approach include the focus on integrated knowledge, a perspective that relates systems at various levels of integration, and actions that are ecological with respect to integrated knowledge and perspective, anticipatory in terms of preventing detrimental outcomes, and ethical by respecting other natural systems [3].

In the ecosystem approach, flood management planning in river catchments should be connected with land use planning and address not only the traditional concerns of flood defense, but also the need to protect water resources in their entirety, both spatial and functional. Spatial considerations require the development of management plans for the whole river catchment and functional considerations should respect ecological functions of the river-catchment systems. Such concerns were often neglected in traditional flood management based on the benefit-cost analysis that addressed direct flood damages only [4].

The primary purpose of this overview was to establish an overall framework for the discussion of major topics of the NATO Advanced Research Workshop on floods, which was held at Malenovice, Czech Republic, in May 1999. These major topics included flood events, flood analysis, flood impacts, and flood damage mitigation.

J. Marsalek et al. (eds.), Flood Issues in Contemporary Water Management, 1–16.

2. Flood Events

Floods are hydrological events characterized by high discharges and/or water levels leading to inundation of land adjacent to streams, rivers, lakes, and other water bodies. Such events are caused and/or exacerbated by intense or long-lasting rainfall, snowmelt, failure of dam or levee systems, earthquakes, landslides, ice jams, high tides, storm surges, and also by human activities, including operation of flood control systems.

Flood severity is described by the average frequency of occurrence (e.g., a flood with a 1:100 year frequency). In most locations, flood risks and impacts are modified by anthropogenic activities, largely by changing the catchment response by land development, and possibly by changing the precipitation and runoff/snowmelt regimes by climate change. While the catchment response changes are relatively well understood, the impacts of climate change on floods are subject to speculation, as discussed in Section 8. Both types of changes of flood regimes weaken a basic assumption of flood frequency analysis – that flood series are stationary [5].

Three types of floods are recognised: (a) naturally occurring floods (without anthropogenic modifications), (b) modified natural floods, and (c) human-caused floods (e.g., by failures of man-built structures, or by errors in operation). Experience further indicates that floods have distinct temporal and spatial properties. High-intensity storms occurring over relatively small catchments may produce flash floods characterised by sharp-peak flood hydrographs. In larger catchments, with longer response time, flood hydrographs are characterised by a slower build up of flows and a very gradual flow decline. Generation of floods in large catchments allows for higher hydrological abstractions and more storage in the catchment system, thus, producing floods with smaller discharges per unit catchment area.

Flood management usually requires three steps: analysis of flood characteristics, estimation of flood impacts, and mitigation of flood impacts. These steps are discussed in the following three sections.

3. Flood Analysis

Flood analysis combines two interconnected components – estimates of flood characteristics required in flood management, and advancing the understanding of hydrologic processes leading to floods. Depending on the significance of the flooding problem addressed and the nature of the available supporting data, estimation of flood discharges is accomplished by means of one or more of the following five techniques: (a) empirical formulas, (b) frequency analysis, (c) regional flood analysis, (d) Probable Maximum Flood (PMF) methods, and (e) conceptual modelling [6].

Empirical formulas for estimation of flood flows were the first tools available to hydrologists. Perhaps the best known was the Rational Method, which dominated certain fields of design practice (e.g., floods in urban areas) as recently as 30 years ago [5]. Empirical formulas were easy to use, because of their computational simplicity and the general availability of supporting data. On the other hand, they suffered from too many oversimplifications and unrealistic assumptions, and currently, their use is limited to a narrow class of design problems.

Frequency analysis methods are statistical techniques used to develop the best estimates of flood flows of specified return periods from the available flood data. This is facilitated by fitting one of the known distributions to the available flood data and extrapolating the observations to the design range of low frequencies of occurrence (1:100, or smaller). These methods are well established in operational hydrology and considered reliable, if the historical flood records are sufficiently long to keep sampling errors small [7].

On the other hand, frequency analysis of short systematic flood records, combined with extrapolations to long return periods, suffers from such shortcomings as uncertainties associated with the errors in flood discharges, the selection of a distribution model and the estimation of model parameters. Consequently, frequency analysis can benefit from the use of non-conventional data, including historical flood data (before recording started) and paleoflood data obtained by analysis of proxy data. Many methods are used to derive these non-conventional data, including historical flow reconstruction, sediment regime-based reconstruction, botanical methods (e.g., tree scars, ring anomalies, age distribution), flow competence methods relating sediment characteristics to flow parameters and paleostage indicators (erosional or depositional) [8].

In regional flood analysis, flood estimates in ungauged catchments are derived by combining all the data available within this region. Regionalization models use the concept of hydraulic similarity by associating basin flood characteristics with climatic and geomorphologic factors. This analysis is usually conducted in three stages, first estimating the flood distribution and its shape parameters, then estimating scale parameters based on the physical meaning of flood distribution parameters, and finally estimating the mean annual flood. In the final stage, the variability from basin to basin clearly dominates over the sampling errors and is not affected as much by the climate as by the geomorphology of the basin. The use of conceptual models accounting for various causative factors is extremely useful for estimating floods at ungauged sites [9].

The concept of the Probable Maximum Flood (PMF) is used in design of major structures, including dam spillways, nuclear power stations and major bridges. While the principles used for determination of PMF are similar to those used in other hydrological methods, meteorological processes and the catchment response are maximized with respect to producing the maximum, physically feasible flood flow. When the PMF is applied to structures classified as representing medium or low hazards, only a percentage of the PMF discharge can be used [5].

Determination of flood flows by conceptual modelling, with explicit soil moisture accounting, is probably the most important method of flood analysis at present. Towards this end, many conceptual models were developed and used for flood estimation, prediction and forecasting [10]. Typically, these models account for three groups of processes, rainfall and snowmelt flux over the catchment, computation of rainfall and snowmelt excess by accounting for hydrological abstractions, and routing the rainfall and snowmelt excess over the catchment [11]. Such computations are done either for individual events, or in a continuous mode encompassing both wet and dry periods.

In single event modelling, the selection of the precipitation input data is of critical importance. For this purpose, rainfall data are subject to intensity-duration-frequency analysis and corrected for areal reduction. Eventually, various design rainfall events are constructed or adopted from historical records and used in flood analysis. Single-event modelling requires assumption of the initial catchment conditions, particularly with respect to soil moisture [11].

Modern conceptual models, which dominate the current hydrologic practice, are physically based and therefore well suited for modelling changes in catchment parameters, for example, those caused by land development, [12], or changes in climatic inputs studied in climate change impacts [13]. Many benefits were brought about by progress in computer hardware and computer science, including the use of database systems, graphic and image processing tools, knowledge-based expert systems, artificial intelligence (particularly neural nets), digital elevation models (DEM), and Geographic Information Systems (GIS) [14]. GIS makes it possible to integrate efficiently not only topography, but also geomorphology, soils types, vegetation, and land use characteristics of the basin with hydrologic models of the catchment. The contemporary hydrologic models require calibration for specific catchments.

4. Flood Hydraulics

Mathematical hydraulic models are used extensively in flood analysis to estimate flood hydrograph propagation to downstream reaches, the impacts of new construction (e.g., bridges) on floods, determination of effects of flood defence works (e.g., embankments), studies of flood control structures, and flood mapping. The leading models are based on St. Venant's equations written in one or two dimensions [15]. One dimensional models are particularly well developed; their applications require the selection of several model constants, input data for cross-sections typically spaced from 0.1 to 1 km, and boundary conditions. Even though 1D models only address the flow in the river channel, they can account for flow spillage onto flood plains by connected storage cells. In complex cases with multiple channels, flows in channel networks can be modelled [16].

More recently, 2D hydraulic models were introduced. They are used in studies of shorter river reaches and often for addressing problems other than just flood analysis. Their advantages include simple use, no distinction between the minor bed, major bed and storage cells, and easier calibration (compared to 1D models). However, they suffer from shortcomings associated with variations in cross-sectional geometry and the resulting complexities of local flows [16].

Both types of models need to be calibrated against measured water levels. Modelling results are affected by the discretization of the river/flood plain system, and proper accounting for many man-made structures interfering with flood flow. Additional challenges arise when dealing with significant river bed changes during floods (e.g., caused by landslides, or deep-cut erosion) and in estimating channel roughness. Finally, it should be acknowledged that physical hydraulic models can be also used for flood routing, but in terms of costs, they cannot compete with mathematical models. However, physical models may offer some advantages when addressing boundary conditions, or local flows through special hydraulic structures.

5. Flood Impacts

Floods cause many different types of impacts which can be appropriately discussed under the three ecosystem components, the society, economy, and environment. In the assessment of impacts, the first two components are usually combined and that approach was adopted here.

5.1. FLOOD IMPACTS ON THE SOCIETY AND ECONOMY

Flood impacts on the society and economy can be classified as tangible and intangible, where the first category includes the loss of human life and both direct and indirect flood damages (property damage, disruption of economic activities). Statistics reported by insurance companies usually deal with direct damage costs only. The Munich Reinsurance Bureau's statistics indicate that such flood damages have been rising during the last decade. In 1998, direct damage costs for all weather related disasters (i.e., storms, floods, droughts and fires; estimated at US $138 billion) exceeded those for the entire 1980s (Fig. 1). The costliest was the 1998 Yangtze River flood resulting in death of more than 3000 people, dislocation of 230 million people, and direct damages of more than $45 billion [17].

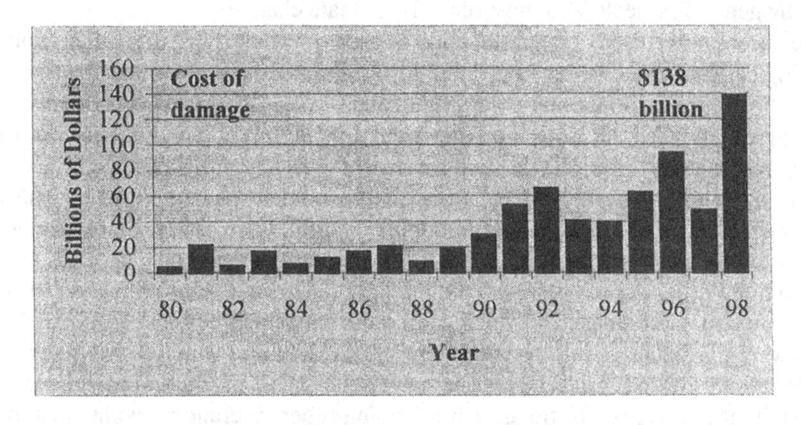

Figure 1. Direct costs of damages caused by weather-related disasters 1980-98 [17].

Direct economic costs, including property damages, income losses, and emergency costs, are well understood and can be evaluated by the conventional techniques of welfare economics. It should be noted, however, that some economic flood effects can be positive, as building business increases during the recovery period and prices may increase as well. Another potential benefit is the improvement of services after reconstruction. Indirect economic costs can be caused by changes in business practices, reduced employment, and the resulting impacts on other sectors [18].

Intangible flood impacts on people are generally poorly understood and difficult to account for in the conventional benefit/cost analysis. In general, intangible impacts include the impairment of human health (by stress, worries about future events, reduced sanitation), loss of memorabilia and non-marketed goods, disruption of life and evacuation, and loss of familiar surroundings and recreational opportunities

[19,20]. Techniques for assessing intangible damages are not well developed; it was suggested to estimate such damages as a percentage of direct costs [21].

When evaluating economic effects of flooding, the inherent risk and uncertainty, and their variation with flood severity, have to be considered. This can be done by developing a frequency-damage curve, and estimating uncertainties associated with this curve. As shown by McBean *et al.* [22], the duration of inundation should be also considered in estimating the magnitude of damages. Benefits of flood control projects are assessed by comparing two situations – the development of the economy with and without the project. Another possibility is to use a detailed questionnaire to determine clients (i.e., occupants of the floodplain) willingness to pay for the project [23].

5.2. FLOOD IMPACTS ON THE ENVIRONMENT

Environmental impacts of floods and flood defence measures have only been addressed relatively recently. From the environmental point of view, floods are natural events, which bring both adverse and beneficial environmental effects. Seasonal flooding of the environment is a natural feedback mechanism serving to replenish floodplains and sustain their ecosystems. However, in most major river basins, this natural feedback has been modified by humans, through catchment development, implementation of flood management projects, and most recently, climate change.

The best recognised environmental impact of floods is the alteration of morphological processes, which contribute to forming the natural habitat for flora and fauna in flood plains. Floodplain flora and fauna serve as key links in many food chains. Flood induced changes in fish, animal, and insect populations are often reported immediately after floods, but are rarely well documented, because of the transient nature of these phenomena. It can be concluded that the flood regime is a direct determinant of the riverine and floodplain natural environments, and changes in their ecosystems are likely to affect an area well beyond the river corridor [24].

Even more pronounced environmental impacts may result from mitigation of flood problems by structural measures, which exert their impacts throughout the year. These impacts were particularly well described for large reservoirs, which serve many other purposes besides flood protection. The adverse aspects of these projects include increases in the severity of floods (through improper operation maximising other benefits), ecological destruction of inundated land, uprooting of the population that lived in the area flooded by the reservoir, interference with sediment transport, and some additional impacts on human health [25]. Embankments contribute to increased river discharges and thereby export flood risks and damages to the downstream areas [24].

In the ecosystem approach, these flood management schemes must be planned in conjunction with runoff source controls throughout the catchment (particularly in urban and agricultural areas), and considerations of water quality, water resources conservation, and aquatic ecosystem protection. Environmental protection and sustainability require long-term planning horizons and cooperation of all levels of government in land use and river use planning. Public engagement is important to create the feeling of public ownership of, and involvement in, flood mitigation projects, and to ease their implementation [21].

6. Flood Management

Throughout history, various societies coped with floods as well as they could within the constraints of their wealth and technology. General strategies for coping with floods comprise learning to live with floods (without attempting to control flood magnitude or to protect the areas susceptible to flooding), protecting life, land and property, and reducing flood peaks and volumes. Specific measures employed in these strategies include: the "do nothing" option, non-structural measures and structural measures [26]. This terminology was adopted here for the sake of consistency with the literature on floods; the description of this first class as "do nothing" is somewhat inaccurate and has a somewhat negative connotation. A classification of flood management measures is shown in Fig. 2.

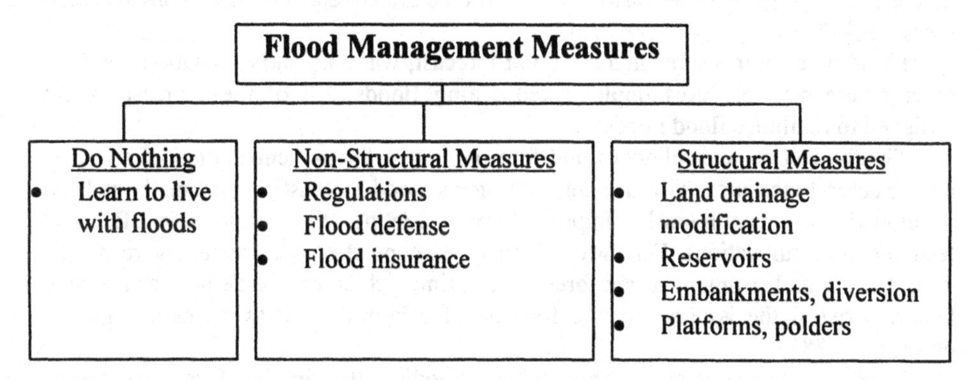

Figure 2. Classification of flood management measures.

6.1. DO NOTHING OPTION

The "do nothing" option is based on the acceptance of floods as inevitable natural events which the society needs to learn to live with. Towards this end, it is helpful to learn as much as possible about flood characteristics and their impacts, and use this knowledge to minimise losses caused by floods, while maximising flood benefits. This generally means to avoid occupancy of floodplains and more or less restrict their use to wildlife habitat, land conservancy, some agriculture activities, and cattle grazing. The "do nothing" approach does not create any new stresses on the environment, but where flood plains are already occupied, it produces tangible as well as intangible flood impacts and damages [21].

6.2. NON-STRUCTURAL MEASURES

Non-structural measures comprise regulations and policies, flood defence and flood insurance. These measures strive to reduce flood impacts, without altering flood characteristics, and as such, focus on policies and emergency measures with low capital investments. Essentially, these measures increase threshold flows, for which no damage occurs.

Regulations control the human use of floodplains and inundation-prone lands. The implementation of these regulations is based on flood risk maps, which are essential tools for land use planning in flood-prone areas. These maps combine hydrological, geomorphological, hydraulic, and land use information related to floods, and serve for assessing flood damages and the feasibility of non-structural (e.g., flood proofing) control measures [11]. Flood proofing regulations specify which structures may be built in flood-prone areas, and how they should be constructed or modified to withstand floods up to a certain frequency, without sustaining substantial damage. This usually requires raising the main floor above the 100 year flood stage. Where the associated costs are too high, relocation out of the floodplain may be a feasible solution. It should be recognised that the effectiveness of these regulations is only as good as their enforcement in land use planning and granting building permits. Weak or inconsistent enforcement seriously undermines the effectiveness of these non-structural measures [27].

Flood defence measures include flood forecast, warning, and evacuation, and any other emergency measures implemented during floods. All of these measures are designed to minimise flood impacts.

Timely, robust, and reliable flood forecasts serve for implementation of all other non-structural measures. Towards this end, many flood forecasting systems have been developed and continuously improved by adopting the latest scientific and technological innovations. Forecast system types depend on the catchment response, described by its lag time, and the forecast lead time, which can be defined as the time lapsed between the issuance of the forecast of a hydrological phenomenon and its occurrence [28].

In small catchments susceptible to flash flooding, the simplest forecast systems, with the shortest lead times, include real-time measurements of streamflow and transposition of this information to the key points of interest by routing or correlation techniques. In more sophisticated systems, streamflow measurements and routing/correlation techniques are enhanced by local precipitation correlation. Longer lead times can be achieved with the systems incorporating real-time measurements of precipitation (using radar), streamflow, and temperature, and using these data as inputs to a watershed model simulating flood flows. Finally, the longest lead times are achieved in systems employing meteorological models and weather forecasts, with frequent updating of the forecasted information [28].

In large catchments, with long lag times, the need for weather forecasting diminishes. In these circumstances, even the simple forecasting methods (transposition of upstream flows by routing or correlation) may produce satisfactory flood forecasts. In general, lead times between 24 and 48 hours are considered optimal for most forecasting purposes [24].

The need for expediency in flood related data acquisition is reflected by the type of instrumentation used (radar, airborne instruments, satellite based instruments) and the data transfer equipment (telephone lines, radio frequency or microwave, satellite, and meteorburst). Many of these elements are incorporated into special integrated systems, which besides flood forecasting also disseminate flood warning. One of such systems is the ALERT (Automated Local Evaluation in Real Time) system, which was developed by the U.S. National Weather Service, and by 1994, was operated in more than 650 U.S. locations. All flood forecasts include uncertainties, which of course increase with the lead time [29].

Flood forecast systems are used before and during floods to direct flood defence activities and emergency operations. Flood defence activities include construction of temporary structures (e.g., earthen or sandbag dikes), operation of flood diversion structures, repairs/fortifications of embankments (levees) and dikes, pumping water from protected areas and other activities. An impressive example of a temporary flood defence structure was the 24 km long Brunkild dike built in 72 hours during the 1997 Red River Flood in Manitoba [27]. General catchment models can be used for dealing with emergencies arising from breaches of embankments, either accidental, or intentional (to increase flood plain water storage and reduce downstream flooding). Some parts of these forecast systems (GIS databases, satellite pictures of inundated areas) are also useful in conducting post-flood rehabilitation and reconstruction.

Flood forecasts are used for issuing flood warnings. However, the success of such systems depends on people's response and cooperation. Warning systems collect information about an impending emergency, communicate it to those who need it, and facilitate correct decisions and responses by those in danger. The management subsystem integrates and evaluates the risk information and warns the public. For effectiveness, people must receive the warning, understand and believe it, and then they can decide what to do. There is a need for confirmation. Factors at play include sender factors (a familiar, credible, official source), and receiver factors (environmental, social, psychological, and physiological aspects) [30].

The last non-structural measure discussed is flood insurance, which is practised in many ways, ranging from public disaster relief to government or private insurance programs. A primary example of such programs is the U.S. National Flood Insurance Program, which was created in 1968 and further modified by the Flood Disaster Protection Act of 1973. This program has played a critical role in fostering and accelerating the principles of flood plain management. Currently, more than 18,000 communities are participating. This program ensures that flood damages to individual properties are paid only once; damage payments require flood-proofing the insured structures for the base flood (1:100 years) [31].

While the non-structural flood mitigation methods have been known and applied for many years, major progress has been accomplished in hydroinformatics support of such activities. This support includes radar and satellite sensing of precipitation; rainfall-runoff, routing and flood forecasting methods, which utilise both physically-based models and artificial intelligence; and, a GIS supported modelling environment. This technology improves the speed of processing, production, and transmission of hydrometeorological data used in flood management for flood warning and emergency operations. Finally, such information is made widely available, on the Internet.

6.3. STRUCTURAL MEASURES

Structural measures aim to achieve two objectives – reduce flood volumes and peaks, and thereby protect people and property against flood damages. Numerous such measures have been developed and classified into various categories, e.g., as point, linear and spatially distributed structures. The first two categories are also called the intensive measures, and the last category, the extensive structural measures [26]. The extensive measures reduce flood generation potential and typically are implemented for a variety of reasons, not just flood reduction. Extensive structural measures strive to delay runoff, increase infiltration, and minimise erosion by reshaping land surface (terracing, contour plowing) and implementing erosion protection (soil conservation, restoring vegetative cover, reforestation).

Intensive flood management structures include storage facilities, enhancement of river bed conveyance capacity, and earthen platforms and polders in flood plains. Storage reservoirs store flood waters and release them when downstream capacity allows it. Well-sized reservoirs can practically eliminate smaller and intermediate floods, and may reduce large floods. However, this protection is achieved at large environmental and ecological costs – loss of land to water storage space, loss of habitat, reduced biodiversity, interference with fish passage, introduction of nuisance species, diseases, loss of land regeneration in flood plains, and creation of a new risk associated with a potential dam failure [25]. Perhaps the least obtrusive are smaller reservoirs built just for flood control and drained to low levels immediately after floods pass [26].

River bed flow capacity can be increased by embankments (levees), dikes, and flood walls. These measures protect land in upstream reaches from flooding, but by eliminating flood plain storage, they increase river discharge and may export flood problems downstream. When these measures fail (e.g., through an accidental or intentional breach), flooding occurs as if they did not exist, and sometimes, it can even be worse [24]. During floods, embankments require protection (reinforcement, repairs, particularly after extended exposure to high water), and land behind embankments has to be drained by pumping. Other concerns arise from reduced low flows in the summer and a gradual buildup of river bed elevations in reaches with embankments. Through this process, river beds become super-elevated with respect to the surrounding land and may flood large areas when embankments fail. Other measures in this category include diversion channels/tunnels, which reduce flows in the protected river reaches [27].

Platforms and polders are earthen structures providing for flood protection of facilities (e.g., industrial or utility plants) located in floodplains. Because of this location, platforms and polders interfere with flood flow and may adversely impact the environment [26].

6.4. SELECTION OF FLOOD MANAGEMENT MEASURES

The selection of flood management measures is based either on the traditional benefit/cost analysis, or on multicriterion ranking. In the benefit-cost analysis, benefits include reduction in costs of floodplain occupancy, reduced flood emergency measure costs, and improvements in floodplain land use. These benefits are compared with the costs of flood management measures [23]. However, economic valuation is not a complete tool because of its limits, uncertainties and poor consideration or neglect of environmental and social aspects. These shortcomings are somewhat mitigated by the multicriterion ranking, which is based on criteria such as low costs, large benefits, limited environmental impacts, limited adverse social effects, and desired social effects. Special techniques for multicriteria assessment are available [32].

Finally, it should be emphasised that the successful flood mitigation schemes are likely to include a mixture of various measures, carefully selected to meet the water management needs of a particular river reach and the entire catchment. In some areas, the do nothing alternative may be the preferred choice, causing the least interference with local ecosystems. Somewhat similar goals can be achieved with non-structural measures serving to increase the flow limit for which no damages occur. Proactive conservation measures, reducing runoff and erosion over large parts of the river catchment, are gaining on prominence in the contemporary land use planning and may be required to compensate for the ongoing land development. Finally, intensive structural measures may have their place in multipurpose projects with low environmental impacts.

The implementation of flood management measures should serve the overall objective of supporting sustainable development of the river catchment and maximising the economic efficiency of the catchment use. This is generally achieved by conjunctive land use and water management, integrated water cycle management, and the use of appraisal methods, which include economic analysis of all impacts of the available options [21].

7. Post-Flood Recovery

For the implementation of recovery and reconstruction, planning before the disaster strikes is important. It gives an opportunity to influence future development, learn from experience of other communities, and capitalise on the fact that the support for mitigation is highest right after the disaster. Flood recovery costs involve the direct, indirect, and secondary losses incurred, and the costs of supporting rescue and relief operations and recovery management. Physical planning should include analysis of disaster vulnerability [33].

Recovery activities start with restoration of essential services, including water utilities. The next in line are job-generating or economic base projects, followed by reconstruction of housing and rehabilitation of agricultural land. Social and human health aspects are also addressed by providing counselling and support for both flood victims and emergency workers in the early phases of recovery. There are also some long-term negative effects of floods on families and communities [19]. Keys to successful recovery include strong and knowledgeable local leadership, and good intergovernmental relations, as various forms of support come from different government and private agencies [27].

8. Discussion

Flood defence is an old issue that has been addressed by many societies and civilisations. While significant progress has been achieved in planning, designing and implementing flood management measures, it is impossible to ignore the fact that, in spite of significant expenditures on flood defence, flood damages are on the rise, as documented earlier in Fig.1 and by other sources [34]. There may be several explanations for this trend, including increasing trends in flood magnitude, deterioration of structural flood management measures, and growing populations and continuing development in and near flood-prone areas. Increases in flood magnitudes may result from climate change and changes in land use and cover.

The literature on climate change impacts on flood generation is largely speculative and includes studies of trends in the historical streamflow data, applications of global circulation models to large river basins, and the modelling of streamflow sensitivity to selected climatic changes.

Recognising that releases of greenhouse gases into the atmosphere have been increasing for some time, it is of interest to examine, whether the climate change impacts can be detected in hydrometeorological records, including flood flows. So far, the confirmation of trends in data has been somewhat successful for specific weather patterns [34], precipitation [35], the start of snowmelt spring runoff [36], and low streamflows [37]. Similar analyses for flood flows yield inconclusive results. In this connection, Bronstert [34] pointed out that extreme floods and the associated damages result from unfavourable combinations of four factors: meteorological inputs, state of the river catchment, state of the river bed, and potential risk to human life an property on flood plains. Thus, in most well populated catchments, the climate change influences on flood flows are likely to be masked by the other factors.

Analyses of almost 400 U.S. historical streamflow records from climate-sensitive flow gauging stations (i.e., the stations without other anthropogenic influences) indicated that only 10–13% of annual peak daily flows exhibited significant trends, and for the longest records (80 years), the numbers of records with upward or downward trends were about equal [36]. Other records are likely to be influenced by historical changes in land use and catchment storage, as demonstrated in a study of four paired catchments by Changnon and Demissie [35]. Although these catchments experienced significant upward trends in precipitation over a period of 50 years, about two thirds of fluctuations were explained by drainage changes in both urban and agricultural catchments.

For the Rhine River, Engel [38] reported an upward trend in precipitation and annual peaks, but no change in extreme peaks with return periods greater than 10 years. For the Elbe River, flood flows with return periods from 2 to 200 years are much smaller in this century than in the last century, as a result of reduced incidence of ice jams [34].

Thus, while there are measurable trends in hydrometeorological variables attributable to climatic change, no definite, robust results are available as yet with respect to trends in flood flows. Difficulties with detecting such trends arise from high natural variability of flood flows, masking effects of land use and land cover changes, and the reduced vulnerability of catchments with high storage volumes to flooding. This vulnerability can be described by a measure of storage, defined as S/Q, where S is the maximum catchment storage volume and Q is the total annual mean renewable supply. Most catchments have small relative reservoir storage volume (described by $S/Q < 0.6$) and are more vulnerable to occurrence of flooding [39].

Other studies of climate change impacts on flooding are based on conceptual modelling. Global atmospheric models provide a realistic representation of climatic changes, but suffer from low spatial resolution (e.g., 2 x 2.5°; [40]), smoothed orography, and inadequate modelling of soil moisture [33]. Under these circumstance, other modelling efforts focused on the so-called impact approach, in which the selected changes in climatic inputs are translated into changes in hydrological regimes by modelling. These studies provide information on the sensitivity of water resources in specific catchments to changes in climatic inputs [13], but often with large uncertainties arising from interpolation of aggregated climatic data to relatively small catchments.

The effects of land use, land cover and drainage changes on flood flows are generally well recognised. Such changes typically start in the headwaters, with deforestation, and continue throughout the catchment with conversion of land for agricultural purposes and urban developments. Deforestation has two impacts on floods – high increases in surface runoff, contributing to flooding, and high soil erosion, contributing to siltation of river beds and storage reservoirs [25]. While these processes are somewhat controlled in the developed countries, the same cannot be said for the less developed countries.

Agriculture also impacts flood generation processes, usually by changes in land cover and surface drainage. Reduced vegetation cover and somewhat degraded soils (e.g., by compaction) contribute to increased generation of runoff [34]. In this respect, soil conservation approaches, typified by reduced till or no-till agricultural operations help prevent soil erosion and degradation, and increase rainwater infiltration into soils. Another factor contributing to generation of runoff and enhancement of flood flows is modified land drainage. Several types of modifications occur, including the construction of new surface drainage ditches and channels, subsurface drainage and elimination of wetlands with significant storage capacity. All of these modifications increase both the volume and speed of runoff and thereby contribute to increased flood flows [35].

The last land use change discussed is urbanisation. The effects of urbanisation on the hydrological cycle have been extensively described in the literature [1]. Fast concentration of runoff on impervious surfaces, with minimal hydrologic abstractions, and fast transport of runoff in gutters, sewers, and drains, result in high runoff rates. This tendency toward generation of high flows was in the past further supported by the straightening, deepening, and lining of natural river beds and channels in urban areas. The final result was flooding within the urban area itself and also in the downstream reaches of rivers with highly urbanised catchments.

Recognising that conventional urban drainage with fast removal of surface waters was not sustainable, urban drainage practices were changed in many jurisdictions by the introduction of stormwater management in the late 1960s. Stormwater management comprises both non-structural and structural measures mitigating progressing urbanisation impacts in the form of increased runoff flows, volumes, and the associated sediments, pollutants, and bacteria [1].

The planning, design, and implementation of stormwater management recognise the concept of minor and major drainage, and the need for sustainability of urban water systems. Specific management measures include source controls (reducing impervious surfaces, preserving natural drainage, slowing down runoff and encouraging on-site infiltration) and structural controls (runoff storage and controlled release, infiltration facilities). While the effectiveness of stormwater management in control of smaller flows is well documented, the performance of such facilities in high flows, or rainfall/snowmelt events is less well established. Also, there are many older drainage systems, which had been built before stormwater management was introduced.

Flood defence structures, such as dams and embankments, may weaken with age and lose their effectiveness in flood protection. Worldwide statistics indicate that a significant number of small dams fail annually, with the majority of these incidents occurring in less developed countries [25]. The resulting floods may result in high losses of human life and property damage. Proper inspection and maintenance need to be applied to all structures serving for storage of water. Cases of embankment failures are also common, particularly during long-lasting floods.

The final reason for increasing flood damages are societal choices leading to a growing population in and near flood-prone areas, and continuing development in flood plains. Lack of land available for development, combined with growing population pressures and sometimes a misplaced reliance on flood defence structures, lead to a continuing encroachment on flood plains. In fact, many flood management projects induce new development and thereby increase future risks. Successful flood plain management requires an accurate delineation of areas at risk in the form of flood maps, and a strict enforcement of flood plain land use regulations. Where land use regulations are not strictly enforced, increased future risks can be expected [27].

9. Summary

Flood management will remain to be one of high priorities in water management. Ever increasing flood damages call for further advances in the understanding of flood generation processes and their modelling, flood impacts, and in designing and implementing flood management measures. Further advances are needed in flood forecasting, in order to further increase lead times for emergency measures of flood management. This objective should be supported by technological innovations including the remote sensing of meteorological data, weather forecast modelling, GIS databases and information technology tools. Future structural measures should focus on extensive measures serving to reduce flood risk potential over large areas. An emerging issue, requiring further research, is the impact of climate change on floods. Finally, in the planning of flood management, floods should be considered within the catchment ecosystem, with clear understanding of connections among the various system components and implications of flood management measures.

10. References

1. Marsalek, J. (1998) Challenges in urban drainage, in J. Marsalek, C. Maksimovic, E. Zeman and R. Price (eds), *Hydroinformatics tools for planning, design, operation and rehabilitation of sewer systems*, NATO ASI Series, 2. Environment, Vol. 44, Kluwer Academic Publishers, Dordrecht/Boston/London, pp. 1-23.
2. Lindh, G. (1986) Water resources management in urban areas - an ecosystem approach. *Vatten* **42**, 3-9.
3. Marsalek, J. (1995) The ecosystem approach to water management in urban areas. *New World Water* **1995**, 27-30.
4. Gardiner, J.L., Dearsley, A.F., and Woolnough, J.R. (1987) The appraisal of environmentally sensitive options for flood alleviation using mathematical modelling. *J. IWEM* **1(2)**, 171-184.
5. Chow, V.T., Maidment, D.R. and Mays, L.W. (1988) *Applied hydrology*, McGraw-Hill Book Co., New York.

6. Rossi, G. (1994) Historical development of flood analysis methods, in G. Rossi, N. Harmancioglu and V. Yevjevich (eds.), *Coping with floods*, NATO ASI, Series E: Applied Sciences, Vol. 257, Kluwer Academic Publishers, Dordrecht/Boston/London, pp. 11-34.
7. Cunnane, C. (1989) Statistical distributions for flood frequency analysis, World Meteorological Organization, Oper. Hydrol. Rep.33, WMO Publ. 718, Geneva.
8. Salas, J.D., Wohl, E.E. and Jarrett, R.D. (1994) Determination of flood characteristics using systematic, historical and paleoflood data, in G. Rossi, N. Harmancioglu and V. Yevjevich (eds.), *Coping with floods*, NATO ASI, Series E: Applied Sciences, Vol. 257, Kluwer Academic Publishers, Dordrecht/Boston/London, pp. 111-134.
9. Cunnane, C. (1988) Methods and merits of regional flood analysis. *J. Hydrology* **100**, 269-290.
10. Todini, E. (1988) Rainfall runoff modelling: past, present and future. *J. Hydrology* **100**, 341-352.
11. Watt, W.E., Lathem, K.W., Neill, C.R., Richards, T.L., and Rousselle, J. (1989) Hydrology of floods in Canada: a guide to planning and design. National Research Council Canada, Publication No. 29734, Ottawa.
12. Ng, H.Y.F. and Marsalek, J. (1989) Simulation of the effects of urbanization on basin streamflow. *Water Resources Bulletin* **25(1)**, 117-124.
13. Nemec, J. and Schaake, J.C. (1982) Sensitivity of water resources systems to climate variation. *Hydrol. Sci. J.* **27**, 327-343.
14. Zeman, E., Vanecek, S. and Ingeduld, PO. (1998) Tools for data archiving, visualization and analysis, applied in master planning, in J. Marsalek, C. Maksimovic, E. Zeman and R. Price (eds), *Hydroinformatics tools for planning, design, operation and rehabilitation of sewer systems*, NATO ASI Series, 2. Environment, Vol. 44, Kluwer Academic Publishers, Dordrecht/Boston/London, pp. 169-178.
15. Ingeduld, P. and Zeman, E. (1998) Modelling receiving waters, in J. Marsalek, C. Maksimovic, E. Zeman and R. Price (eds), *Hydroinformatics tools for planning, design, operation and rehabilitation of sewer systems*, NATO ASI Series, 2. Environment, Vol. 44, Kluwer Academic Publishers, Dordrecht/Boston/London, pp. 357-392.
16. Labadie, G. (1994) Flood waves and flooding models, in G. Rossi, N. Harmancioglu and V. Yevjevich (eds.), *Coping with floods*, NATO ASI, Series E: Applied Sciences, Vol. 257, Kluwer Academic Publishers, Dordrecht/Boston/London, pp. 291-323.
17. The Toronto Star (1998), Toronto, Nov. 28, 1998.
18. Moser, A. (1994) Assessment of the economic effects of flooding, in G. Rossi, N. Harmancioglu and V. Yevjevich (eds.), *Coping with floods*, NATO ASI, Series E: Applied Sciences, Vol. 257, Kluwer Academic Publishers, Dordrecht/Boston/London, pp. 291-323.
19. Green, C.H. and Penning-Rowsell, E.C. (1986) Evaluating the intangible benefits and costs of a flood alleviation proposal. *J. IWEM* **40(3)**.
20. Simonovic, S. (1998) Criteria for social evaluation of flood management decisions. An unpublished paper, Natural Resources Institute and the Department of Civil and Geological Engineering, University of Manitoba, Winnipeg, Manitoba (6 p.).
21. Gardiner, J.L. (1990) River catchment planning for land drainage and the environment. *J. IWEM* **4(5)**, 442-450.
22. McBean, E.A., Gorrie, J., Fortin, M., Ding, J., and Moulton, R. (1988) Adjustment factors for flood damage curves. *Journal of Water Resources Planning and Management* **114(6)**, 635-646.
23. Moser, A. (1994) Economics of selection of flood mitigation measures, in G. Rossi, N. Harmancioglu and V. Yevjevich (eds.), *Coping with floods*, NATO ASI, Series E: Applied Sciences, Vol. 257, Kluwer Academic Publishers, Dordrecht/Boston/London, pp. 585-596.
24. Arnold, M. (1976) Floods as man-made disasters, *The Ecologist* **6(5)**, 169-172.
25. Goldsmith, E., and Hildyard, N. (1984) *The social and environmental effects of large dams*, Wadebridge Ecological Centre, Cornwall, United Kingdom.
26. Yevjevich, V. (1994) Classification and description of flood mitigation measures, in G. Rossi, N. Harmancioglu and V. Yevjevich (eds.), *Coping with floods*, NATO ASI, Series E: Applied Sciences, Vol. 257, Kluwer Academic Publishers, Dordrecht/Boston/London, pp. 573-584.
27. Morris-Oswald, M., Simonovic, S. and Sinclair, J. (1999) Efforts in flood damage reduction in the Red River Basin: practical considerations. Report to the International Joint Commission, University of Manitoba, Winnipeg, Manitoba (78 p.).
28. Nemec, J. (1986) *Hydrological forecasting*, D. Reidel Publishing Co., Dordrecht/Boston.
29. Feldman, A.D. (1994) Assessment of flood forecast technology for flood control operation, in G. Rossi, N. Harmancioglu and V. Yevjevich (eds.), *Coping with floods*, NATO ASI, Series E: Applied Sciences, Vol. 257, Kluwer Academic Publishers, Dordrecht/Boston/London, pp. 445-458.

30. Mileti, D.S. (1994) Public response to flood warning, in G. Rossi, N. Harmancioglu and V. Yevjevich (eds.), *Coping with floods*, NATO ASI, Series E: Applied Sciences, Vol. 257, Kluwer Academic Publishers, Dordrecht/Boston/London, pp. 549-564.
31. Federal Emergency Management Agency (FEMA) (1998). Floodplain management summary, FEMA Web Page (www.fema.gov), updated Sept. 14, 1998.
32. Ko, S.K., Fontane, D.G. and Labadie, J.W. (1992) Multiobjective optimization of reservoir systems operation. *Water Resource Bulletin* **28(1)**, 111-127.
33. Gruntfest, E. (1994) Flood disaster relief, rehabilitation and reconstruction, in G. Rossi, N. Harmancioglu and V. Yevjevich (eds.), *Coping with floods*, NATO ASI, Series E: Applied Sciences, Vol. 257, Kluwer Academic Publishers, Dordrecht/Boston/London, pp. 723-732.
34. Bronstert, A. (1995) River flooding in Germany: influenced by climate change? *Phys. Chem. Earth* **20(5-6)**, 445-450.
35. Changnon, S.A. and Demissie, M. (1996) Detection of changes in streamflow and floods resulting from climate fluctuations and land use-drainage changes. *Climatic Change* **32**, 411-421.
36. Burn, D.H. (1994) Hydrologic effects of climatic change in west-central Canada. *J. Hydrology* **160**, 53-70.
37. Lins, H.F. and Slack, J.R. (1999) Streamflow trends in the United States. *Geophysical Research Letters* **26(2)**, 227-230.
38. Engel, H. (1995) Hochwasser am Rhein, Ursachen und Entwicklungen - Moeglichkeiten und Grenzen von Vorhersagen, in Proc. AIC Conf. Hochwasser Katastrophen - Ursachen und Praeventivmassnahmen, June 29&30, 1995, Koeln.
39. Gleick, P.H. (1988) Vulnerabilities of United States water systems to climatic change, in J.C. Topping (ed.), *Coping with climate change*, Proc. of the Second North American Conference on Preparing for Climate Change: A Cooperative Approach, Washington, D.C., Dec. 6-8, 1988, 327-336.
40. Van Blarcum, S.C., Miller, J.R. and Russell, G.L. (1995) High latitude river runoff in doubled CO_2 climate. *Climatic Change* **30**, 7-26.

IMPROVEMENT OF FLOOD PROTECTION IN CZECH REPUBLIC

PAVEL PUNCOCHAR, MAREK MATA and DANIEL POKORNY

Department of Water Management Policy, Ministry of Agriculture, Tesnov 17, Prague 1, Czech Republic

1. Recent floods in the Czech Republic

In 1997 the eastern parts of the Czech Republic were hit by floods, which were the greatest in this century. The high water levels caused extensive destruction in many places throughout the basins of the Morava, the Odra and the upper Labe (Elbe) rivers. In total, 538 cities and villages in 34 districts were afflicted and there were 50 casualties. The overall damage amounted to 63 billion Czech crowns. The evaluation of flood data showed [1,2] that the peak flows exceeded 100-year floods for which most of the existing flood control measures and structures were designed. In addition, the duration of the floods due to the exceptional duration of the rainfall contributed to increased damages.

One year later, in 1998, the north-eastern part of the Elbe River watershed was seriously damaged by a flash flood which was caused by a local storm rainfall. In this case the flood was caused by short but very intensive rainfall (reaching 204 mm in few hours) which was followed by extremely fast and huge increase of water levels in small brooks and streams in the mountainous part of the watershed. Hydraulic effects of water velocity in hilly territory with narrow valleys caused serious damages. During the violent flood there were 6 casualties, 23 houses were completely destroyed, 1 322 buildings damaged and the overall damages in the most affected district (Rychnov nad Kneznou) reached nearly 2 billion Czech crowns.

In spite of different durations and courses of floods, the heavy rainfalls were the only cause of their origin as indicated in Fig. 1. Such rainfalls (and consequent floods) have not occurred on the territory of the Czech Republic during the last 100 years. Obviously, the general public has been seriously concerned about this situation and speculations about the possible beginning of unusual climate changes accompanied by the heavy rainfalls and floods were frequently discussed.

The historical record of the extreme annual flow rates observed in the Czech watercourses indicates the occurrence of even higher floods and their uneven distribution in the past (Fig. 2). However, the absence of extreme water levels and flood disasters during the past 100 years influenced the sensitivity as well as attention of the public. As a result of increased flood awareness, the inhabitants of the hilly areas in upstream parts of watersheds, subject to high rainfalls, expect the water management authorities and state administration to improve flood defense measures.

J. Marsalek et al. (eds.), Flood Issues in Contemporary Water Management, 17–24.

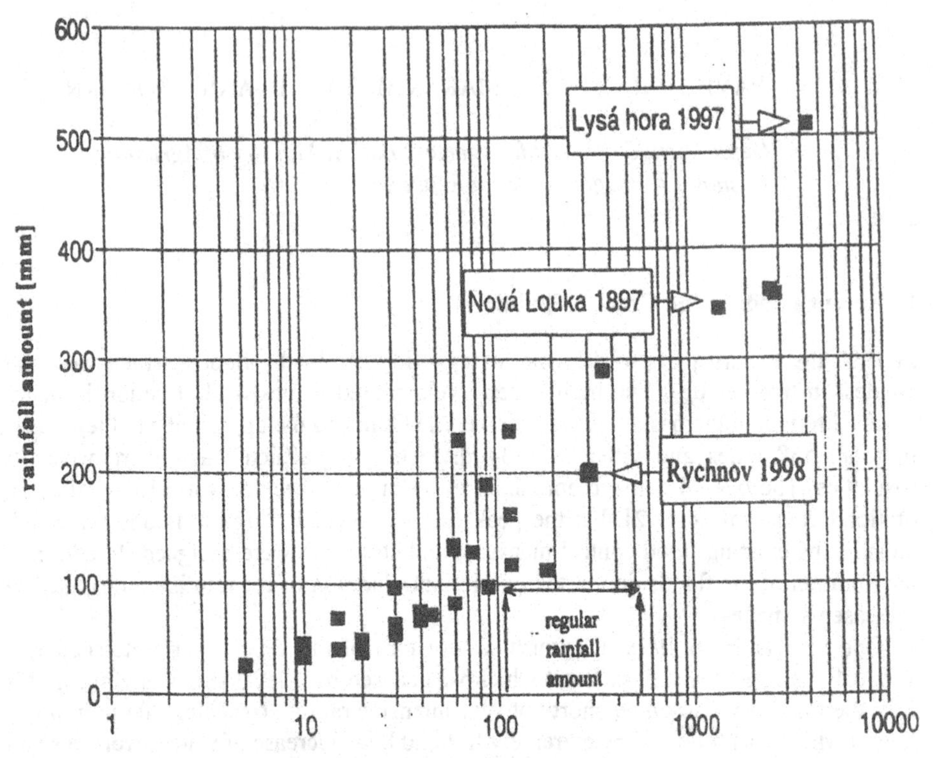

Figure 1. Maximum rainfalls on the territory of the Czech and Slovak Republics.

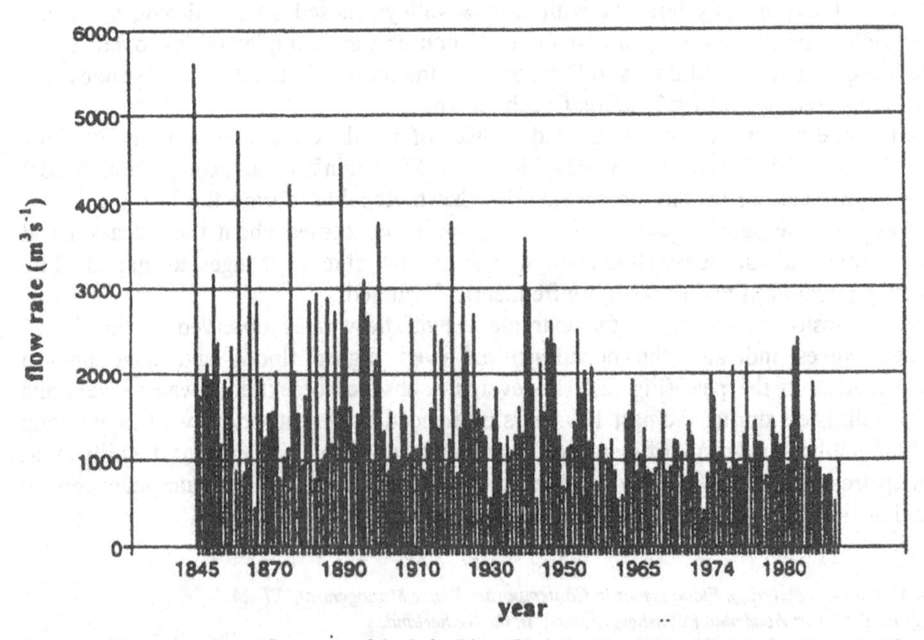

Figure 2. **Annual maximum flow rates of the Labe River (Ustin L., Czech Republic) during the period from 1845 to 1990 (produced by the VUV TGM, Praha).**

Therefore, the responsible bodies in water management, namely the Ministry of Agriculture and the Ministry of the Environment, in co-operation with the Ministry of Interior and the Ministry of Regional Development, have to initiate the appropriate measures and actions for the improvement of flood defense and flood protection, on the basis of analyses of the recent events.

The following paper gives the basic information on increasing activities in the flood protection with focus on preventive activities implementing both the structural as well as non-structural measures.

2. Activities of the state administration in water management

2.1. LEGISLATIVE TOOLS

The first action was the improvement of the Government Decree 27/1975 dealing with the flood defense (responsibilities, duties and activities of the state administration and municipalities) where the new structure and system of the state administration was also introduced after the political-economical change in 1989. However, the main improvement is expected in the new Water Act, which is under preparation and should be introduced by the year 2001. Duties of the property owners as well as the joint funding schemes and the roles of the municipalities together with the state administration, have to be clearly determined, including the role of the newly established regional structure of the country in accordance with the system of European Union countries. Special attention will be paid to the financing of water management, with the most important issue being prevention. The basic concept of the new Water Act was agreed upon recently. The preparation of the above legislative tools is conducted by the Ministry of Environment in co-operation with the Ministry of Agriculture.

2.2. STRATEGY OF FLOOD PROTECTION

On the basis of the evaluation of state administration activities during recent floods and experiences with the restoration of the territory hit by floods, the Strategy of the Flood Protection in the Czech Republic has to be developed by March 2000. This Strategy, being developed by the Ministry of Agriculture, will identify the priority catchments for the implementation of relevant flood protection activities, including the time schedules for their implementation. Simultaneously, program proposals of the preventive flood protection measures will be prepared in co-operation with the Ministry of Environment and the Ministry of Regional Development, which is responsible for land use planning.

2.3. WATER MANAGEMENT PLANS

In 1998, the introduction of the new system of water management planning began by undertaking five Pilot Plans. The flood protection measures and their relation to the relevant land use planning form one of the most important parts of the planning procedure. The Pilot Plans are related to the hydrological sub-catchments (an area size of 2-3 districts) and are located in the watersheds exhibiting various characteristics of the hydrological regime, socio-economic situation, etc. The main goal of the Pilot Plans is to develop and prove the methodology for introducing the new concept of water management master plans, which will originate from a broader participation of stakeholders (i.e. a "bottom-up" process, instead of the "top-down" approach typical for planning before 1989).

3. International projects

Some of the European Union countries offered special help to the Czech Republic, in addition to the direct support during the floods, in the form of the transfer of expertise in the improvement and innovation in flood defense. Recently, two such joint projects have started in the Czech Republic and their outputs will be used in the above mentioned state administration activities.

3.1. FLOOD MANAGEMENT IN THE CZECH REPUBLIC

This project, funded by the Danish Environmental Agency (DEPA, 4 mil. DKK), has started under the above title in 1998. The main goal is the implementation of the simulation model MIKE 11 in the Morava River catchment to help decision makers in the selection of preventive flood protection measures. The main partners are the Danish Hydraulic Institute (Horsholm, Denmark) and the Hydroinform Co. (Prague), together with the Morava River Board. A very important part of the project is the transfer of skills in using this software to other River Boards and also to several other important water management agencies of the Czech Republic. The project is running well with the data supplied from a national project [1]. Several information exchange workshops and study stays have been held in both countries. Some of the recent results are also presented in this volume [3] and confirm a successful implementation of the MIKE 11 model. The first comparison of four different scenarios of possible flood protection measures in the concerned watershed has been submitted for testing. There is no doubt that the project results will facilitate the comparison of the effectiveness of different protection measures. This information together with an economic evaluation will serve to reach optimal decisions on the type of flood protection measures. The project will be finished by the end of 1999.

3.2. FLOOD DAMAGE RESTORATION – FLOOD DEFENSE IN THE CZECH REPUBLIC

This project, which is funded by the Netherlands government (about 1 million NLG) in the framework of the PSO Program (i.e. program for the support of the Central and Eastern European countries), started in January 1999.

It deals with three topics:

- transfer of know-how in the water management planning in the Netherlands to the Czech Republic (public participation, co-ordination of users, structure and financing etc.)
- implementation of a runoff model in the chosen Czech watershed

- supply and installation of the advanced ultrasonic flowmeters in the chosen Czech watershed for increasing the prediction possibilities.

The main partners are the ARCADIS Heidemij Advies together with Delft Hydraulics on the Dutch side, and the Czech Hydrometeorological Institute, the Labe River Board and the Ministry of Agriculture on the Czech side. The first part of the project has already been finished and 25 Czech water managers attended a one-week course on water management planning with a special focus on flood defense in the Netherlands. The information produced in this project is used in the Pilot Plans mentioned earlier.

The implementation of the runoff model started in the upper part of the Orlice River watershed (where the severe flood occurred in 1998) and the installation of the first flowmeter (Fa. Instromet) will be finished in September 1999. Thus, the project will support both the transfer of know-how (licensed software) and the improvement of the flow rate monitoring network in the chosen watershed. The project will be finished by the end of 1999.

4. National programs and projects for the improvement of flood defense

4.1. IDENTIFICATION OF FLOOD PRONE AREAS ALONG IMPORTANT WATER COURSES

The program was established by the decision of the Czech government in 1999 and the main aim is to identify the flood prone areas along the main Czech watercourses. The current situation is not satisfactory, because so far, flood flows and flood prone areas have been determined for only one third of the length of the main rivers (important water management watercourses). This determination will start in those parts of watersheds, which possess the highest danger of flood occurrence.

4.2. STUDIES OF FLOW CONDITIONS IN IMPORTANT WATER COURSES

The main aim of these Studies is to summarize and evaluate various available and reliable proposals of flood protection measures under consideration and their integration into individual watercourse catchments.

The studies will address the following:

- assessment of recent flood protection in individual catchments
- compilation of various proposals of flood protection measures (structural and non-structural)
- detailing and evaluation of combined alternatives of flood protection measures with respect to their efficiency and feasibility
- proposals of complex (integrated) systems of flood protection for given watercourses and their catchments.

The Studies will be incorporated into the Water Management Plans and, therefore, they will be included in land use planning of the concerned territory (through co-operation with the Ministry of Regional Development). This program is closely related to the determination of flood prone areas and is to be executed over a period of more than 5 years, because of high costs of the data required (a digital model of the terrain might be commonly needed).

4.3 IMPLEMENTATION OF FLOOD PROTECTION MEASURES

This program started in 1998 and covers the realization of the most urgent actions for enhancing flood protection measures. It supports the activity of the bodies responsible for the management of water courses (River Boards, State Reclamation Agency, Forests of the Czech Republic and also municipalities) which also co-sponsor this program.

The first phase of the program is to be completed by 2002 and the following phases will focus on the realization of findings from the Studies of flow conditions in the important watercourses mentioned earlier.

The program contains mainly structural (technical) measures (e.g. enlargement of embankments, construction of embankments and platforms, reconstruction of embankments, land drainage modification, etc.) which have been shown as the most urgent and also suitable for fast implementation.

5. National Research Programs

5.1. RESEARCH AND DEVELOPMENT OF FORECASTING MODELS FOR THE IMPROVEMENT OF FLOOD DEFENCE

This project had been proposed by the Ministry of the Environment and approved by the Government Board for the Research a Technological Development. It started in 1999 and was planned for 4 years. The main coordinator is the Czech Hydrometeorological Institute (Prague). The main goal is the development and pilot application of hydrological modeling techniques suitable for flood warning and forecast of river stages and flow rates at chosen sites in real time. A special attention will be paid to snowmelt, which is very important on the territory of the Czech Republic. The developed and calibrated models will be ready for a limited number of watercourses; for the rest, general guidelines for the estimation of watershed response to the measured precipitation will be used to produce the expected runoff. For some important watersheds, an information system will be proposed and will include the expected consequences of the forecasted flood situation.

5.2. EXTREME HYDROLOGICAL CONDITIONS IN CATCHMENTS

This 3-year project started in 1999, and is funded by the Grant Agency of the Czech Academy of Sciences. It mainly focuses on the analysis of interactions within the system atmosphere-watershed-watercourse-affected flood areas along the streams, and their use for an integrated flood defense. The major topics addressed include the following:

- Methods and models of hydrological forecasts in relation to extreme flow conditions
- Functions and possibilities of water courses, flow regulation structures and water management systems for the management of flood conditions
- Functions and possibilities of landscape for the management of extreme hydrological conditions.

The Technical University Prague is the coordinator of this project.

5.3. DEVELOPMENT OF METHODS FOR ESTIMATION OF EXTREME FLOODS

This project was initiated by the ICOLD (International Committee on Large Dams) and started under the coordination of the Czech Hydrometeorological Institute in 1997. The main expected output is the assessment of the safety of large Czech dams under extreme floods (probability: 1:10 000 years).

6. Activities of River Board Companies

6.1. OVERVIEW AND PLANNING OF FLOOD PROTECTION PROPOSALS IN THE FIVE CATCHMENTS

Inventories of the current flood protection were produced by the River Boards for the five main Czech watersheds, within a year after the catastrophic 1997 flood. These reports were prepared in co-operation with other bodies responsible for the management of waters in the Czech Republic and also included references to small watercourses and forest streams. The description of the recent situation in flood protection is accompanied by proposals of the most urgent and effective actions/measures. These activities were submitted for detailed and careful discussions by the municipalities located on or near floodplains of the respective watercourses. Simultaneously, some of the municipalities (and also ecological associations) proposed various improvements in flood protection along the watercourses, namely in the watersheds affected by recent floods. Thus, the program of Studies of flow conditions in the important watercourses, mentioned above, followed from these River Board materials.

7. Summary

The overview of various activities concerning flood defence and flood protection improvements in the Czech Republic is given. Preparation of the new water management legislation and the national strategy of flood protection are essential activities of the state administration. These steps have to be accompanied by the determination/identification of the flood areas along two thirds of the total length of the important Czech watercourses and by the development of an integrated system of flood protection measures in individual catchments.

8. References

1. Hladný, J., ed., (1998) Vyhodnocení povodňové situace v červenci 1997 (Evaluation of the floods in July1997 – In Czech) Souhrnná zpráva projektu MŽP ČR (Praha) 163 p.
2. Kubát, J. (1999) 1997/98 Floods in the Czech Republic. (*This volume*)
3. Zeman,E. and Biza P. (1999) Modeling Tools – the Key Element for Evaluation of Impacts of Anthropogenic Activities on Flood Generation, with Reference to the 1997 Morava River Flood. (*This volume*).

1997/1998 FLOODS IN THE CZECH REPUBLIC: HYDROLOGICAL EVALUATION

J. KUBÁT

Czech Hydrometeorological Institute
Na Šabatce 17, 143 06 Praha 4, Czech Republic

1. Hydrological aspects of floods in the Czech Republic

The Czech Republic is located in the temperate zone with regular cycles of air temperatures and precipitation. This long term variation cycle is also perturbed by short term fluctuations, which are caused by passing atmospheric fronts, separating warm and cold atmospheric masses, and producing precipitation. The precipitation pattern is rather continental, with maximum monthly totals occurring from May to August, and the minima in February and March. Frequent storms with intensive rains of relatively small areal extensions occur during the summer months. Orographic effects are important for both the precipitation formation and its areal patterns.

The snow cover usually lasts from mid-December to mid-April, but in mountains, it often stays until May. Snowmelt events are not very regular; those causing floods may appear from December until April.

The floods in the Czech Republic are generated almost exclusively on the Czech territory, with a few exceptions in the catchments of the rivers Ohře and Dyje.

Large floods are the most severe and frequent natural disasters in the Czech Republic, causing immense damages and a large number of deaths. According to the historical hydrological records, which have been kept for the River Elbe since 1845, the frequency of floods was rather high during the second half of the last century. After this period a decline in the occurrence of large scale floods was observed and in the second half of the twentieth century the occurrence of such floods was exceptionally low. This trend changed recently, when a few very high floods occurred, including the flood of 1997, which by its areal extent and damages was the largest flood of the 20th century.

2. July 1997 Floods in the basins of the rivers Morava, Odra and Upper Labe (Elbe)

The floods of July 1997 were generated by two heavy regional rainfalls of long duration over Eastern Bohemia and Northern Silesia and Moravia. These rainfalls caused flooding in all rivers in these regions. The flood waves proceeded downstream in both rivers Morava and Odra and resulted in the inundation of an area of 1250 km^2.

J. Marsalek et al. (eds.), Flood Issues in Contemporary Water Management, 25–39.

The resulting damages were severe [1]: 50 deaths, 5400 damaged or destroyed houses, evacuation and emergency housing of 10,000 persons, partial or total destruction of 1850 km of roads and 950 km of railtracks, destruction or heavy damage of 48 road and 26 railroad bridges, flooding of several hundreds of industrial and agricultural facilities, and damages of structures along 800 km of rivers. The total cost of flood damages is evaluated at 62.6 billion Czech korunas (CZK).

2.1. METEOROLOGICAL ORIGINS

This flood was caused by two periods of heavy rains from July 4 to 21, 1997, resulting from a low pressure system proceeding from the Atlantic to South-East, which became stationary over Central Europe with an influx of warm and humid air. The mountains in the north of the country lifted this flow of air and thus accentuated the rain field formation.

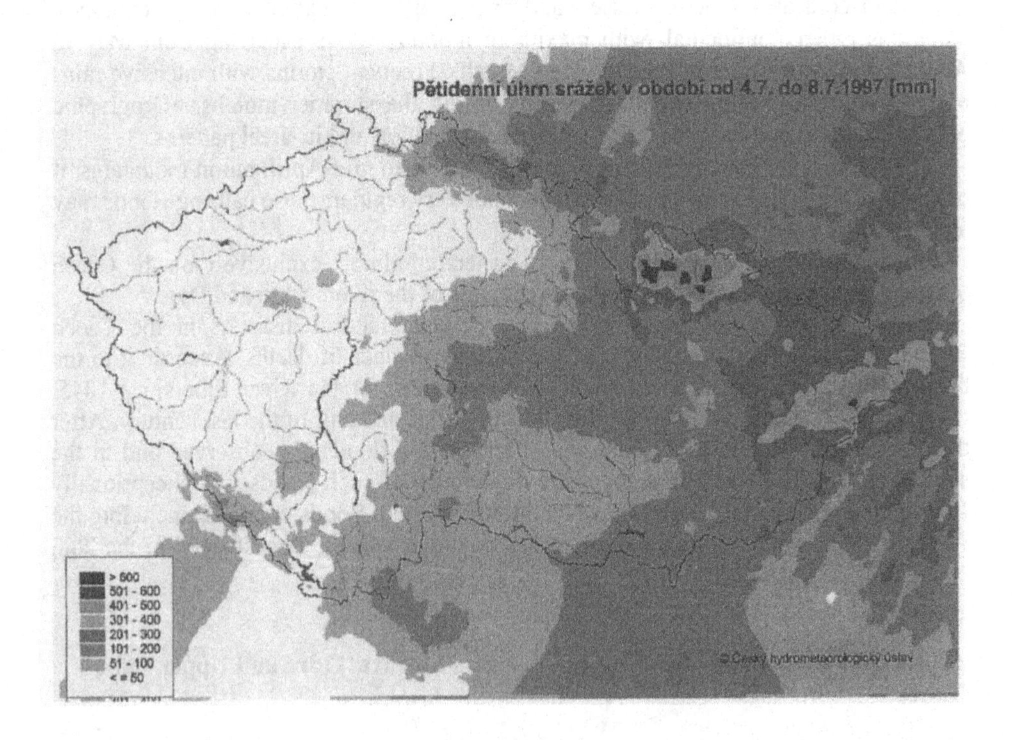

Figure 1. Five-day rainfalls [mm] from July 4 to 8, 1997.

From the synoptic analysis it appeared that this "flood causing" cyclonic pattern over Southern Poland, lasting five days, is rather unusual, the first in this century. It was also concluded that the saturation of the air by water was exceptionally high. The combination of these two phenomena caused the extreme rainfall. The first episode resulted in 100 to 200 mm of rainfall over the North-Eastern part of the country, in

North Moravia even more than 300 mm. The maximum five-day rainfalls in the Moravian mountains were over 500 mm (Šance 617 mm, Lysá Mountain 586 mm, the Jeseník Mountains 512, Praděd 455 mm). Such values are most exceptional for these locations.

The second episode, with 50 to 100 mm of rain, covered East Bohemia and all of Moravia. The maximum was in the Krkonoše Mountains, over 200 mm.

2.2. EVALUATION OF THE FREQUENCY OF THE OBSERVED PRECIPITATION

Compared to the period 1961-90, the monthly rainfall totals in July 1997 in mountainous regions were greater than 50% of the annual average, and in some cases, even more than 80%. In particular the first episode was very extreme. Using the longest available precipitation record (since 1890), the return periods of such rainfalls were from 200 to 500 years in the catchment of the River Odra, 500 to 1000 years in the upper part of the Morava river catchment and 200 to 500 years on tributaries of the upper Morava.

2.3. HYDROLOGICAL EVALUATION

The discharge at many river gauging stations was the highest on record. In agreement with the rainfall pattern, two flood waves occurred, with the first being the most severe. The computed APIs (Antecedent Precipitation Indices) in the areas of concern ranged from about 20 to 60 mm. The highest ones were noted in the Beskydy and Jeseníky Mountains and in the Orlice river catchment. Thus it was concluded that even before the first flood wave, the catchment storages were quite saturated.

The evaluation of flood peak flows was difficult. In most cases, they were above all the data within the span of rating curves. Typically, flood waters spilled over floodplains outside of the measuring cross-sections and the river beds also changed in most cases, and some gauges were damaged or destroyed. However, 85 complete and about 40 auxiliary discharge measurements were completed. After the flood, the gauging cross-sections were newly surveyed and the discharge values were recalculated using hydraulic methods. The calculated results were cross-checked and verified in consultations with the River Basin Authorities. Some pertinent data are presented in Table 1.

TABLE 1. Evaluation of the flood peak flows at selected sites

River Catchment	Gauging station	Catchment area (km^2)	Date and time	Max. Q (m^3/s)	Unit q (m^3/km^2)	Runoff coefficient
Labe	Špindler. Mlýn	52.92	19.7. 08.45	155	2.93	0.84
Labe	Debrné	476.80	19.7. 17.00	180	0.38	0.46
Úpa	Horní Maršov	81.63	7.7. 14.00	78.4	0.96	0.65
Třebovka	Hylváty	174.23	8.7. 04.15	66.2	0.38	0.51
Tichá Orlice	Malá Čermná	689.96	9.7. 04.00	251	0.36	0.54
Orlice	Týniště n.O.	1590.75	8.7. 18.45	497	0.31	0.41
Odra	Svinov	1615.12	8.7. 12.00	688	0.43	0.50
Odra	Bohumín	4662.33	8.7. 14.00	2160	0.46	0.61
Ostravice	Ostrava	822.74	9.7. 05.00	898	1.09	0.85
Opava	Krnov	370.50	7.7. 14.00	375	1.01	0.64
Opava	Opava	929.65	7.7. 16.00	647	0.70	0.61
Opava	Děhylov	2039.11	8.7. 00.00	744	0.36	0.47
Opa vice	Krnov	175.98	7.7. 10.00	175	0.99	0.50
Bělá	Mikulovice	222.24	7.7. 06.00	335	1.51	0.83
Olše	Věřňovice	1068.00	9.7. 05.45	673	0.63	0.67
Desná	Šumperk	241.16	8.7. 05.45	191	0.79	0.58
Morava	Moravičany	1558.82	8.7. 15.30	625	0.40	0.52
Morava	Olomouc	3322.07	9.7. 19.00	760	0.23	0.47
Vset.Bečva	Vsetín	505.78	7.7. 05.00	302	0.60	0.58
Rožn. Bečva	Krásno	253.32	7.7. 01.00	489	1.93	0.87
Bečva	Dluhonice	1598.79	8.7. 00.45	838	0.52	0.56
Morava	Kroměříž	7014.44	10.7. 10.00	1034	0.15	0.45
Morava	Strážnice	9.146.92	14.7. 05.00	901	0.10	0.44
Dřevnice	Zlín	311.99	7.7. 04.00	282	0.90	0.63
Olšava	Uherský Brod	401.23	8.7. 09.00	140	0.35	0.30
Svitava	Letovice	419.31	8.7. 08.00	96.9	0.23	0.30

Table 1 lists the peak flows of the first wave, except for the upper Elbe, where the second wave peak was larger. The unit discharges indicate the usual decrease with the increasing contributing area. In regions with higher mountains (Krkonoše, Jesníky and Beskydy Mountains) the unit discharge reached values as high as 3 m^3/km^2, while in the lowland reaches, q values were much smaller, about 0.1 m^3/km^2. The coefficients of runoff were extremely high, in some cases over 0.8, indicating that during these rainfall events the retention capacity of the catchments was entirely filled.

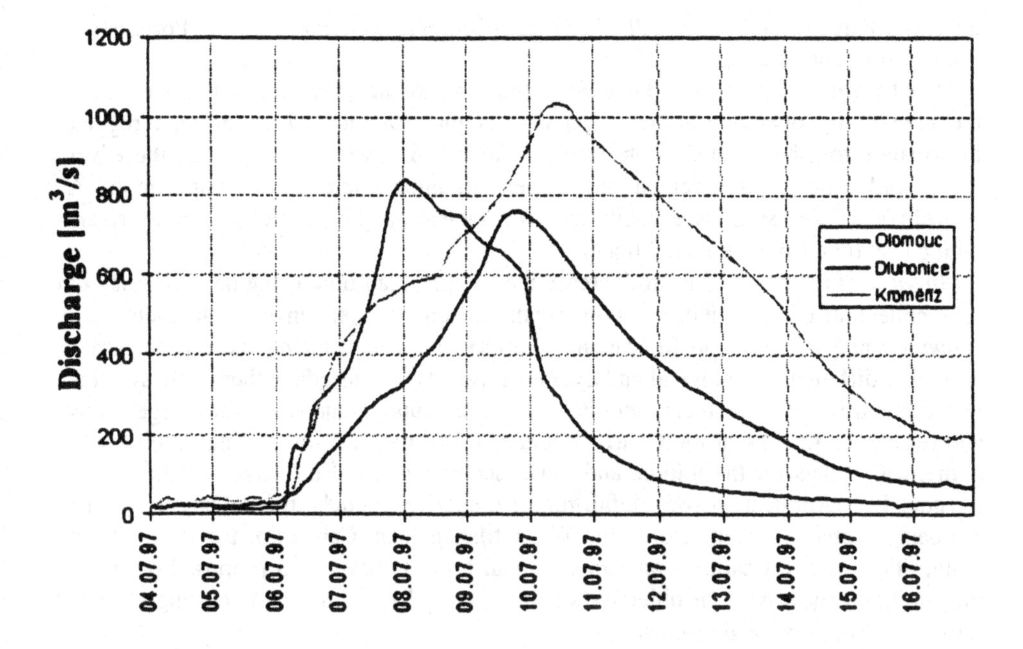

Figure 2. The hydrographs of the flood waves on rivers Morava and Bečva.

The flood waves were modified and retained by 24 larger reservoirs with a total retention capacity of 124.2 10^6 m^3. Some of them fulfilled the retention task they were designed for and reduced substantially the peaks. These reservoirs were emptied before the arrival of the flood wave. Others did not play such an important role by retention of the flood volume, but delayed the flow peaks so as not to coincide with the peaks from catchments without reservoirs.

In the evaluation of the 1997 flood peaks, the corresponding return periods were calculated. These periods decreased in the downstream direction at all concerned rivers. Thus, for the Opava River the return period of the maximum peak at Krnov was 700 years, at Opava 500 yr and Děhylov 300 yr. For the Morava River at Raškov, it was 800 yr, at Olomouc 500 yr, at Kroměříž 300 yr and at Strážnice only 100 yr. For the rivers Odra, Bečva and those in Eastern Bohemia, the return periods were not more than 200 yr. The standard $Q_{100\ yr}$ flow was recomputed for the whole region taking into consideration the new data including the July 1997 flood, and it was noted that it was necessary to correct the Q_{100} for several upper reaches of the Morava and Bečva rivers.

3. The perspectives for flood forecasting in the Czech Republic

The flood forecasting and warning service in the Czech Republic has been entrusted by law (No.458/1992) to the Czech Hydrometeorological Institute (CHMU) in cooperati;on with the River Basin Authorities. The flood warnings are issued in cooperation with several other institutions in accordance with the flood regulations, plans and instructions of the bodies in charge of emergencies. The CHMU established a Central Forecasting Office (CPP) in Prague-Komořany and six Regional Forecasting Offices at its subsidiaries.

The hydrological service of the CHMU monitors the actual situation on the rivers in the country by operating some 150 river gauging stations, which provide regular information together with the data from the Water Management Centres of the River Basin Authorities on flow regulation in reservoirs impacting on flood propagation. A hydrological forecast is issued daily for 18 principal gauging stations on main rivers with a lead time from 6 up to 24 hours.

In view of the possibilities to increase the forecast lead time using new methods of data collection of precipitation, snow depth and cover extent in the catchment, and snowmelt and precipitation forecasting, the CHMU is introducing these procedures, including different hydrological and hydraulic models. To introduce these new systems in the hundreds of catchments monitored in the country requires a very large effort and resources. At present the CHMU operates a hydrological forecasting system with numerical models for the middle and lower sections of the Elbe River on the Czech territory [2]. Applications of rainfall-runoff numerical models in smaller catchments are being presently tested also by the Water Management Centres of the River Basin Authorities. Several national and international projects have been proposed or are in their initial stages, with an objective of introducing hydrological forecasting models into the daily operational routine.

TABLE 2. Routine forecasting profiles of the CHMU

River	Forecasting location	Forecast lead time (hours)	Issuing office
Labe	Jaroměř	3	RPP Hradec Králové
Metuje	Jaroměř	3	RPP Hradec Králové
Labe	Přelouč	14	RPP Hradec Králové
Jizera	Bakov n.J	6	CPP Praha
Labe	Brandýs n.L	12	CPP Praha
Vltava	Orlík (inflow)	7	RPP České Budějovice
Berounka	Beroun	12	RPP Plzeň
Vltava	Praha	6	CPP Praha
Labe	Mělník	12	CPP Praha
Ohře	Karlovy Vary	15	RPP Ústí n.Labem
Ohře	Nechranice (inflow)	22	RPP Ústí n.Labem
Ohře	Louny	8	RPP Ústí n.Labem
Labe	Ústí n.L.	24	CPP Praha
Labe	Děčín	24	CPP Praha
Odra	Bohumín	6	RPP Ostrava
Morava	Olomouc	12	RPP Ostrava
Bečva	Dluhonice	12	RPP Ostrava
Morava	Strážnice	24	RPP Brno

3.1. EVALUATION OF THE QUANTITATIVE PRECIPITATION FORECASTING

In order to gain longer lead times for hydrological forecasts in the Czech Republic conditions, it is very important, if not indispensable, to have a good Quantitative Precipitation Forecast (QPF). The hydrological forecasting can then produce reliable, long-term forecasts and at least make a good guess of the implications for flooding [3].

However, the forecasting of rain fields in time and space is one of the most difficult problems of the meteorological forecasting. The forecasters at the CHMU use four numerical meteorological models for forecasting rainfall quantity twice a day.

TABLE 3. Numerical meteorological models that produce QPF used by forecasters of the CHMU

Name of Model	Country of origin	Surface grid size (km)	Time step (hours)	Lead time (hours)
EUROPA Model	Germany	50	6, 12, 24	78
DEUTSCHLAND Model (DWD)	Germany	15	6	48
LAM Model	United Kingdom	50	3, 6	36
ALADIN Model	France Czech Republic	15	3, 6, 12, 24	48

The outputs of these models were compared for the 1997 flood with raingauges read at synoptic stations of CHMU [4]. From the comparison it appeared that the least number of errors was made by the DWD Model. The DWD was particularly more successful than ALADIN for lower elevations; for higher elevations and mountain locations the number of errors produced by DWD and ALADIN appeared similar. ALADIN underestimated the rainfall quantity more often than DWD. LAM Model also underestimated rainfall quantity almost systematically to the extent that experienced meteorological forecasters took this into consideration.

It appears that even with relatively short lead times, the errors in numerical model forecasts may be substantial for both areal distribution and rainfall quantity. The simultaneous use of several models permits to some extent to eliminate the errors of meteorological models and thus increase the reliability of the QPF.

3.2. EVALUATION OF THE ACCURACY OF HYDROLOGICAL FORECASTS

During the 1997 flood, a number of river gauges were damaged or malfunctioned and this reduced the amount of information on upstream flows. Thus, the reliability of forecasting at downstream stations was hampered. Despite the above difficulties, the accuracy of the forecasts was good to very good, except for the absence of forecasts for some of the peak segments of the flood waves.

It was noted in forecasts for the final profile of the Odra river on Czech territory in Bohumín that the average deviations of the measured stages from the forecasted ones were between 5 to 13 cm (except for the first wave peak). In terms of flows, the errors were less than 10 to 15%. Ninety percent of the forecasts were within the acceptable ±20% error band and more than half of them contained errors less than 5% of the discharge.

The quality of forecasts for the upper Morava river in Olomouc and lower Bečva river in Dluhonice was similar to that above, although not as accurate. In view of the lack of input data, a good number of forecasts were missing. An average deviation of most forecasts was less than 25 cm for stages, and less than 20% for flows. More than half of forecasted flows were within 5% of the measured flows. The average dispersion of errors was however larger than for other locations, particularly for the first flood wave.

Forecasts for the lower Morava River at Strážnice - the absence of reliable stage and flow data from upstream was the main reason for errors in these forecasts. Furthermore the rating curve for this location was proven unreliable and a large quantity of flood flow was outside of the bed or lost elsewhere because of embankment breaks. Under such circumstances, the lag time as an important element of hydrometric forecasting did not serve any purpose. Taking these points into consideration, a majority of forecasts at Strážnice were not bad. It is important not to consider an artificial peak caused by an embankment break as a forecast error which was the case here. In those circumstances, the evaluation of stage forecast errors cannot be connected with that of flow forecasts, as these two basic elements of water movement are no longer related. The stages were forecasted with an average error of 20 cm and the flows with a deviation of up to 17%. Seventy percent of forecast errors were below 5%.

It appeared that accurate and successful hydrological forecasts can be made for floods up to a 20-year flood. For the time being, satisfactory results cannot be achieved in forecasting extreme values of hydrological variables, while working with insufficient or inaccurate hydrological and meteorological inputs.

4. The flood of July 1998 on the Orlice River

This was a summer flood, caused by a heavy storm rainfall on a relatively small area of 500 km^2, which resulted in [5] 6 deaths, 40 totally destroyed houses, hundreds of flooded or damaged structures including 18 road bridges, tens of damaged crossings over smaller rivers or brooks, tens of industrial and agricultural structures flooded, and 60 km of damaged water courses. The total damage was estimated to be about 2 billion CZK.

4.1. METEOROLOGICAL SITUATION

On July 22nd 1998, a cold front was moving South-East across Bohemia. The front stopped in the evening, when it encountered very warm tropical air masses; behind the front, cooler oceanic air was flowing. The front resulted in a heavy storm rainfall activity. On the leeward side of the Orlice Mountains the ascending air developed additional heavy storm rainfall. The total daily rainfall measured at Deštná was nearly 204 mm and other gauges in the Czech Republic and in Poland read between 100 to 200 mm. This amount of rain is very rare; only two of these stations recorded similar rainfalls in this century, 101 mm at Deštná (in 1910) and 136 mm at Šerlich (1956). A larger rainfall total was recorded in Bohemia in 1897 (Nová Louka) – 345 mm/day.

4.2. HYDROLOGICAL EVALUATION

This was a typical "flash flood" on several water courses; the most severely hit were the right-hand tributaries of the Orlice River. The flood developed very fast. The time of rise in the upper catchment was only 2 hours. The celerity of the flood wave was 2-3.5 m/s. The narrow valleys were filled with water which eroded the valley slopes and the bed, and carried away all obstacles it encountered. The eroded material was

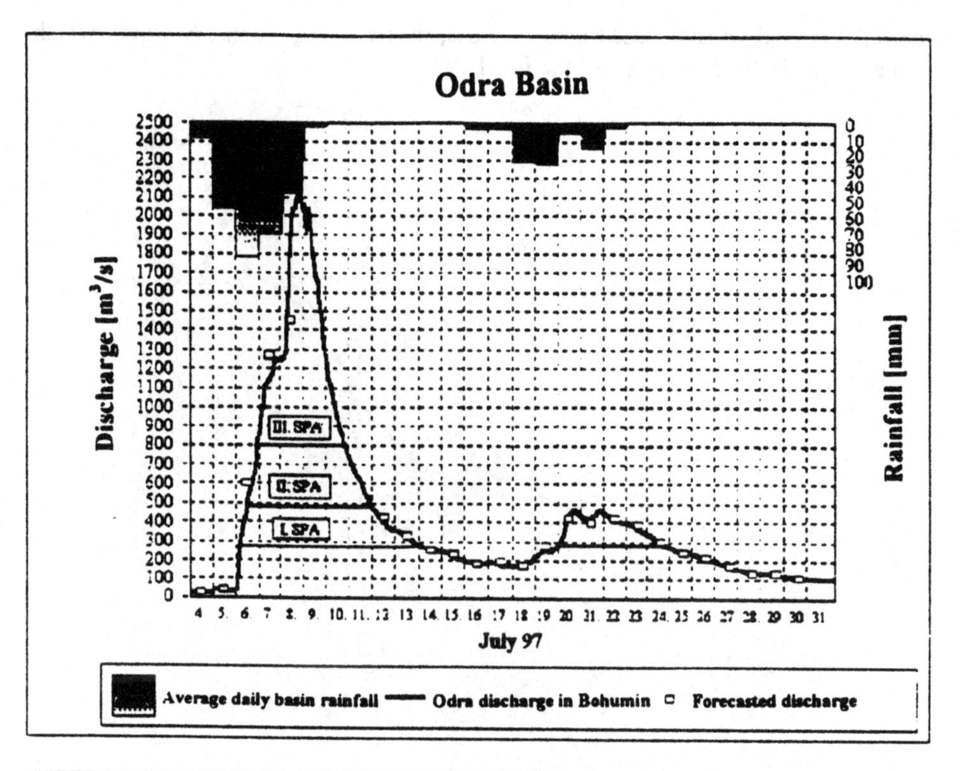

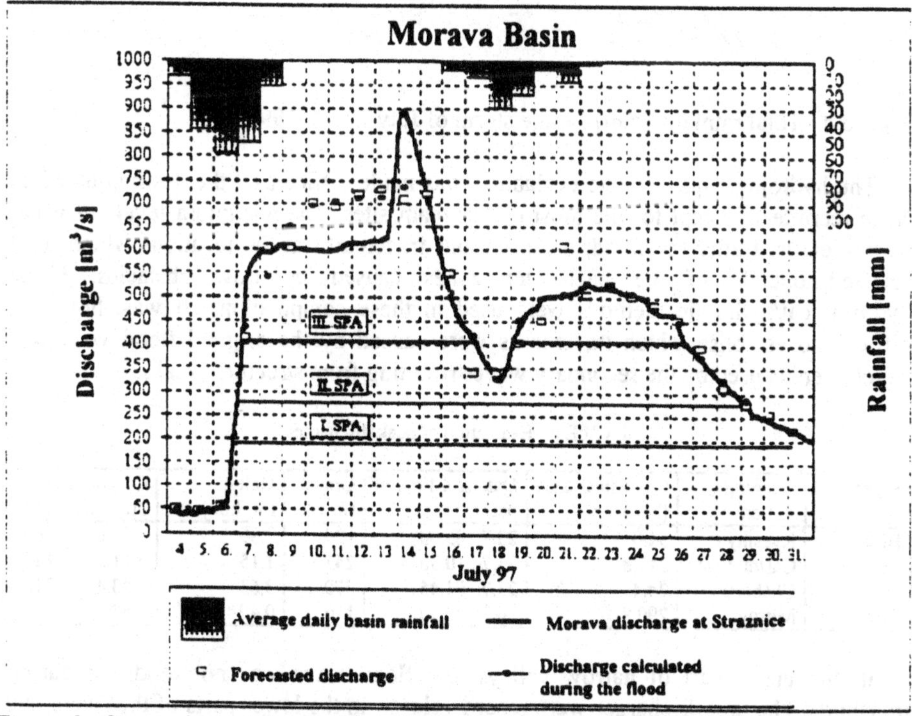

Figure 3. Comparison of forecasted and observed flows in the rivers Odra and Morava

deposited at bridges and caused their collapse, thus creating secondary waves which increased the destructive energy of the flood wave.

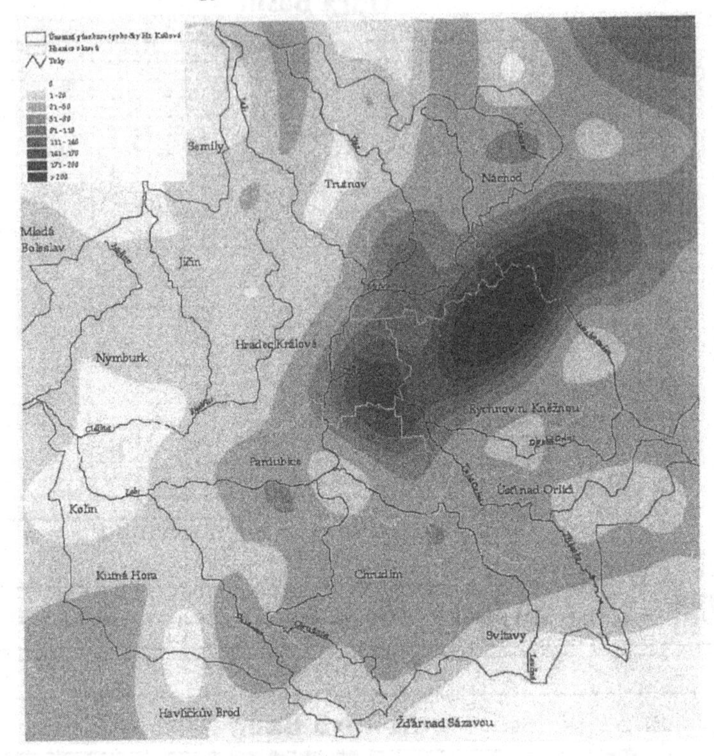

Figure 4. Total rainfalls [mm] for the storm of July 22-23, 1998.

The hydrologic quantitative evaluation was made difficult by the destruction of all stage recorders, except for the most downstream sites. The maximum levels of water were measured after the flood, using various tracemarks at the locations which were selected outside of the backwater zones caused by flow obstacles. Usual hydrological/hydraulic methods were used in these computations. It was, however, difficult to establish which tracemarks were caused by the original flood wave and which were caused by the secondary waves resulting from structure breaks.

TABLE 4. Evaluation of peak discharges

River	Gauging station	Catchment area (km^2)	Date and time of peak	Qmax (m^3/s)	Unit q (m^3/km^2)	$Q_{100year}$ (m^3/s) previous	new
Bělá	Kvasiny	54.0	23.7. 05.00	129	2.39	61.0	76.8
Bělá	Častolovice	213.8	23.7. 10.30	247	1.15	114	120
Dědina	Chábory	74.4	23.7. 03.45	270	3.63	43.4	71.4
Dědina	Mitrov	290.8	24.7. 06.30	116	0.40	68.5	100

In the upper part of narrow valleys, the flood wave was not modified during transport. The unit discharges were record values. In the lower parts of the rivers, the spread of the water over floodplains caused significant reductions of flood peaks. The frequency analysis of the flood peaks was not performed, as the period of observations on smaller streams was rather short, about 50 years, and the values of discharges contained large computational errors. Yet the computed discharges were several times larger than those obtained previously on these rivers for hydrologic design purposes by analogous methods.

5. Perspectives of forecast in flash floods situations

The forecasting of flash floods, caused by local storm rainfalls, seems rather unreliable. The meteorological weather models used in precipitation forecasting are mostly applicable to frontal rainfall, but cannot correctly forecast the timing and location of the storms with a precision sufficient for hydrological purposes. The situation of July 22, 1998 is a good example of this state of affairs. The weather forecast models were unsuccessful in this case. The weather model of the German Meteorological Service forecasted the precipitation in this area to be about 30 mm. The ALADIN model operated by the Czech meteorologists predicted hardly any rain in the area for this night.

On the basis of analysis of the actual synoptic situation, the meteorological forecaster on duty is usually able to formulate only general comments and/or warnings with respect to heavy storm rainfall, without indications of the rainfall quantity or its precise spatial extent. Thus, the weather report for July 27, 1998 of the Czech Hydrometeorological Institute included at the end the following: “Advisory: today and tomorrow storms may produce heavy rains and possibly sparse hail.”

There is also a problem with recording the rainfall over small areas. The raingauge network of the CHMU includes about 800 stations, which represents one station per less than 100 km^2, on average. However the number of stations which provide information in real (or quasi-real) time to the forecasters on duty and can be used for issuing flood warnings is limited. The Institute is gradually installing precipitation gauges with automatic data transmission and negotiates with the River Basin Authorities a mutual exchange of real-time data. It is expected that in the future, the number of automatic real-time reporting stations may reach 300, e.g. one station per 250 km^2. This density may prove sufficient for monitoring regional rainfalls. However, localised rainfalls and, in particular their core intensities, will probably be never monitored by a surface observation station network accurately enough for hydrological forecasting.

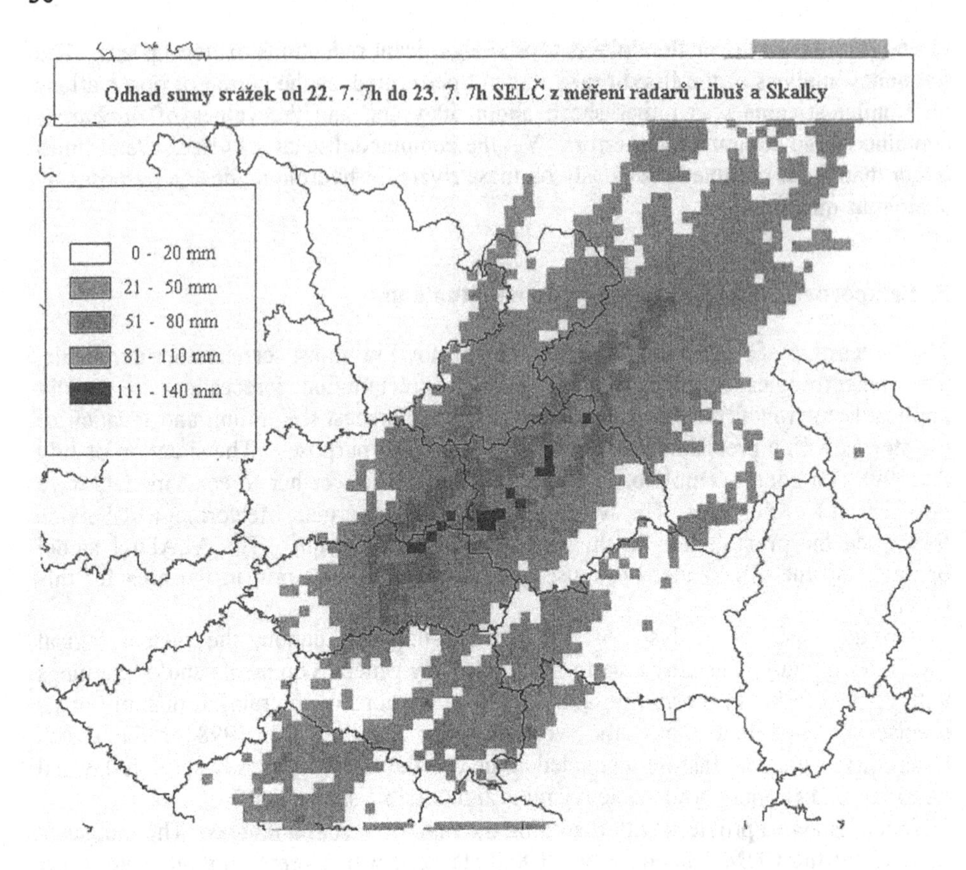

Figure 5. 12-hour rainfalls [mm] indicated by the meteorological radar during the night of July 22, 1998.

As done in many other countries, the Czech Hydrometeorological Institute is increasing the exploitation of data provided by meteorological radars. Two such radars are operated by the Institute, one in Prague and the second one in central Moravia. Together with the data available from radars in the neighbouring countries, the Czech territory is well covered and the data are available at 15 minutes intervals. Thus the rainfall field of the night of July 22, 1998 was monitored by this observational instrument.

While the radar in most cases monitors the areal extent and dynamics of rainfall fairly well for the needs of meteorological analysis of the atmospheric processes, the quantitative information on the rainfall depth at the ground surface is not always reliable and can be in error by several orders of magnitude.

Advanced hydrometeorological services, such as the National Weather Service (NOAA) in the USA and others, provide an hourly evaluation of the data received from the ground surface network, meteorological radars, satellites and numerical models in the form of the so-called composite rainfall field, which is presented as grid maps in digital mode. The Czech Hydrometeorological Institute intends to explore this development in order to provide more reliable input data for forecasting in general and for the flood forecasting and warning purposes in particular.

Another direction which is being explored in the Czech Republic is the establishment of local warning systems for small catchments to be operated by local authorities, either on a county or city basis. These systems known in the USA and other countries under the acronym ALERT include automatic observation sensors, transmission and simple warning computerised devices, which are located at local emergency centres to alert the authorities and the general population in the case of flood danger. Thus, local warning systems can appropriately complement the efforts of country-wide organisations such as the Hydrometeorological Services to provide disaster prevention assistance.

6. Conclusions with respect to flood forecasting services

The floods of 1997 and 1998 thoroughly tested the level of flood prevention and the system of flood emergencies in the Czech Republic. This system functioned well in some places, and not so well, or not at all at other places. All agencies involved in this system, including the CHMU, have been analysing their performance with the intention of improving it.

6.1. INFORMATION ON RAINFALL AND ITS FORECAST (QPF)

- The present network of raingauges is sufficient for the monitoring of regional precipitation amounts. It is necessary to accelerate the automatic collection of data from real-time reporting stations and to add stations in the mountains.
- Local heavy rains can be monitored only by radars. The new radar in Moravia will be supplemented by a new radar in Prague. A system of radar data calibration by telemetering rainfall stations has to be yet introduced.
- As indicated above, the QPF is provided on the basis of information from several European numerical weather models and the ALADIN model operated in Prague. It is up to the meteorological forecaster on duty to decide which model data to use for the QPF forecast.

6.2. FLOW AND FLOOD FORECAST DATA

- The present flow data collection covers the main rivers and will need to be supplemented in some regions. The flow measuring cross-sections need to be stabilised, so that they remain functional during floods. It is necessary to adjust the rating curves for high flows and complicated flow conditions with the help of hydraulic models.
- Automatic collection of data from flow measuring stations is implemented by different telecommunication means linking them to the Forecasting Office of CHMU and Water Management Centres of the River Basin Authorities. These systems will need to be interconnected, supplemented and in some places backed up

so that important information arrives even in extreme situations. Eventually, an integrated system should be installed for all important sources of flood data.

- Methods of flood forecasting from ground measured precipitation have been already introduced in the forecasting practice and models were tested on various catchments. The problem is to use these models countrywide and to operate them under the conditions of the forecasting services. It is necessary to select the models to be used and calibrate them for the different catchments so that they can be used routinely.
- To increase the lead time of hydrological forecasts, which in the Czech Republic is limited by short times to peak on short rivers, the models should use QPF as input and forecasted temperatures as well as other reliable data pertinent to snowmelt.
- Hydrological and water management interpretation of meteorological warnings derived from the fallen or forecasted precipitation depend at present on the professional experience of the forecasters rather than on an objective procedure. The use of hydrological models will not be possible in all catchments even in the distant future particularly on small rivers subject to flash floods. Therefore, it is necessary to classify the flooding vulnerability of individual areas for various meteorological events and use this information in issuing flood warnings.

6.3. DISSEMINATION AND USE OF FLOOD FORECASTS

- The Central Forecasting office of the CHMU is able to disseminate information on floods to the emergency flood prevention institutions and other users at the national and regional levels (including the River Basin Authorities); the Regional Forecasting Offices can do the same at the county level and for large cities. Problems remain with reaching the final user, and at the decision making and execution level where the information is immediately used. To improve this situation, it is necessary to examine the whole system of warning transmission, by involving local authorities and coordinate it with the flood regulations and emergency plans. Some information can be disseminated by public media (television and radio broadcasting).

- The system of classification of flood danger and the associated emergency activity is convenient, because of its simplicity and universality at all levels of flood prevention services. After each larger flood; however, it is necessary to review the degrees of flood prevention activities at all locations for which warnings are issued and relate them to the activities to be executed according to the flood prevention plans. It is also necessary to update at regular intervals the "Regulations for flood warning service at regular intervals".

- The legal documents establishing the forecasting and warning services should be reviewed together with the forthcoming revision of the Water Law. Simultaneously it is necessary to review and streamline the structure of flood prevention authorities, because of the preparation of a new law on Crisis Management and the Integrated Emergency System.

The experience from several countries proves that by early warning and operation of well functioning systems of emergency preparedness, it is possible to substantially reduce flood damages and eliminate or limit the loss of lives during floods. The immense amounts of damage caused by the recent floods clearly support the need of priority of all flood prevention steps even under the present financial constraints.

7. References

1. Hladný, J., Blažek, V., Dvořák, V., Kubát, J., and Švihla, V. (1998) Vyhodnocení povodňové situace v červenci 1997 - Souhrnná zpráva projektu. (Evaluation of flooding in July 1997 - Final project report). ed. CHMU. Praha.
2. Řičicová, P., Zezulák, J. 1996. Využití modelu Aqualog v předpovědní praxi (The use of Aqualog model in forecasting practice). Hydrologické dny, Stará Lesná.
3. Kubát, J. 1997. Předpovídání povodní v České republice (Flood forecasting in the Czech Republic). Planeta, 11/1997.
4. Kubát, J., Wolek, M. 1997. Činnost Českého hydrometeorologického ústavu za povodňových situací (Activities of Czech Hydrometeorological Institute during floods). Meteorologické zprávy, 6/1997.
5. Pozler, R., Šiftař, Z. et.al. (1998) Zpráva o povodňové situaci v červenci 1998 v povodí Orlice (Report on floods in July 1998 in the Orlice river catchment). ed. CHMU. Hradec Králové.

REGIONAL ESTIMATION OF DESIGN FLOOD DISCHARGES FOR RIVER RESTORATION IN MOUNTAINOUS BASIS OF NORTHERN SLOVAKIA

S. KOHNOVÁ and J. SZOLGAY

Dept. of Land and Water Resources Management
Slovak University of Technology, Radlinského 11, 813 68 Bratislava, Slovak Republic

1. Introduction

The estimation of design floods in ungauged small and medium basins in Slovakia was often performed by simple regional flood formulae derived for various geographical regions. Envelope curves were also applied in order to get an acceptable regional safety factor for design purposes. Usually the 100-year flood was estimated from the basin area; correction factors were used to account for deviations from the average of local runoff forming conditions in the region. Floods with shorter return periods were computed by regional frequency factors. Comparisons of design floods derived from such formulae with at site values in 260 small and mid-sized basins [1,2] showed a case-sensitive, highly variable safety factor in these schemes. They are not generally applicable for river training and restoration since they tend to overestimate the design discharges in almost all basins of a particular region. For these tasks other regional approaches, which do not include the regional safety introduced by the envelope curve concept, but allow for site specific over- and under-estimation resulting from the regional average concept, seem to be more appropriate. Some of the recently introduced concepts of regional homogeneity as e.g. in [3,4,5,6,7] were suggested to be used together with regional flood formulae composed of catchment characteristics other than the basin area only. To eliminate the heterogeneity in the annual flood data series due to the diverse genetic origin of flood events, it was proposed to consider annual maximum peak discharges separately for two seasons - the summer season with rainfall-induced discharges and the winter season with floods originating from snowmelt and mixed events.

2. Methodology

For a site with no flow records within a homogeneous region, the mean annual maximum rainfall induced flood (Q_{pr}) and its standard deviation (S_d) were proposed to be estimated by a regional regression equation of the form:

$$Q_{pr} = k.\, A^a.\, B^b.\, C^c \ldots \tag{1}$$

J. Marsalek et al. (eds.), Flood Issues in Contemporary Water Management, 41–47.

$$S_d = i.\, U^u.\, V^v.\, W^w \ldots \quad (2)$$

where *i, k, a, b, c* ... and *u, v, w* ... are regional parameters, and *A, B, C* ... and *U, V, W*...are climatic and physiographic catchments characteristics.

Formulae of this type were previously derived for floods computed from the log-normal distribution in the flysh region of Slovakia in [8,9] yielding not fully satisfactory results. In this paper, therefore, different concepts for the delineation of hydrologic regions and design flood computation were tested in catchments from the high core mountain region of Slovakia.

Design discharges for return period N = 2, 5, 10 years, which are usually needed in channel design for river training and restoration, are suggested to be computed from the Gumbel EV1 distribution:

$$F_{Q_{max}} = e^{-e^{(Q_{max} - \mu)/\alpha}} \quad (3)$$

It was found as one of the applicable distributions in Slovakia by a L-moment diagram analysis of 250 catchments [10]. Following the work of Burlando et al. [11], the method of moments was applied to the estimation the parameters of the cumulative distribution function:

$$\mu = Q_{pr} - 0.5772.\alpha \quad (4)$$

$$\alpha = 0{,}779.S_d \quad (5)$$

Homogeneous regions for flood frequency analysis were defined in the following ways. Firstly, in method (I) the whole study area was considered to be homogeneous by tradition and for comparison. Secondly, in method (II), it was attempted to divide it into geographic sub-regions based on subjective analysis of the geomorphologic properties of the landscape. Next the idea of geographical regions was abandoned, and in the third approach (III) it was attempted to group catchments into regional types with similar variability of flood frequency curves of the individual sites. In method (IV), physiographic properties of basins and flood runoff characteristics were used as variables in cluster analysis (K-means clustering with Euclidean metrics after Hartigan [12]) to define homogeneous regional types. The number of regions in each method chosen had to be less than four in order to get a comparable number of basins in each group and acceptable estimates of parameters in the regional formulae.

Since no independent data set was available in the study area to test the performance of the proposed approach, for each regionalisation method the relative differences *(X-Y)/Y* between the values of Q_N computed from the regional formulae and statistically from the site data were compared at the sites analysed. Values computed from the envelope curve approach by the traditional formula were included into the comparison as method (V):

$$q_{max} = A / (F+1)^n \quad (6)$$

where q_{max} is the specific annual maximum discharge with a return period of 100 years in $m^3.s^{-1}.km^{-2}$, *F* is the catchment area in km^2 and *A, n* are regional parameters. Floods with shorter return periods were computed with the help of regional frequency factors from the 100-year discharge:

$$Q_N = a_N . Q_{100} \quad (7)$$

where Q_N is a design flood discharge with a return period of N years, a_N is a regional frequency factor, and Q_{100} is a 100-year design discharge given by the regional formula.

3. Data collection

The region of high core mountains was selected as a test site for the proposed approach. It forms the inner side of the West Carpathian belt and comprises the East, West and Low Tatras. The base of this unit consists prevailingly of granitoids and metamorphic rocks with several tectonic slices. Extreme summer floods are significant for this region. The Tatras Mountains belong to the highest mountains in Slovakia, with altitude rising from 800 to 2655 m above sea level. The long-term mean annual precipitation amount varies from 900 to 2000 mm.

Flood records from 48 small and middle-sized basins with catchment area in the range 10 to 500 km^2 were used in the analysis. The length of observation ranged from 15 to 60 years. Annual maximum peak discharges from the summer season with rainfall induced discharges were selected. Because of the high altitude of the region, the origin of each selected event was separately analysed to exclude floods of mixed origin, rain and snowmelt, in spring and autumn.

Several climatic and physiographic catchment characteristics influencing the runoff formation process were derived from a set of digitised maps and a raster digital elevation model DEM with the help of a GIS: the catchment area F (km^2), the length of the river network L (km), the mean catchment slope IP (%), the mean stream slope IT (%), the catchment shape coefficient ($A2$) defined as the ratio of the catchment size and the square of its length, the mean aspect *OPRIEM* (degrees), the mean catchment elevation *HPR* (m a.s.l.), the percentage of forested area in the catchment LS (%), the long-term mean annual runoff for the period 1930-1960 qa ($l.s^{-1}.km^{-2}$) and an index of the infiltration capacity of soils (*INDEXPOD*).

The maximum daily precipitation for the period 1901-1980, *MAXD* (mm), was taken from nearby stations and the time of concentration TK (h) of each basin was computed according to [13,14,15].

4. Results

The geology and morphology of Slovakia are significant for their extreme heterogeneity. Several classifications of the geomorphology have been reported in the literature. The geomorphological classification of Slovakia from Mazúr and Lukniš [16] divides the territory of Slovakia into 84 geomorphologic units, with some of the units containing a more detailed subdivision. In order to select homogeneous regions of rainfall-induced flood formation, geomorphologic units in the high core mountains were combined with a hydrogeological classification by Fusán et al. [17] and the corresponding atmospheric pattern. The regionalisation according to method (II) led to three regions, the West, East and Low Tatra regions.

Selection of catchments according to similar values of the coefficient of variation of the annual rainfall flood series in method (III) formed three groups of geographically discontinuous basins with Cv values ranging from 0.4 to 0.6, 0.6 - 0.7 and 0.7 and more, respectively.

Regionalisation using cluster analysis in method (IV), which was based on several combinations of catchment characteristics with strong influence on Q_{pr} and S_d and little correlation among each other, resulted in a number of possible regional types. An analysis of the cluster profile plots revealed, that the site characteristics strongly overlap between clusters in each case. A typical cluster profile plot, which shows the grand mean across all variables, the mean within each cluster and one standard deviation above and below the mean, is given in Fig.1.

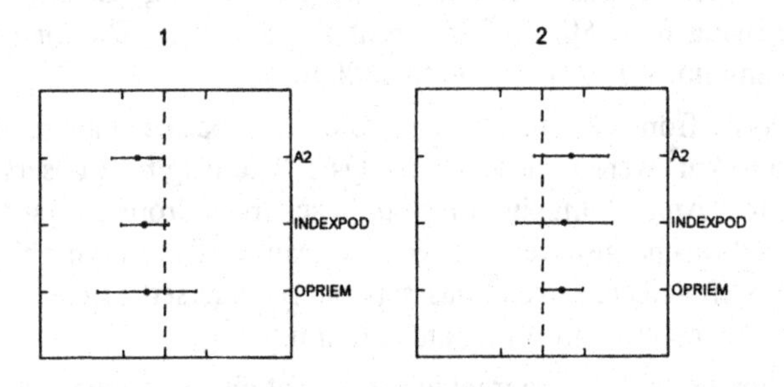

Figure 1. One of the typical cluster profile plots for the high core mountain region.

This would make the assignment of ungauged sites to a regional type difficult. Homogeneity of the regions according to Hosking and Wallis [18] was therefore not tested. Similar problems were reported in [10], where it was suggested to cluster catchments using the distribution parameters *(Q_{pr}, S_d)* instead in order to get similar flood frequency curves in the derived regions. Mapping of the areal distribution of the frequency curve parameters enabled in [10] to assign ungauged catchments into regional types. Three regions with similar flood frequency curves were therefore selected in method (IV) using Q_{pr} and S_d as clustering variables.

Stepwise multiple regression was used to seek the most convenient relationship between the climatic and physiographic basin characteristics and the dependent variables in the regions. Attention has been paid to the choice of predictors with low mutual dependence in order to minimise the effect of multicollinearity. Several subjectively chosen and hydrologically reasonable starting combinations of the independent variables were used as seeds in the discrimination process. The number of predictors was restricted between two to four in each formula for computational reasons. No unique set of predictors was found to be optimal for all catchment groups neither within nor across the regionalisations. The values of the multiple correlation coefficient ranged from 0.6 to 0.95 for the resulting combination of predictors. In

Fig.2, the comparison of Q_{pr} values estimated from the regional approach and at-site data are presented.

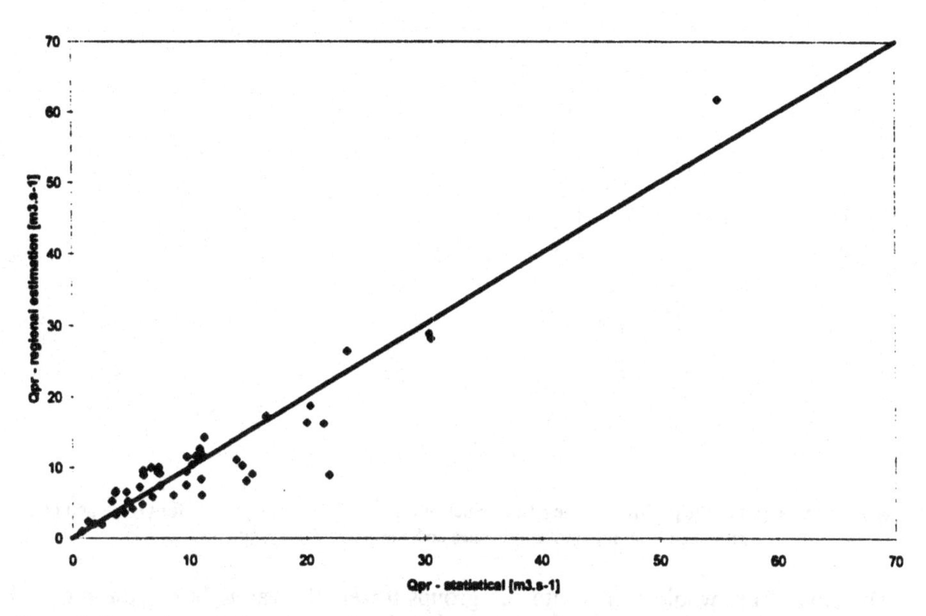

Figure 2. Comparison of Q_{pr} values estimated from the regional approach in method [III] and computed from at site data.

Since no independent dataset was available to test the derived formulae, it was decided to observe their performance in the design discharge estimation. Design discharges for return period of N=2, 5, 10 years, which were usually used in channel design for river training and restoration, were computed from the Gumbel EV1 distribution with parameters estimated by the regional formulae. The relative differences $(X-Y)/Y$ between Q_5 computed using the EV1 distribution (X) and from the regional formulae (Y) are compared in Table 1. Averages of the positive and negative values of relative differences for Q_2, Q_5 and Q_{10} and all methods are shown in Fig. 3.

TABLE 1. Statistics of the relative differences $(X-Y)/Y$ between Q_5 estimated from regional approaches and at site data.

Regionalisation Method	(I)	(II)	(III)	(IV)	(V)
Minimum	-0,614	-0,500	-0,567	-0,623	-0,280
Maximum	1,589	1,172	0,765	1,378	4,327
Mean	0,090	0,063	0,072	0,063	0.852
Std. Deviation	0,472	0,384	0,334	0,391	0.993

As expected, the mean values of the relative differences are close to zero when using the derived formulae in methods (I-IV), whereas in the traditional approach (V) they are biased with most regional estimates lying over their reference values and only a few above them. However the distribution of the relative differences seems to be slightly skewed also in methods (I-IV); this suggests the need for further analysis of the choice of the form of the formulae and the estimation of their parameters.

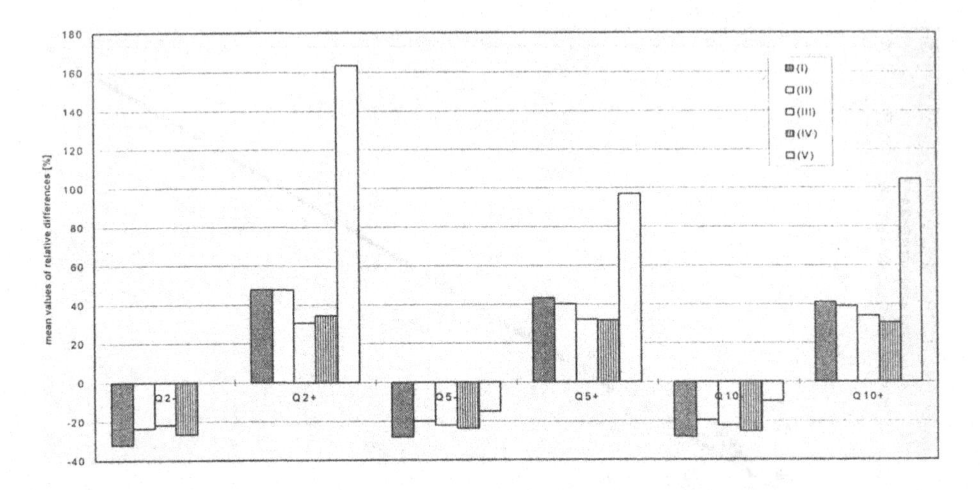

Figure 3. Averages of the positive and negative values of relative differences $(X-Y)/Y$ for Q_2, Q_5 and Q_{10} estimated from regional approaches and computed from at site data.

Division of the whole region into sub-groups lowers the variability of the regional estimates over their reference values. The differences between methods (II–V) are negligible. Therefore, no further analysis was made towards the possibility of mapping the values of C_v, Q_{pr}, and S_d in the region.

5. Conclusions

The methods (I - IV) are valid in the region, on the average. Typically, underestimation of the design values in approximately one half of the sites and overestimation in the other half occurs. Therefore, regionalisation approaches with the lowest variability over the regional reference values should be preferred in a practical application. However, no preference can be stated regarding the choice of the regionalisation method. They are generally applicable for river training and restoration since they do not tend to overestimate the design discharge in all basins of a particular region as the traditional approach. The traditional approach has to be preferred in tasks where safety of design is required, since under-design cannot be compensated by over-design.

Several questions remain to be investigated such as the preference of a particular regionalisation approach, aspects under which the concept of regional homogeneity can be used for design purposes involving safety, and the applicability of the methods in other regions. Regional frequency analysis following principles of Hosking and Wallis [18] could lead to the selection of more appropriate regions than subjective regionalisation methods relying on the concept of geographical regions. However, with the climatic and physiographic catchment parameters used in this paper it can lead to difficulties, when assigning ungauged catchments to regional types. Inclusion of other variables controlling the variability of flood formation, describing the intensity of overland flow formation, upper layer permeability, catchment storage, etc., should be further investigated in order to test the practical applicability of the concept of regional homogeneity based on physiographic characteristics.

6. Acknowledgement

The Slovak Grant Agency supported the research presented in this paper under the Project No. 2/6008/99. This support is gratefully acknowledged.

7. References

1. Kohnová, S. and Szolgay, J. (1996) The regional estimation of maximum specific discharges in small catchments of Slovakia (in German). *Zeitsch. für Kulturtechnik und Landentwicklung*. **28**, 116-121.
2. Kohnová, S. and Szolgay, J. (1995) On the use of the Dub regional formulae on the territory of Slovakia (in Slovak), *Vodohspod. Casopis*, **43**, 28-56.
3. Acreman, M. C. and Sinclair, C. D. (1986) Classification of drainage basins according to their physical characteristics; an application for flood frequency analysis in Scotland. *J. Hydrol*., **84**, 365-380.
4. Meigh, J. R., Farquharson, F. A. K. and Sutcliffe, J. V. (1997) A world-wide comparison of regional flood estimation methods and climate. *Hydrol. Sci. J*., **42** (2), 225-244.
5. Zrinji, Z. and Burn, D. H. (1994) Flood frequency analysis for ungauged sites using a region of influence approach. J. Hydrol. **43**, 153, 1-21.
6. Mosley, M. P. (1981) Delimitation of New Zealand hydrologic regions. *J. Hydrol*., **49**, 173-192.
7. Wiltshire, S. E. (1985) Grouping basins for regional flood frequency analysis. *Hydrol. Sci. J.*, **30** (1), 151-159.
8. Kohnová, S. and Szolgay, J. (1998) Regional estimation of the mean annual summer flood and its variability in the flysh region of Slovakia. *In: XIXth Conference of the Danube Countries*, Osijek, June 1998, pp. 215-222.
9. Kohnová, S. and Szolgay, J. (1999) Regional estimation of design summer flood discharges in small catchments of northern Slovakia. IAHS at IUGG Birmingham, Symposium 1. Hydrological extremes. IAHS Press Wallingford (in press).
10. Čunderlík, J. (1999) *Regional estimation of design discharges in Slovakia*, Bratislava, PhD. Thesis, 120p.
11. Burlando, P. et a.l. (1998) Materials from Postgraduate course "*Flood management*", ETH Zurich 1998, 233p.
12. Hartigan, J. A. (1975) *Clustering Algorithms*. New York, John Wiley and Sons.
13. Kirpich, Z.P. (1940) Time of concentration of small agricultural watersheds. Civ. Engng., **10** (6), p. 362.
14. Nash, J.E. (1960) A unit hydrograph study with particular references to British catchments. Proc. Inst. Civ. Engin., **12**, 249-282.
15. Hradek, F. (1989) The maximum runoff at a small experimental catchment (in Czech), VŠZ, Praha, p. 90.
16. Mazúr, E. and Lukniš, M., et al. (1986) *Geomorphological division of ČSSR and SSR* (in Slovak), Part I, maps.
17. Fusán I., et al. (1980) *Geological division of SSR*, (in Slovak), Atlas SSR, SAV, SÚGaK, Bratislava.
18. Hosking, J.R and Wallis, J.R (1997) *Regional Frequency Analysis. An Approach based on L-moments*. Cambridge Univ. Press, 220p.

FLOOD FLOW REGIONALIZATION BASED ON L-MOMENTS AND ITS USE WITH THE INDEX FLOOD METHOD

JURAJ ČUNDERLÍK

Department of Land and Water Resources Management
Faculty of Civil Engineering, Slovak University of Technology
Radlinského 11, 813 68 Bratislava, Slovak Republic

1. Objectives

Regional flood frequency analysis is currently one of the most often applied methodologies for flood estimation at ungauged sites worldwide. The performance of any regional analysis is very dependent on regionalization - which is its most important step. Currently, prevailing methods of regionalization usually base the formation of flood regions on the similarity of river catchments in physiographic characteristics. This approach is derived from the concept that catchments similar in physiographic characteristics have consequently similar flood response. This concept cannot be guaranteed in general, particularly if the spatial variability of catchment characteristics is large. A usually significant variance of sites' L-moment ratios *L-Cs* and *L-Ck* in a region derived from similarity in physiographic characteristics tends to be explained by their sampling variability. Naturally, one can ask, to what extent this variance may be regarded as sampling variability? If the variance is not caused by sampling variability, then such a region is heterogeneous, having significant differences between sites' probability distributions.

This paper presents an alternative approach to flood flow regionalization based on the similarity in sample scale and shape distribution parameters of annual maximum floods (Q_{max}^{AN}), which are in a strictly homogeneous region constant. The rationale behind this approach is that certain probability distribution parameters can be seen as a quantitative expression of a catchment hydrologic response. The reliability of such regionalization is dependent on how well the sample distribution represents the true distribution of the population. A method is proposed, which will allow to use this regionalization also for ungauged sites. This alternative approach is tested on catchments which were not used in the regionalization analysis.

J. Marsalek et al. (eds.), Flood Issues in Contemporary Water Management, 49–59.

2. Method

2.1. FORMATION OF FLOOD FLOW REGIONS

Cluster analysis is a commonly used tool for grouping river basins into flood flow regions. It is objective and the most practical method, recommended by many authors [1,2,3,4,5]. In this study the hierarchical and nonhierarchical (K-means) methods of clustering were applied. Distances among data vectors and cluster centers were computed by the Euclidean distance measure. The L-moment ratios *L-Cv, L-Cs* and *L-Ck,* which are in a strictly homogeneous region constant, were chosen as regionalization's parameters. These parameters were standardized and equally weighted. Both methods of clustering were performed for different pre-defined number of clusters ranging from 10 to 20. The average values of within-region variance of the regionalization's parameters were calculated in terms of standard deviation according to the following equation [6]:

$$\overline{W}rv_{LCvsk} = \frac{\sum_{i=1}^{N} \sqrt{\frac{\sum_{j=1}^{n_i} \left(LCv_{i,j} - W_{LCv^i}\right)^2 + \left(LCs_{i,j} - W_{LCs^i}\right)^2 + \left(LCk_{i,j} - W_{LCk^i}\right)^2}{n_i}}}{N-1} \quad (1)$$

and between-region variance was quantified by:

$$\overline{B}rv_{LCvsk} = \sqrt{\frac{\sum_{i=1}^{N} \left(W_{LCv^i} - \overline{W}_{LCv}\right)^2 + \left(W_{LCs^i} - \overline{W}_{LCs}\right)^2 + \left(W_{LCk^i} - \overline{W}_{LCk}\right)^2}{N-1}} \quad (2)$$

where W_{LCvi}, W_{LCsi} and W_{LCki} are the regional averages of L-moment ratios in *i*-th region, $\overline{W}_{LCx}$ are the total averages of W_{LCvi}, W_{LCsi} and W_{LCki}, N is the number of clusters (regions) and n_i the number of sites in *i*-th region. The optimal result was chosen from all clustering results as the one with the minimum within, and maximum between, regional variance of *L-Cv, L-Cs* and *L-Ck.*

2.2. SELECTION OF A REGIONAL DISTRIBUTION OF QMAX

For every derived region the most appropriate distribution was determined by means of the L-moment ratio diagram and Hosking's goodness of fit test [5,7]. The L-moment diagram is a widely used tool for graphical interpretation and comparison of L-moment ratios *L-Cs, L-Ck* of various probability distributions with the estimated sample L-moment ratios. Regional averages *L-Cs* and *L-Ck* were weighted proportionally to the sites' record length and plotted on the diagram. The distribution which was closest to the regional average value was chosen. In this decision procedure the scatter of individual sites' L-moment ratios in a given region was also taken into account.

Hosking's goodness of fit test is based on the judgment of the quality of a distribution fit by the difference between the L-kurtosis *L-Ck* of the fitted distribution (τ_4^{DIST}) and the regional average *L-Ck* (t_4^R). To assess the significance of this difference Hosking suggests comparing it with the sampling variability of t_4^R. The goodness of fit measure Z^{DIST} is then:

$$Z^{DIST} = \frac{\left(\tau_4^{DIST} - t_4^R + B_4\right)}{\sigma_4} \tag{3}$$

where B_4 is the bias of t_4^R and σ_4 is the standard deviation of t_4^R calculated as:

$$B_4 = \frac{1}{N_{sim}} \sum_{m=1}^{Nsim} (t_4^{[m]} - t_4^R) \qquad \sigma_4 = \sqrt{\frac{\sum_{m=1}^{Nsim} \left(t_4^{[m]} - t_4^R\right)^2 - N_{sim} B_4^2}{N_{sim} - 1}} \tag{4-5}$$

where $t_4^{[m]}$ is the regional average *L-Ck* of the *m*-th simulated region and *Nsim* is the number of simulations of the original region with the same number of sites and the same site record lengths. Simulated regions have kappa distribution as their frequency distribution. The fit is adequate if $|Z^{DIST}| \leq 1{,}64$.

2.3. THE ESTIMATION OF N-YEAR FLOODS

The index flood method developed by Dalrymple [2] is currently one of the most frequently used method for quantile (Q_N) estimation [8]. Its key assumption is that the sites form a homogeneous region and that the frequency distributions of the sites are identical apart from a site specific scaling factor, the index flood. N-year quantiles $Q_{N,i}$ in a *i*-th catchment are derived by [9]:

$$Q_{N,i}(F) = \mu_i q(F), \quad i = 1,\dots,M \tag{6}$$

Here μ_i is the index flood, *q(F)* is the dimensionless regional growth curve, and *M* is the number of catchments in a region. Regional parameters needed for derivation of the regional growth curve can be computed as the averages of at-site parameters weighted by the sites record length. The index flood was estimated by the sample mean of the $Q_{\max}^{AN}$ ($\overline{Q}_{\max}^{AN}$). In the case of ungauged sites, the sample mean is usually estimated from catchment characteristics. Contours of the relative error δ can be expressed as:

$$\delta = \frac{\overline{Q}_{\max}^{AN} - eq\overline{Q}_{\max}^{AN}}{eq\overline{Q}_{\max}^{AN}} \tag{7}$$

where $\overline{Q}_{\max}^{AN}$ is the original sample mean and $eq\overline{Q}_{\max}^{AN}$ is the mean estimated from catchment characteristics using regression, which was plotted on a map. The final estimation of $\overline{Q}_{\max}^{AN}$ ($est\overline{Q}_{\max}^{AN}$) can then be corrected by the relative error δ inferred from the map:

$$est\overline{Q}_{\max}^{AN} = eq\overline{Q}_{\max}^{AN}(1+\delta) \tag{8}$$

In the regionalization based on at-site data one must define a method for allocating ungauged sites to the derived regions. If the correlations between regionalization parameters and catchment characteristics are not significant, the alternative way is to estimate them from maps. For this purpose, at-site values of *L-Cv* and *L-Cs* from 243 sites were attributed to the catchment gravity centers and drawn on contour maps using kriging interpolation method. An ungauged site is then attributed to the specific region by the values of *L-Cv* and *L-Cs* inferred from the maps.

3. Study catchments

For this analysis only catchments with areas less than 400 km^2 were selected. Sites influenced by dams were excluded from the analysis. All sites had at least 15 years of continuous records of annual maximum discharges. The average record length was 33 years, the longest 77 years. In total, 266 river catchments representing all physiographic regions of Slovakia were included in the database. Another 10 selected catchments, which were not included in the analysis, were used for testing purposes. Table 1 gives a short summary of catchments according to their area and record length.

TABLE 1. Summary of the 266 selected river catchments (1) and 10 testing catchments (2)

Type of catchment	Catchment area [km^2] 0-100	100-200	200-300	300-400			
1	181	47	24	14			
2	2	2	2	4			
	Record length [years] < 15	15-20	20-30	30-40	40-50	50-60	> 60
1	3	24	118	70	19	7	25
2	1	0	2	3	0	2	2

4. Results

The optimal result of any regionalization is such that minimizes the within-region variance of used regionalization parameters and maximizes their between region variance. Figure 1 shows within-region variances in terms of standard deviations of *L-Cv, L-Cs* and *L-Ck* for different numbers of clusters derived by the hierarchical and splitting (nonhierarchical) methods. For both cluster methods the within-region variance continuously decreases towards the realizations with higher numbers of clusters. At 11 clusters, a weak local minimum can be observed. In the case of between-region variance in Figure 2, again there is a local maximum at 11 clusters. The splitting method of K-means shows better results in both figures. Taking into account also the size of regions, the result of the method of K-means with 11 clusters was chosen.

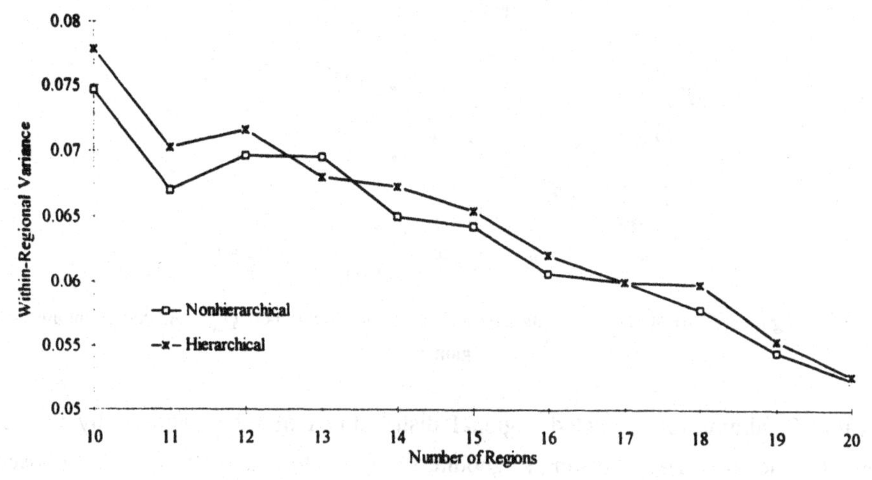

Figure 1. Within region variance of *L-Cv*, *L-Cs* and *L-Ck*.

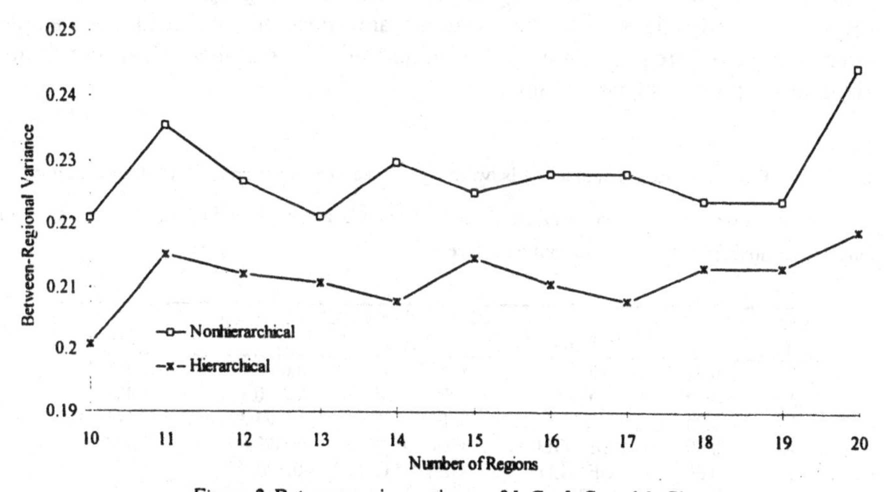

Figure 2. Between region variance of *L-Cv*, *L-Cs* and *L-Ck*.

The discordancy measures defined by Hosking a Wallis [5] were calculated for sites in all regions; none of the sites had this measure greater than the defined limits. Tests of regional heterogeneity were not applied, because in this case one would judge the degree of heterogeneity by means of the same statistics, which were used for formation of regions. Surprisingly good correlation between the average annual maximum flood $\overline{Q}_{\max}^{AN}$ and the catchment area was found in all regions, with the coefficient of determination (r^2) ranging from 0,52 to 0,91. Figure 3 shows an example of this correlation in region 4.

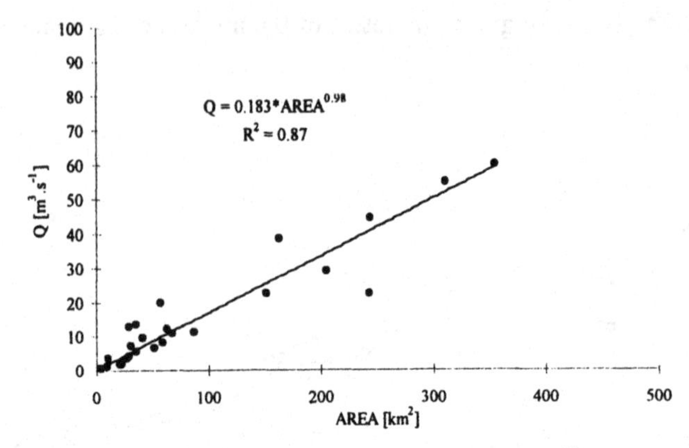

Figure 3. Correlation between the average annual maximum flood $\overline{Q}_{max}^{AN}$ and catchment area in region 4.

Table 2 summarizes selected regional distributions of Q_{max}^{AN} chosen by the L-moment ratio diagram, Hosking's goodness of fit test and finally the chosen distribution. Both methods identified the same distributions. The L-moment ratio diagram for all regions is shown in Figure 4. Regions are not geographically compact, however, some indications of spatial relations are apparent. Dimensionless sample distributions in every region can be well summarized with one theoretical probability distribution function as shown in Figure 5.

TABLE 2. Coefficients of determination (r^2) between $\overline{Q}_{max}^{AN}$ and catchment area, regional distributions of Q_{max}^{AN} chosen by: Ldiag = L-moment diagram, Test = Hosking's goodness of fit test, Z^{DIST} = values of Hosking's test statistics, Reg = final distribution choice.

Reg id	r^2	Choice of a regional distribution Ldiag	Test	Z^{DIST}	Reg
1	0.71	GPA	GEV/GPA	0,69/0,87	GPA
2	0.52	GPA	GPA/GEV	0,22/0,88	GPA
3	0.74	LN3	GEV/LN3	1,12/1,32	LN3
4	0.89	GEV/GLO	GLO/GEV	0,56/0,89	GEV
5	0.69	GEV/LN3	GEV/GLO	-0,49/0,67	GEV
6	0.76	GLO	GLO	1,13	GLO
7	0.91	GEV	GLO/GEV	0,55/-0,57	GEV
8	0.80	GEV/LN3	GEV	1,44	GEV
9	0.77	GPA	GLO/GPA	-0,82/1,51	GPA
10	0.57	GLO	GLO	-0,64	GLO
11	0.86	LN3	LN3/PE3	0,39/-0,78	LN3

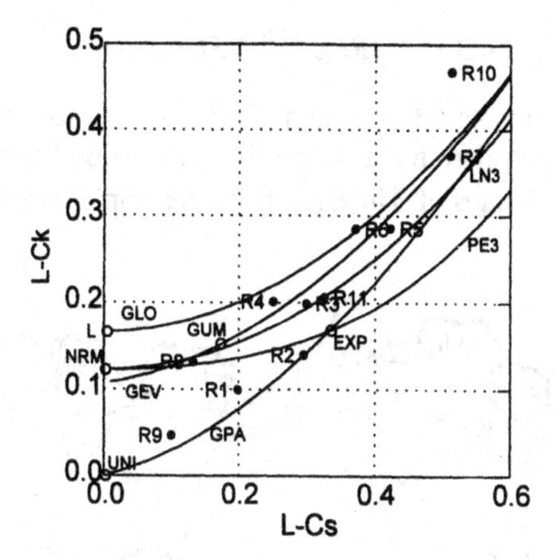

Figure 4. L-moment ratio diagram for derived regions (EXP-exponential, GUM-Gumbel, LO-logistic, NRM-normal, UNI-uniform, GEV-generalized extreme value, GLO-generalized logistic, GPA-generalized Pareto, LN3-log-normal, PE3-Pearson III. R1-11 = regional average values of *L-Cs* and *L-Ck* weighted proportionally to the site record lengths).

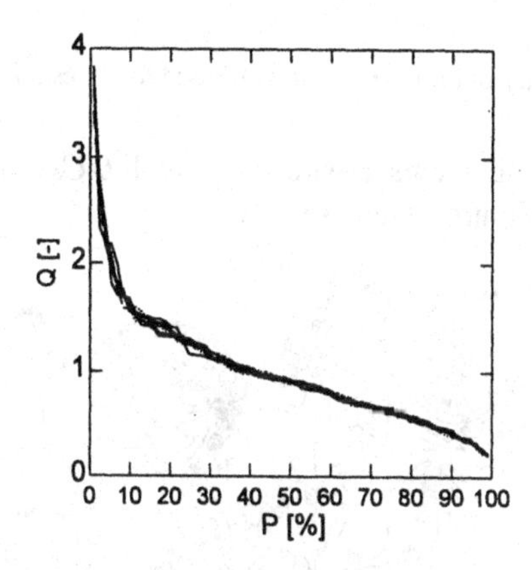

Figure 5. Dimensionless empirical exceedance curves of Q_{max}^{AN} in region 1.

Several regression models were defined comprising basic physiographic characteristics including catchment area, main river length, drainage density, average catchment slope and elevation, average annual rainfall and 24-hour extreme rainfall. The best result in terms of the correlation coefficient was achieved using the equation:

$$\overline{Q}_{max}^{AN} = 0{,}393 * AREA^{0{,}791} * ECRANGE^{0{,}11} \tag{9}$$

where *AREA* stands for catchment area and *ECRANGE* is the catchment elevation range. The coefficient of correlation *r* is 0,767. The index flood was then estimated by equations 8 and 9 with the values of δ inferred from the map shown in Figure 6.

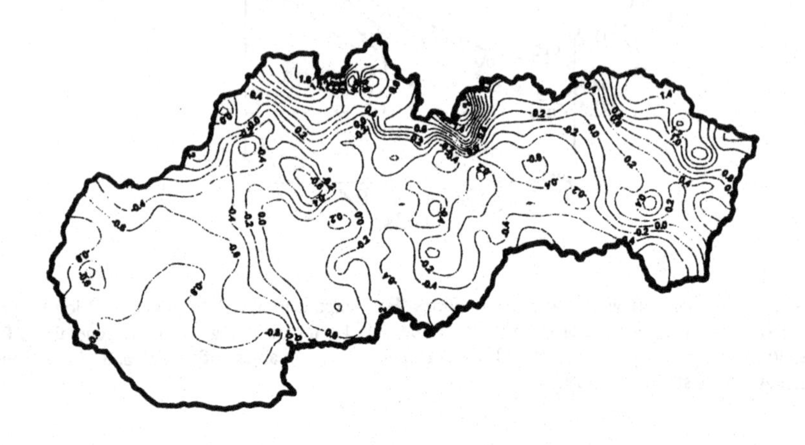

Figure 6. Contour map of the relative errors (δ) used for the estimation of $\overline{Q}_{max}^{AN}$.

For the selected test catchments at-site *L-Cv* and *L-Cs* were inferred from the contour maps displayed in Figures 7 and 8 below.

Figure 7. Contour map of the *L-Cv*.

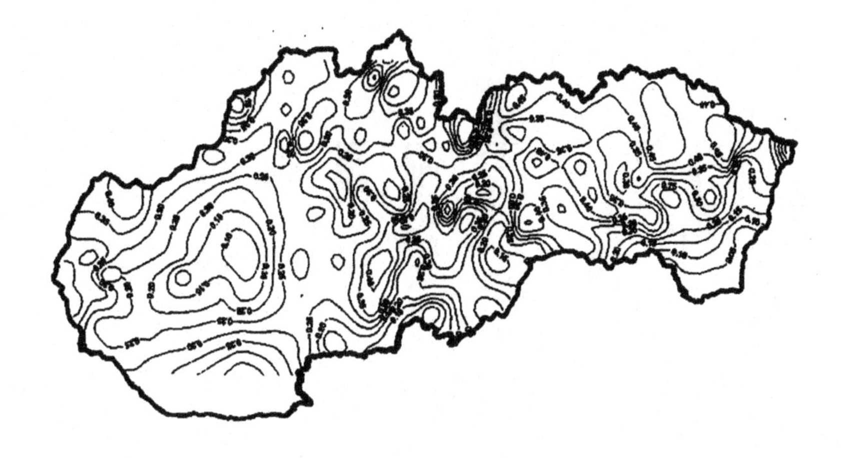

Figure 8. Contour map of the *L-Cs.*

Relative errors of *L-Cv*, *L-Cs* and the index flood, calculated for all 10 test catchments, are shown in Figure 9. Except one case (27%) the relative errors are rather low, within ± 20%.

In the case of *L-Cv*, relative errors are quite small, ranging from -18,5% to 14,2%. The relative errors of *L-Cs* are higher, from -34,8 to 45,2%. N-year floods (Q_N) were estimated by combination of the proposed regionalization with the index flood method. Figure 10 shows relative errors of Q_N derived from the comparison of the estimated quantiles and the quantiles calculated from at site data. It follows from Figure 10, that the suggested approach gives relative errors of Q_{10} within ± 20%, Q_{50} less than 25% and Q_{100} not exceeding 35%.

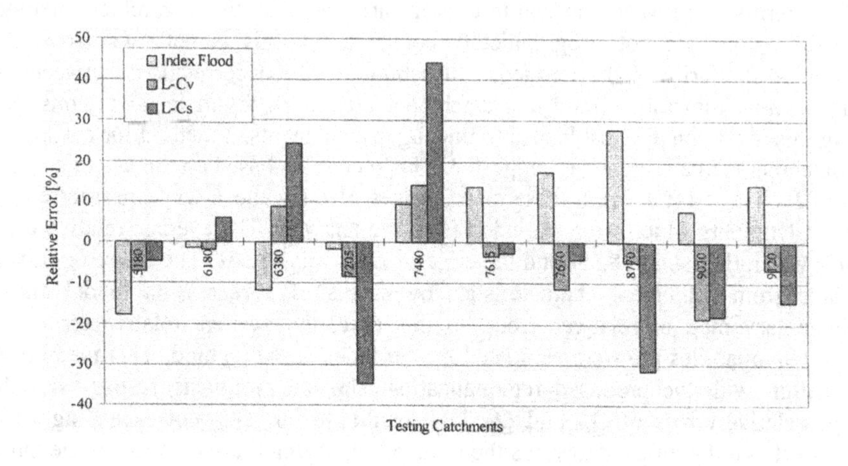

Figure 9. Relative errors of $\overline{Q}_{\max}^{AN}$, *L-Cv* and *L-Cs* for 10 test catchments.

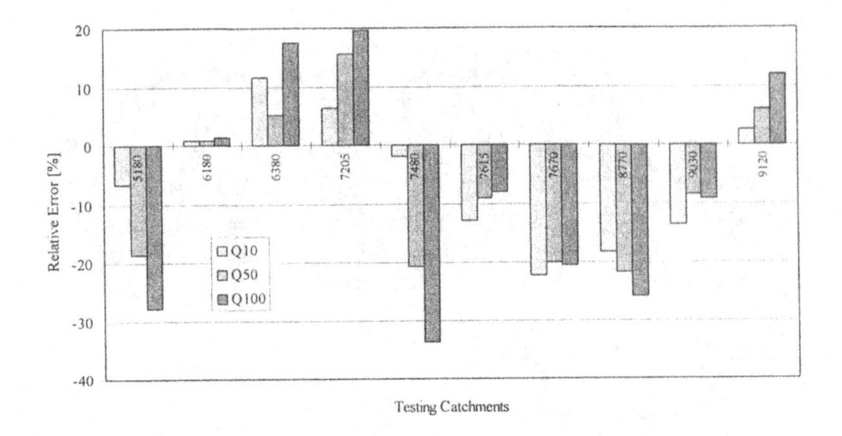

Figure 10. Relative errors of Q_N estimation for 10 test catchments.

5. Conclusions and recommendations

The method introduced in this paper is an alternative approach to flood flow regionalization. A homogeneous region is defined as a group of catchments with equal sample distribution of Q_{max}^{AN} except of the location parameter. Reliability of this concept is dependent on how well the sample distribution represents the true distribution of Q_{max}^{AN}. Over 260 selected river catchments were grouped into regions by means of hierarchical and splitting (nonhierarchical) cluster analysis methods according to *L-Cv*, *L-Cs* and *L-Ck*, which are in a strictly homogeneous region constant. The K-means splitting method with Euclidean distance metrics gave better results in terms of between and within cluster variances than the hierarchical method. Acquired regions are not geographically compact, but this is not regarded as an indicator of similarity of the frequency distribution. Good correlations between the average annual maximum flood and catchment area were found in all regions. For making this approach applicable also to ungauged catchments, a method for attributing such catchments to a region was suggested. This method is based on the use of contour maps of *L-Cv* and *L-Cs*. From these maps, values of *L-Cv* and *L-Cs* were inferred for 10 test catchments which were excluded from the analysis. The average relative error of *L-Cv* was rather small, 8.5% and the error of *L-Cs* was 18.9%. The index flood was estimated from catchment characteristics by means of a regression model and a supplementary map of relative errors. In this case, the average relative error was 12.3%. The quantiles Q_N were estimated by the index flood method. This method, in combination with the proposed regionalization, showed promising results with the average relative errors of Q_{10} and Q_{50} less than 13% and Q_{100} not exceeding 15%. Future work will be directed towards the issue of improving the estimation of the index flood and the coefficient of linear variation, which can be estimated from catchment characteristics. Regionalization can be improved by including the flood and rainfall seasonality measures.

6. References

1. Burn, D.H. and Boorman, D.B. (1992) *Catchment Classification Applied to the Estimation of Hydrological Parameters at Ungauged Catchments*, Report **118,** Institute of Hydrology, Wallingford, 71 p.
2. Dalrymple, T. (1960) *Flood frequency analyses*, Water Supply Paper 1543-A, U.S. Geological Survey, Reston.
3. Flood Estimation Handbook Notes. (1997), Institute of Hydrology.
4. Gordon, A.D. (1981) *Classification*, Monograph on Applied Probability and Statistics. Chapman and Hall, London.
5. Hosking, J.R.M. and Wallis, J.R. (1997) *Regional Frequency Analysis: An Approach Based on L-moments*. Cambridge University Press, London.
6. Čunderlik, J. (1998) Probability distributions of annual maximum discharges in selected river catchments of Slovakia, *Slovak Hydrometeorological Institute Studies* **59**, 41-56.
7. Hosking, J.R.M. and Wallis, J.R. (1993) Some statistics useful in regional frequency analysis, *Water Resources Research* **29-2**, pp. 271-281.
8. Bobeé, B. and Rasmussen, P. (1995) *Recent advances in flood frequency analysis*, U.S. National Report to International Union of Geodesy and Geophysics 1991-1994. Washington, D.C., 1995.
9. Stedinger, J.R., Vogel, R.M. and Foufoula-Georgiou, E. (1992) Frequency analysis of extreme events, in D.R. Maidment (ed.), *Handbook of Hydrology*, New York.

THE FIRST USE OF HEC-5 IN FLOOD ANALYSIS IN TURKEY

OSMAN N. OZDEMIR

Civil Engineering Department
Faculty of Engineering and Architecture
Gazi University - Maltepe 06570 Ankara-Turkey

1. Introduction

The Seyhan River Basin is located within the bounds of 36°30' to 39°15' North latitude and 37°00'East longitude in Southern Turkey. The Seyhan River Basin climatologically expands across two regions, namely, the Central Anatolia Region and the Mediterranean Region. It is mainly composed of the Zamanti River, Goksu River, Seyhan River, and other tributaries after the confluence of the Zamanti and Goksu. The total catchment area of the Seyhan River basin at the base of the Seyhan dam is estimated to be 19400 km^2. The maximum river length from the Seyhan Reservoir's highest water level is estimated to be 420 km.

The Mediterranean climate is dominant in the Mediterranean Region and the continental climate is dominant in the Central Anatolian Region. In the Mediterranean climate, winters are warm and rainy, summers are hot and dry.

In the coastal area of the Seyhan River basin, precipitation is about 800 mm annually. At higher elevations, it increases to 1000 mm, and in the northern parts of the basin the corresponding value diminishes to 400 mm. Much of the precipitation, 50% of the annual total, falls between December and March. The mean annual precipitation of the basin is 590 mm. According to the climatic features of Turkey, the precipitation falling in winter over the areas above 1000 m elevation is generally in the form of snow, and produces snow cover on the ground. Snowmelt starts around the end of winter or at the beginning of spring, when air temperatures rise.

The City of Adana, which is the fourth largest city in Turkey with approximately one million inhabitants, is the political, commercial, industrial, and cultural centre of the region. Most of its economic activities, including industrial, commercial, and agricultural activities are significantly concentrated in the lower area of the city. In the City of Adana and its vicinity, many vital educational, cultural, entertainment, health and social welfare organizations and facilities exist.

The flood season in this area spans from November to April. Rainfall increases in this season and melting snow boosts up the river discharge. Some annual peak runoffs are recorded at the tributaries of the Seyhan River basin in the summer season. The 1980 flood was the largest flood causing the highest damage. Flood damage took place not only on the Cukurova plain but also in the upstream area and tributaries of the Seyhan River. Around the Feke district, floods hit in 1979 and 1980. The 1980 flood

J. Marsalek et al. (eds.), Flood Issues in Contemporary Water Management, 61–65.

destroyed 21 buildings and damaged 76 buildings. After the flood, the river bed was widened from 16-17 m to 30 m. This flood was equivalent to a 100-year flood as indicated in [1].

A feasibility study for the Seyhan Dam was carried out from 1949 to 1951, and the dam was constructed from 1953 to 1956. The dam spillway has a capacity of 2500 m^3/s; the power plant design discharge is 231 m^3/s and the total amount of water used for irrigation is 33 m^3/s.

The Catalan Dam was described in the Lower Seyhan Master Plan Report in 1980. Flood damages in 1980 contributed to an urgent implementation of the Catalan Dam construction, starting in February 1982. The Catalan and Seyhan Dams are able to protect the downstream area against the 500-year flood.

2. Flood Experience

An intensive rain and high temperature observed during the last week of March 1980, especially on March 27 and 28, caused the Seyhan reservoir lake elevation to rise after 1600 hours on March 27 [2,3]. The thick snow cover in the area and temperature rise in the basin alerted the dam personnel to be cautious. There was a spillway discharge from the Seyhan dam reservoir on the day the flood started. Table 1 shows how the Seyhan dam was operated during the flood. The discharge into the reservoir was high on March 27, 1980. Therefore, an immediate action was taken to evacuate the villages downstream of the Seyhan dam.

TABLE 1. Operation of the Seyhan dam reservoir during the three-day flood in 1980

Date	Hour	Reservoir Elevation (m)	Spillway Gate Bottom Elevation (m)	Inflow (m^3/s)	Outflow (m^3/s)
24 March	13	62.45	64.00	300	410
25	13	63.04	64.00	444	812
26	13	63.80	64.00	591	1083
27	16	64.10	64.00	1000	800
27	24	64.92	64.00	3800	800
28	01	65.12	64.00	4178	800
28	03	65.59	64.00	5091	1033
28	05	66.11	64.00	5224	1050
28	07	66.60	64.00	5640	1117
28	09	67.10	64.00	5990	1187
28	10:30	67.45	64.00	6079	1180
28	12	67.75	64.00	4964	1342
28	14	68.15	64.00	5297	1422
28	16	68.50	64.00	4826	1674
28	20	69.01	64.00	4438	2020
28	24	69.33	64.00	3691	2560
29	06	69.60	63.50	3160	2760
29	16	69.72	63.50	2830	2830
29	24	69.62	63.90	2170	2791

The three days in late March (27-29) were the most critical days during the 1980 flood. The critical water level in the reservoir forced high discharge from the reservoir and serious flooding in the region. The safe capacity of flood channel downstream of the

Seyhan dam was estimated as 800 m^3/s. In this event, the flood embankments designed to convey flood discharges up to 1400 m^3/s were insufficient. At several locations, these embankments were swept and demolished causing flooding in the plains of Adana.

The hydrograph of this flood event was simulated using the HEC-5 program package. The observed and simulated reservoir elevations for the Seyhan dam during the flood period are compared in Figure 1.

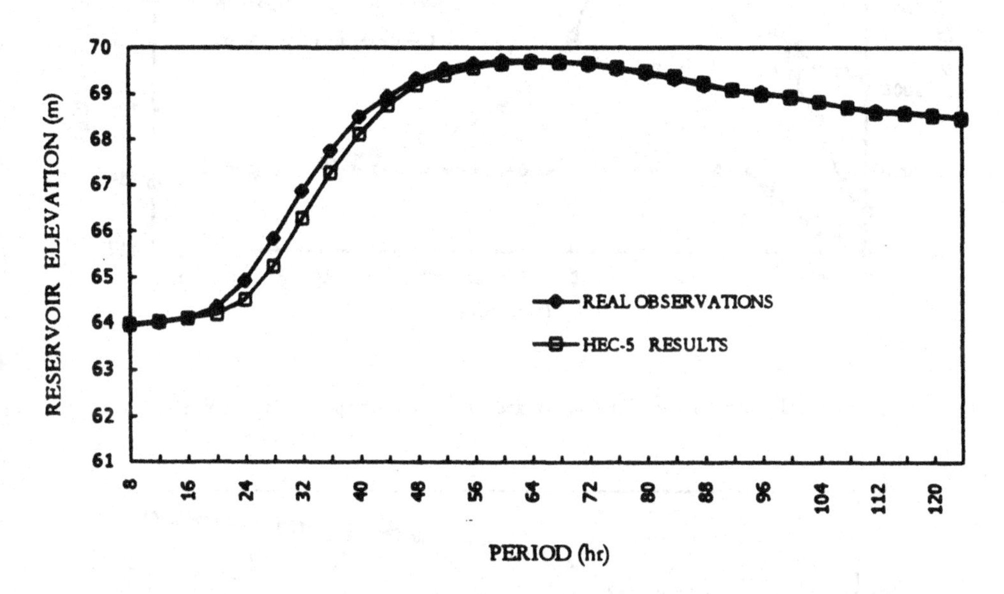

Figure 1. The observed and simulated reservoir elevations for the 1980 flood.

3. Flood Control

The Seyhan dam which was put into operation in 1956 is subject to sedimentation that decreases its active storage volume. Therefore, a new volume-area-elevation curve was prepared for this reservoir. After experiencing the 1980 flood, the construction of the Catalan dam, with a considerable flood control storage, was started just upstream of the Seyhan dam. The two dams together can contain the 500-year flood [2,3].

The 500-year flood was simulated using the HEC-5 [4,5]. Because the embankments with a 1400 m^3/s capacity broke and caused flooding downstream of the Seyhan dam, it was assumed that the flood channel capacity of the Seyhan river below the dam was just 800 m^3/s. The flood season in this region lasts from November until April, and therefore, the appropriate seasonal lake elevations shown in Table 2 were imposed on both Catalan and Seyhan reservoirs. The Catalan and Seyhan dams inflow and outflow hydrographs for a 500-year flood, together with the corresponding lake elevations, are shown in Figures 2 and 3.

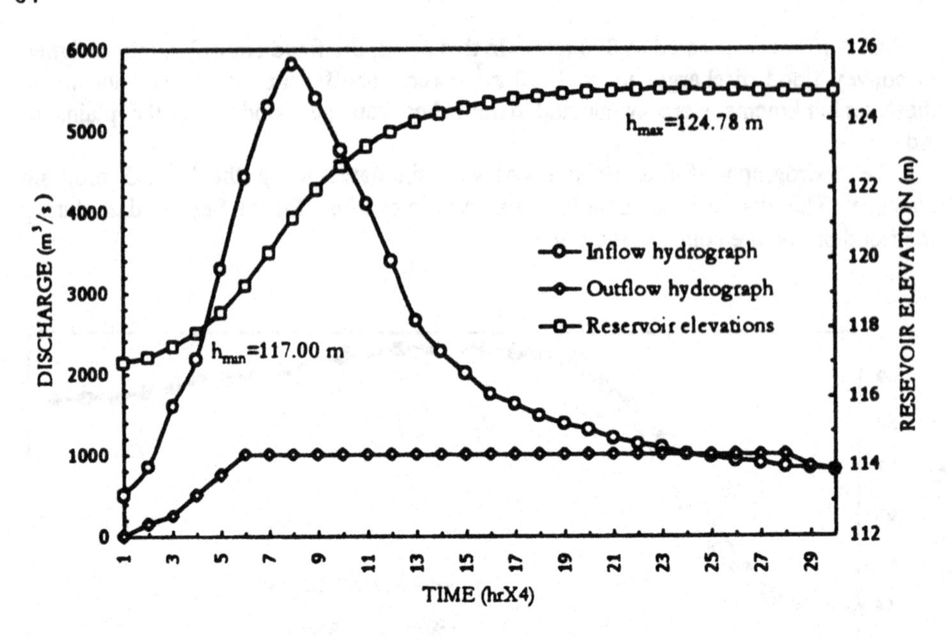

Figure 2. Catalan dam 500-year flood inflow and outflow hydrographs, and reservoir elevations.

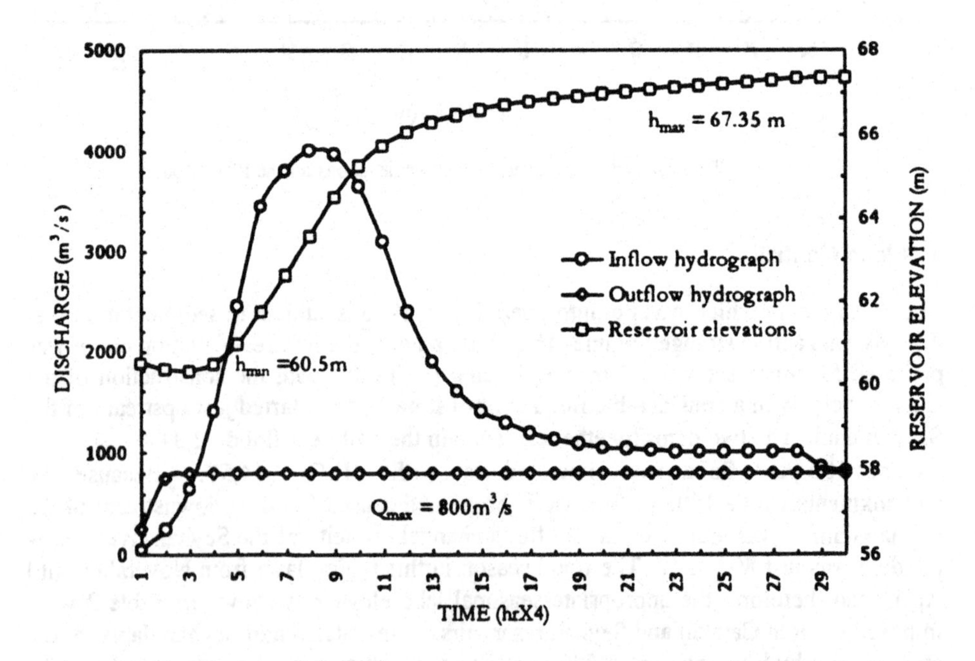

Figure 3. Seyhan Dam 500-year flood inflow and outflow hydrographs, and reservoir elevations.

DSI, the General Directorate of State Hydraulic Works, is responsible for the operation of multi-purpose dams in Turkey. In order to prevent flood damage in the Seyhan River basin downstream of the Adana province, the seasonal maximum reservoir levels are specified for the Catalan and Seyhan dams. Table 2 gives the comparison of the DSI present directive and the HEC-5 simulation for a 500-year flood, without exceeding the flood channel capacity of 800 m^3/s, as performed in this study.

TABLE 2. The comparison of the DSI directive and HEC-5 simulation for a 500-year flood

	Crest Elevation (m)	Maximum Operation Level (m)	Simulation Starting Level (m)		Simulation Ending Level (m)	
			DSI Directive	HEC-5 Simulation	DSI Directive	HEC-5 Simulation
CATALAN	126.50	125.00	118.00	117.00	125.00	124.80
SEYHAN	72.50	67.50	60.50	60.50	67.50	67.35

4. Conclusion

The DSI personnel in the Adana province were able to release the 100-year flood from the small Seyhan dam without a loss of life. The lack of an early warning system forced engineers to use their experience during the flood. Since the construction of the Catalan dam was finished in 1997, the 500-year flood can be released from the two dams without causing any property damage. The use of the HEC-5 program package can be useful for planning flood prevention measures during critical seasons.

5. References

1. *Feasibility Study on Flood control, Forecasting and Warning System for Seyhan River Basin*, Final Report, Volume I, (1994), Japan International Cooperation Agency (JICA).
2. Seyhan Taskin Raporu, '*Seyhan River Flood Report*' (1980). DSI Printing Office, Ankara, Turkey.
3. Taskin Yilligi, '*Flood Report*', Volume V, (1998), DSI Printing Office, Ankara, Turkey.
4. *HEC-5 Simulation of Flood Control and Conservation Systems, Users Manual*, (1982) U.S. Army Corps of Engineers.
5. *HEC-5 Simulation of Flood Control and Conservation Systems*, Exhibit 8 of Users Manual, (1987) Input Descriptions.

THE GREAT FLOOD OF 1997 IN POLAND: THE TRUTH AND MYTH

K. SZAMALEK

National Security Bureau
10 Wiejska st. 00-902 Warsaw, Poland

1. Introduction

In the second half of June 1997, the weather in Poland was influenced by cyclonic precipitation of high intensity and depth, which occurred throughout the country, except in northern and north-western Poland. The heavy rains of July '97 in Poland could be added to the list of the largest known rains in the officially recorded history of the Earth (Fig.1) [7]. The highest precipitation (585 mm) was recorded at Lysa Hora (Czech Republic) from July 5 to 9, while in the Polish drainage basin of the Odra River, it amounted to 484 mm in Kamienica and 455 mm in Miêdzygórze (Fig. 2) [8].

After the first rain storm, two more followed (July 15 - 23 and 24 - 28) [7]. The flood started in the Czech part of the Odra Basin. Following three rain storms, three consecutive stages of the flood were formed. The first stage was characterised by fast runoff, the resulting flood was very dynamic and destructive. The water level in the Upper Odra rose above the highest level ever recorded in the Polish history (Fig. 3). The time lag between rainfall and runoff was short and, because of lack of adequate water storage capacities, there was no way to avoid catastrophic material losses. In the second stage, a huge flood wave was already in the river channel of the Odra and it propagated downstream, where it inundated low-lying towns. At the time of the flood, it was impossible to avoid inundation of towns, but some preparations could have been made to reduce it. To understand the 1997 flood situation fully, besides the flood hydrology, it is necessary to take into account the political situation in Poland at that time.

The 1997 flood was widely discussed and constituted a subject of political controversy and political game played between the opposition and the ruling coalition [5]. The opposition controlled and had a great influence on the majority of the Polish mass media. The media considered the government as being weak and incapable of fighting this flood. It was stressed repeatedly that the actions of public institutions were late and inadequate. At this point, it is necessary to emphasise that the summer of 1997 in Poland was very sunny, dry and hot in the areas not affected by the flood. People visiting the unaffected areas could see very heavy rainfalls, destroyed bridges and roads, and desperate and suffering people only on TV. The mass media blamed the government for these unfortunate events. It was easy to believe this propaganda by people who were resting on beaches, sailing on lakes, etc. in areas without any rain during the two-week

J. Marsalek et al. (eds.), Flood Issues in Contemporary Water Management, 67–74.

holiday period. It is paradoxical that Poland in 1997 was divided into two parts: an extremely wet part in the SW and the dry and hot rest of the country.

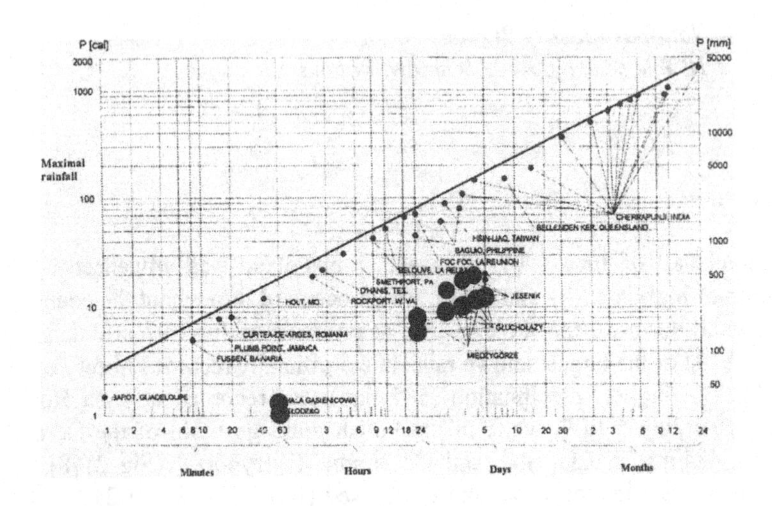

Figure 1. World maximum recorded rainfalls versus the 1997 rainfalls in the Upper Vistula and Odra River catchments (large black circles) [6].

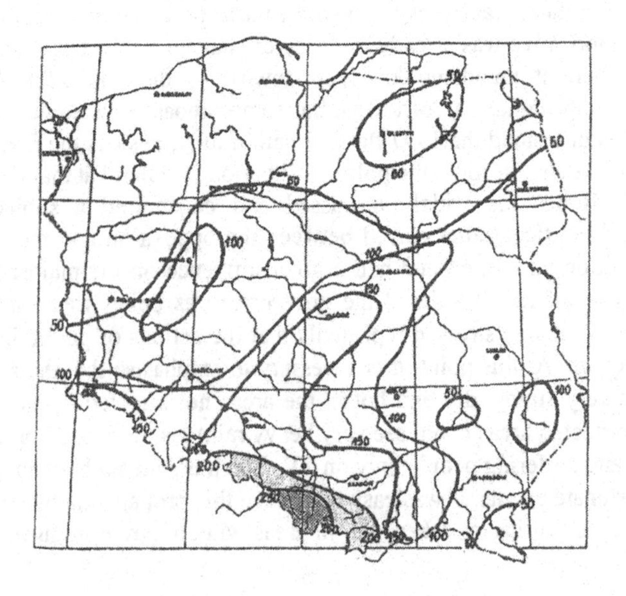

Figure 2. Rainfalls [mm] in Poland from July 4 to 8, 1997 [9].

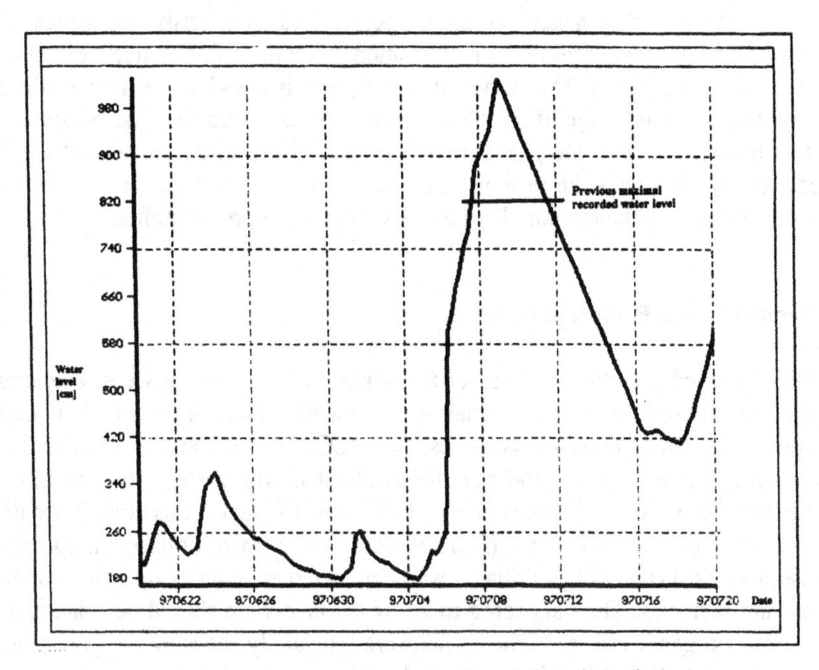

Figure 3. Hydrograph of the Odra River (Miedonia Station).

2. Response Time

The first official warning about intensive rainfall was announced by the Polish Meteorological Institute (IMGW) on July 4, 1997. From July 5 to 7, the first areas were flooded, mostly fields, river valleys and sporadically some houses. The scale of the flood was similar to the previous events. The fire-brigades started their normal activities, upon request. On July 7, at 1430 hours, Prime Minister Cimoszewicz boarded a helicopter to visit the flooded areas (accompanied by the Chief of Civil Defence, deputy minister of interior, and the deputy minister of environmental protection - the author of this paper). Localities Raciborz, Glubczyce, Glucholazy and Prudnik were inspected. The helicopter landed three times and the prime minister had an ample opportunity to talk directly with local residents, as well as with firemen, policemen and other servicemen. There were no signs of panic or nervousness, the situation seemed to be under control. The prime minister returned to Warsaw at 2100 hours. Unfortunately, it was impossible to monitor the situation in the Klodzko Valley. The pilot refused to fly to this region because of extremely heavy atmospheric conditions. At that time, the situation in Klodzko was the worst: there were first destroyed houses and first victims.

As indicated in the post-flood report of IMGW [6] the obvious and direct reason for the flood were heavy rains which started on July 5 and, occurred in two places; from 1000 to 1900 hours in the upper Odra catchment and from 0800 to 2200 in the Nysa Klodzka catchment. Rains continued uninterruptedly during the next three days and nights.

On July 8, the prime minister set up a special state anti-crisis committee (ACC) responsible for fighting the flood. Its composition and terms of reference were presented by Szamalek elsewhere [9]. The author of this paper was the vice chairman of the ACC. To the best of our knowledge of this flood, prime minister's reaction was adequate. The myth that the government, the prime minister and public institutions reacted too late is contradicted by the truth - the response time of the public institutions mentioned above was largely caused by the magnitude of the flood and its unpredictability.

3. Methods Used in Fighting the Flood

The ACC received full powers to co-ordinate work of all state services, agencies and organisations. Its duty was also to analyse the current situation and predict the flood development. In order to fulfil its duties, the ACC informed and warned the media, public services as well as general population about newly arising dangers. The ACC concentrated the means and forces at the state level. However, this state body did not control directly the flooded areas. It was the duty of firemen, policemen, soldiers and administrative staff (including civilians) who conducted emergency operations in towns, villages, and factories. They appeared to know the best what should be done and what were the most urgent measures. The media were repeatedly accusing the government of acting at the central level only, instead of acting in the flooded areas as well. By comparison, in the Czech Republic the state crisis body worked in the flooded areas and very soon appeared to be the unable to act because of lack of communications and proper conditions for monitoring and co-ordinating rescue operations.

The ACC mobilised and monitored the use of the state reserves of food, medical supplies, disinfection means, sand bags, telecommunication systems, etc. The rescue operation was the biggest civil and military operation of that kind in Poland since the World War II. In total 25,000 firemen, 10,000 policemen, 45,000 soldiers and hundreds of thousands of civilians and volunteers were engaged. During the flood of '97 more than 162,000 people were evacuated. A review conducted by the Supreme Chamber of Control [6] after the flood confirmed that the work of the ACC was efficient and well organised.

A flood of this enormous scale was an ultimate test for legal and organisational systems of crisis management. The older procedures, which were complex but proven in previous flood fighting actions, appeared in this case to be largely irrelevant. Undoubtedly, the main reason for this failure was the tremendous magnitude of this natural disaster. Even though the old system worked well during previous actions, this time it was insufficient.

Thus, the myth that the methods used during flood fighting action were wrong and inadequate (the opposition proposed to move the ACC from Warsaw to the flooded areas) is contradicted by the truth - the ACC used all the necessary methods, means and forces at government's disposal.

4. Opposition Views and Suggestions

In July and August 1997, the political opposition assumed that it was necessary to change the legal and organisational system of flood prevention in Poland as soon as possible. Additionally, the opposition announced that it had already proposals and plans for a new system. They criticised the chaotic system of water management administration, the lack of adequate funding for building new water reservoirs, embankment systems, repairing old dikes, etc. In October 1997, the new Polish constitution was introduced. It defined three different states of emergency as follows: (a) the state of natural disaster, (b) the state of emergency and, finally, (c) the martial law. In October 1997, the President of Poland submitted three drafts of the emergency state law - as of May 1999, they were not in force [11]. Thus, the 1999 regulations are exactly the same as in July 1997 (the same bad regulations). In July 1998, in the Klodzko Valley, a new flood occurred on a much smaller scale than in July 1997, and the same problems concerning flood warning and fighting were experienced, though on a smaller scale. Thus the myth that the opposition was aware of deficiencies in the legal and structural system of flood protection and wanted to change it quickly is contradicted by the truth - in late May 1999, the legal and administrative system of flood protection was just as bad as before, and in fact, some dikes were even in a worse state than in 1997 [6].

5. Poland vs. Czech Republic and Germany

In 1997 a kind of comparative analysis was presented in the media. Their popular thesis was that the flood response was better organised in Germany and the Czech Republic. These countries were able to minimise losses, offer people efficient assistance, and control the flood situation. The truth about the magnitude of the flood in the Czech Republic and difficulties in crisis management was released gradually. In the opposition view, only Poland was unprepared for the flood. Reality, however, was different.

The World Disasters Report mentioned "the scale of floods found both countries desperately unprepared, Germany was able to marshal....Czech and Polish governments seemed almost impotent to prevent flooding, organise relief efforts, and mobilise funds to rebuild" [12]. Therefore, it was impossible to avoid the flood of July 1997 and the associated losses. Precipitation which caused the flood was so high (Figs. 1 and 2) that the amount of water significantly surpassed the retention capacities of river basins and retention reservoirs in the Upper Odra catchment [2]. The Polish flood response system was, generally speaking, similar to that used in Western Europe [3].

Destruction, panic and chaos in the concerned areas of Poland in the beginning of the flood were presented by media as well as by the opposition in comparison to the "Ordnung" (order) of the preparatory action on the German side of the border reach and in the Lower Odra. Yet, this was at the time when the flood peak was still far away upstream of the Lower Odra. When high water reached the Slubice/Frankfurt area, it turned out that embankments on the Polish side, subject to massive strengthening efforts, withstood the stress while those on the German side broke in several places (Ziltendorf, Breskow) and contributed to massive inundation and high flood damages in Germany.

Thus, the myth that Germany was well prepared to prevent a flood of any scale and that German troops and civil defence, thanks to their preparations (embankment reinforcement, flood monitoring, good organisation) could cope with floods and avoid heavy damages is contradicted by the truth that embankments were broken and many villages and towns were flooded. This happened in spite of the flood warning from Poland and the Czech Republic, and the previous experience from the 1993 and 1995 Rhine floods, during which some embankments were also broken.

6. Information and Communications

During every flood rescue operation the crucial problem is to inform the public properly, broadly and accurately. The question is who should inform journalists and other media specialists. During the first days of the 1997 Great Flood the access to the public TV station was limited. Among a number of TV stations in Poland, the public TV has the biggest audience. The first information about heavy rains appeared on the public TV station on July 6 [4]. Between July 7 and 13, the public TV heavily criticised public services, especially those provided by the government. Sometimes, one would get the impression that the flood was caused by the government itself. Many journalists described the flood situation, its origins and consequences. It was obvious that their presentations were more emotional than professional. Therefore, the Anti-Crisis Committee requested on July 9 that the head of the public TV station provides a chance for the ACC to present flood information on the basis of accurate data from hydrometeorological institutions and civil defence. The special flood programme, prepared and conducted by the ACC, appeared on July 14 and was presented many times, up to August 4, 1997. The programme appeared twice a day at noon and at 2000 hours. The ACC specialists and ministers and other officials were the guests of this flood program. They explained the existing situation, predicted the flood development, and indicated which towns and villages had to be evacuated.

On July 20, 1997, the ACC flood programme was watched by 37% of the total TV audience and this was a record for the public TV station [4]. Very often the members of ACC had to discuss and correct the misinformation published in newspapers or presented by public and private TV stations. After decades of censorship, the freedom of media became an essential human right in Poland. Yet, during the flood, the absolute freedom of the media did not go together with the underlying responsibility. Chasing sensations did not serve well the rescue operation. In the country, where so many individuals did offer their opinions on flood management in the media, the questioning of individual decisions on flood mitigation measures (e.g. whether or not it was necessary to move amphibious vehicles from central Poland to the flood zone) was not uncommon [5]. Furthermore, the press presented “alternative” weather forecasts, some of which mostly underestimated precipitation causing the second flood, correctly reflected in the official Polish Meteorological Institute forecast. A local TV station in

Warsaw disseminated a news that a flood wave seven metres high (!) was approaching the capital. This misinformation caused chaos.

In tune with public criticism, journalists nearly laughed at statements about damages to embankments done by rodents and small mammals (beavers, otters, muskrats, etc.) [10]. People did not understand why the flood developed so fast, why the embankments were not sufficient to keep water back, why water appeared in streets and inside buildings, even though the embankment system was not broken and was not overtopped by water. The ACC showed on TV sketches illustrating these cases. The flood time was the time to mutually learn about communications. The conclusion which can be drawn from the Great Flood lesson is the necessity to be prepared for this kind of crisis situation. This preparation should include production of animated pictures to be used in TV programs, and the training of well prepared and educated journalists able to explain, and comment on, natural disasters.

The Great Flood was extensively covered by the Polish media. For several weeks, it was the dominating topic in the press and the main subject of cover stories of weekly magazines. TV programmes, including regular news and information from inundated areas and live broadcasts from the Anti-Crisis Committee, were numerous and gained high visibility. Many journalists, politicians, social activists and other public personalities considered it appropriate to share their, typically negative, opinions on the flood management activities via the media. In this respect, the myth that only the journalists informed the public accurately and the media presented an objective flood situation is contradicted by the truth that most media took sides in this public debate of flood fighting activities and were mostly supporting the opposition side.

7. Conclusions

In 1997 people thought that with high enough embankments, numerous water reservoirs of large capacities and weather radar, it was possible to avoid flooding and the associated damages. As understood now, it was a false perception, because the magnitude of the 1997 flood exceeded the previous ones. The Polish weather radar system consists of two stations located near Warszawa and Katowice. During the July 1997 flood, the radar near Katowice appeared to be of little help, because of lack of software and poor telecommunication infrastructure [1]. Unfortunately, this situation has not changed for the time being.

Poland needs a new system of flood protection and management. First of all, it is necessary to adopt a new law describing emergency states. Secondly, one must built as soon as possible a full meteorological radar system, which would provide warning about intensive precipitation. Thirdly, digital maps of towns and villages (eventually covering whole Poland) should be prepared and used during floods to forecast the extent of flooded areas and to guide evacuation. The last but not the least, the society as well as the media should be educated that natural disasters have no political colour. They are independent of all, and during their occurrence, the most important thing is to concentrate all the means and forces on fighting the disaster. It is better to discuss the actions taken after rather than during the rescue operation. It is much easier to determine the institutional and personal responsibilities after the operation has been completed. Fighting the flood is not a political war between different parties and social movements. Therefore, it is immoral and unethical to use this subject as a political gun. Nobody knows which government will have to pass a similar exam in crisis management in the future. Anything that was successful can be repeated during next operation, anything that was wrong can be changed after proper analysis. After all, Noah has managed to survive the Genesis Flood only because he was informed early enough to prepare the Ark. And what is even more important, he believed the messages received...

8. References

1. Bobiński E., Zelaziński J. (1998). Are we more clever after recent floods? (in Polish). In: B. Więzik & M. Maciejewski (eds.) *Natural disaster emergency*(in Polish), *Bielsko-Biala, 14-16 October 1998. Proc. of the Conf.* pp. 43-56.
2. Feluch, W. (1998) Hydrological aspects of flood in July 1997. Warsaw. Unpublished.
3. Rosenthal, U. & Hart P. (eds). (1998) *Flood Response and Crisis Management in Western Europe. A comparative Analysis,* Springer.
4. KRR&TV (National Council of Radio and Television) (1997). Media during flood -Report of the National Council of Radio and Television (in Polish). KRR&TV, Warsaw.
5. Kundzewicz, Z., Szamalek K.,and Kowalczak, P. (1999): The Great Flood of 1997 in Poland.. *Hydrol.. Sci. J.* Vol.**44** (in press)
6. NIK (Supreme Chamber of Control) (1998). Information on results of review of the state flood protection of the country and the course of flood mitigation during the flood in S and W Poland in July 1997 (in Polish), NIK, Warsaw.
7. Ozga-Zielińska, M. (1997). The necessity of determination of maximum likelihood of runoff for Polish rivers caused maximum likelihood rainfalls (in Polish). *The flood'97 Conference. 10-12 September 1997 Ustroń.* Polish Hydrometeorological Institute. Vol.2, pp.1-11.
8. Stachy, J., (1997). Differences between peak flows in the Oder and Vistula river basins (in Polish). *The flood'97 Conference. 10-12 September 1997 Ustroń* .Polish Hydrometeorological Institute, Vol.2, 191-208.
9. Szamalek K. (1998): Oder flood'97 - Lessons learnt in Poland. In: Bronstert A., Ghazi A., Hladny J., Kundzewicz Z. & Menzel L. (Edts). (1998) *The Odra/Oder Flood in Summer 1997: Proceedings of the European Expert Meeting in Potsdam, 18 May 1998,* PIK Report No. 48, Potsdam Institute for Climate Impact Research (PIK), Potsdam, 21-25.
10. Szamalek K. (1998). The Titanic flood syndrome (in Polish), Prz. Tyg. **27** (800), pp. 8-9.
11. Szamalek K. (1999). Legal regulations of emergency states and the role of media (in Polish). In: Maciejewski M. (Editor) (1999), Contemporary problems of extreme environment disasters, IMGW Warszawa. Pp. 99-104.
12. *World Disasters Report* (1998) - International Federation of Red Cross and Red Crescent Societies. Oxford University Press.

ACHIEVEMENTS AND FUTURE NEEDS TOWARDS IMPROVED FLOOD PROTECTION IN THE ODER RIVER BASIN: RESULTS OF THE EU-EXPERT MEETING ON THE ODER FLOOD IN SUMMER '97

A. BRONSTERT[1], Z. W. KUNDZEWICZ[2] and L. MENZEL[3]

[1]*Potsdam Institute for Climate Impact Research, P.O.-Box 601203, 14412Potsdam, Germany*
[2]*Res. Centre of Agricultural and Forest Environment, Polish Academy of Sciences, 60-809 Poznań, Poland*
[3]*Potsdam Institute for Climate Impact Research, P.O.-Box 601203, 14412 Potsdam, Germany*

1. Introduction

This paper offers a synthesis of the discussions during the International Workshop on the Odra/Oder flood. The objective of this expert meeting was to bring together scientists and policy-makers from the three affected countries and additional experts from various EU member states. The topics discussed included meteorology, climate, hydrology, river hydraulics, flood management and socio-economics [1]. After analysis and discussion of the flood and the corresponding circumstances, conclusions were drawn with respect to what has already been achieved and what needs exist for further research towards better and environmentally sound flood protection.

2. The July '97 flood in retrospective

The Workshop clearly demonstrated that all the three countries affected by the Oder flood of the summer of 1997 have undertaken vigorous efforts to evaluate the flood in a broad context.

In the flood-affected countries, work is underway to determine the exceedance probabilities (return periods) of the flows observed along the river. Analysts tend to distinguish three stages of the flood. In the headwaters of the Oder and its mountainous tributaries in the Czech Republic and in Poland (see Figure 1), the flood was a very rare event. During the first stage, in the mountains and headwaters, it was a destructive, killer flood, with very strong dynamic effects in both Poland and the Czech Republic. Numerous villages and towns were inundated or even destroyed. When moving downstream, the flood became less violent, yet high losses were caused by the flooding of urban areas.

After inundating the town of Racibórz (65,000 inhabitants), the Oder river inundated other large towns located downstream, including Opole (131,000) and Wroclaw (700,000). In these cities, high material losses were unavoidable, because their flood protection systems had been designed for much lower design flows. Even

J. Marsalek et al. (eds.), Flood Issues in Contemporary Water Management, 75–83.

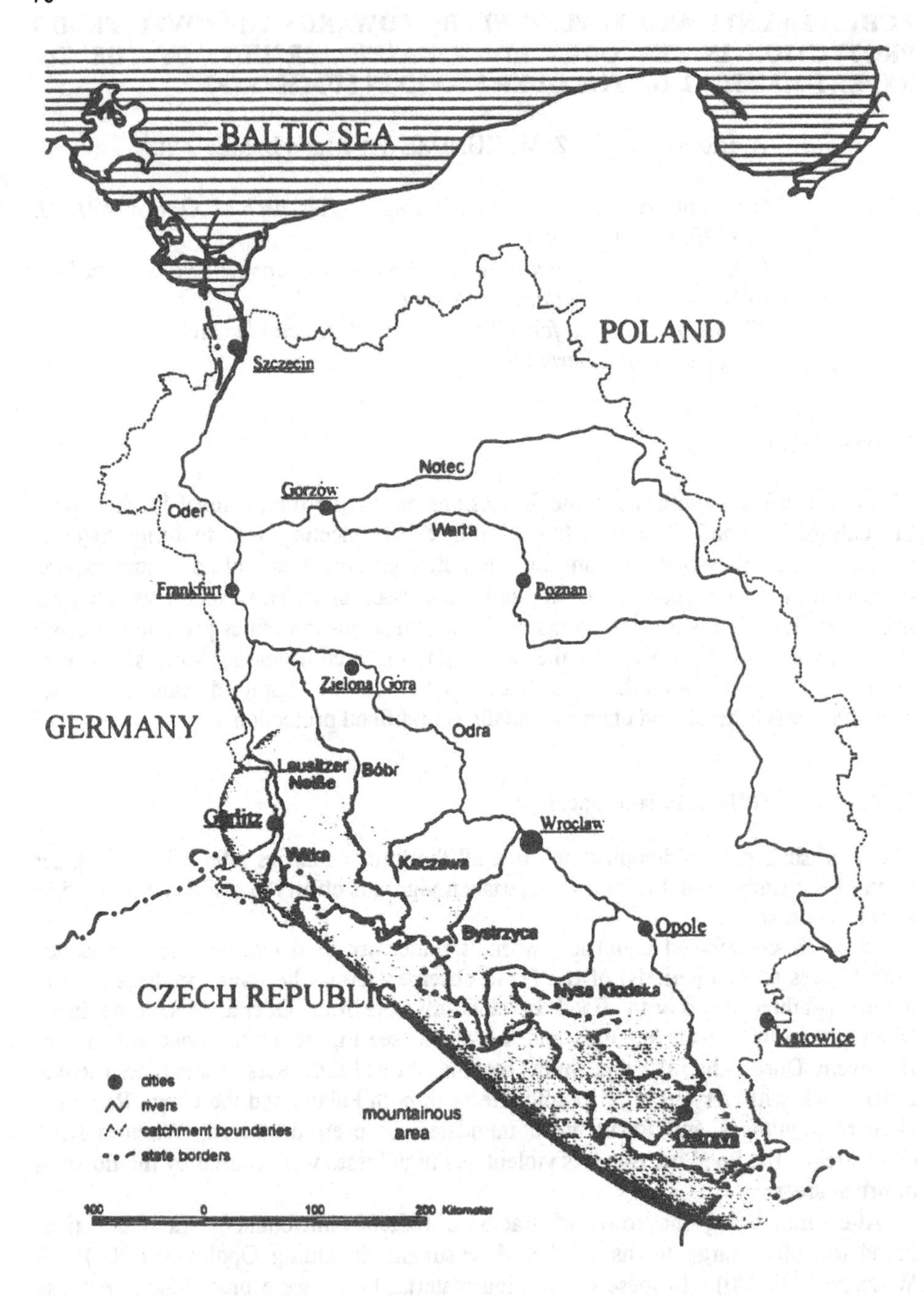

Figure 1. **Map of the Odra/Oder catchment including the major tributaries and cities.**

further downstream, along the Polish-German border reach, the flood return period was further reduced, but still exceeded 100-year flood conditions, on which the flood defense design was based. In this latter phase of the flood, enough time was available for preparations, build-up and strengthening of embankments, evacuation, etc. In addition, the 8-day basic flow forecast data were of good quality. In the lower Oder, the flood was not as record-breaking as upstream, yet in most locations historical maxima were exceeded. Inundation strongly endangered the towns of Eisenhüttenstadt (50,000 inhabitants) and Frankfurt/Oder (80,000 inhabitants).

The 1997 summer flood on the Oder lasted long; alarmingly high water levels were exceeded continually during several weeks. Even the exceedance of historical absolute maximum water levels persisted up to 16 days. Table 1 summarizes flood damages in all three affected countries.

TABLE 1. Summary of changes caused by the 1997 Odra/Oder flood (after [2])

	Czech Republic	Poland	Germany
Fatalities	60	54	--
People evacuated	26,500	162,500	6,500
Economic losses (in US$)	1,950,000,000	2,800,000,000	360,000,000

- Many losses of human lives could have been avoided, if the flood warning would have reached these people and/or if they would have believed this warning and reacted to it; i.e., the majority of fatalities was caused by a partial (local) failure of the warning/information system or by the ignorance of the severity of the flood danger.
- There is a very high uncertainty in quantifying the economic losses. Estimates range from 0.55 to 2.2 billion US$ in the Czech Republic, from 2.5 to 4.0 billion US$ in Poland, and from 0.3 to 0.7 billion US$ in Germany.
- Only a small portion of the total damage (13%) was insured [3], mostly in Germany.
- The losses and evacuations in Germany have been moderate, at least in comparison to those in the Czech Republic and Poland. However, it is highly relevant to mention that water levels in the German reach of the Oder river were strongly reduced because of dike breaks and the huge inundation upstream in Poland. Without this retention (of course highly undesirable from the Polish perspective), the flood would have been catastrophic in Germany, too.

The flood magnitude was unexpectedly high. Over decades, people have become used to floods on the Oder and its tributaries. They knew where the safe places were, and where to find shelter for their animals and mobile property. This time, water entered even the usually safe places. As demonstrated by the catastrophic flood in July 1997, there is a lack of an adequate flood protection system for several towns. It is the urban flooding that caused most of the material losses. Also vast areas of agricultural land could not be adequately protected. Since 1998, there has been an ongoing process of re-assessment, strengthening, modernization and extension of the existing flood protection infrastructure in all the affected countries.

3. Strategies for flood protection and management

The 1997 flood triggered the initiation of a broad discussion on action strategies, which fall into the following three categories:

3.1. MODIFICATION OF FLOOD WAVES

Flood risk may be partly controlled by following the concept of "keeping water where it falls". The necessity of watershed management and environmental conservation in flood risk areas includes an appropriate land use planning serving to modify water storage in, and the proportion of the inundated area of, the landscape. Examples might be the re-establishment of wetlands along water courses, conservation of structural landscape elements, stormwater management in urban areas, changes in agricultural practices or reforestation. The dying of forests in the Oder headwaters was said to contribute to fast runoff formation and high peak flow levels [4].

- River engineering measures can help modify the depth, volume, and velocity of the flood wave (reinforcement of embankments, dikes, dams, reservoirs etc.). The influence of linear structures, such as roads, railways or pipelines, and the discharge capacity of bridges should be examined.
- Effects of reservoirs in upstream areas can be very important. The reduction of the peak flow of the first Oder flood wave in the Czech Republic proved to be very effective [5]. For many years there have been plans for the Racibórz reservoir (close to the Polish-Czech border), which would become the largest reservoir in the entire drainage basin, if it were built. The presence of this reservoir would have helped reduce the losses during the 1997 Oder flood. At least in Poland, this flood brought a significant change in the public opinion towards a higher acceptance of water storage reservoirs [6].

3.2. MODIFICATION OF SUSCEPTIBILITY TO LOSSES

- A robust warning system, risk awareness of the population and their preparedness for evacuation are a must to preserve human lives.
- Recommendations for the prevention of high losses include the demand for flood forecasting and alert centres to coordinate flood forecasting and warning dissemination, and evacuation of the population. It should be clarified before the flood, who will be involved in the decision making processes during the flood. Reducing the probability of high losses includes land use regulations for frequently inundated areas, e.g. defining the zones free of settlements.

3.3. MODIFICATION OF CONSEQUENCES OF UNAVOIDABLE FLOOD LOSSES

Awareness building is one step towards a culture that lives with risks of natural disasters. The people should realize that in extreme situations, damages caused by natural disasters are unavoidable and that in such cases the common interests stand above those of individuals. This involves the resettlement of people in some extreme cases. On the other hand, a reliable flood insurance system with franchises should be implemented to guarantee fast post-flood recovery [3].

4. Structural or non-structural measures?

What is really needed is a best mix of all three components (3.1)-(3.3) of the above strategy, including both structural and non-structural methods. The 1997 flood unveiled the many weak points of the existing flood protection systems and where improvements were badly needed. Among the immediate needs are: Working out a flood management strategy, including repair and modernization of technical infrastructures (embankments, dikes, dams, polders, flood plains) and non-structural flood protection measures. Indeed, every link in the chain of operational flood management (observation - forecast - response - recovery) needs strengthening.

There have been numerous floods in historic times along the Oder, both summer rain-caused floods and winter floods, and consequently, the need for an adequate flood protection system has arisen. It consists of dams, reservoirs (including dry flood protection reservoirs), embankments and relief channels along the Oder and its tributaries, and a system of polders. Yet, the existing system proved to be inadequate for an extreme flood. Efforts to make the upper part of the Oder navigable will conflict with the proposed flood protection measures.

Structural flood protection measures such as embankments, dikes and dams, give indeed an illusion of complete safety. However, an embankment designed to withstand a 100-year flood would clearly fail when a 1000-year flood arrives, despite the design safety margin and all the emergency strengthening/build-up efforts during the flood.

The Oder River needs vast flood retention areas in order to accommodate future huge flood volumes. Yet, a problem remains: how to create such flood storage areas? At least, from the perspective of flood protection, the last remaining natural flood plains should be preserved. How to relocate people living there? This requires a broad general understanding of what has to be done for flood protection. The burden of flood protection should be borne by the whole population. Relocation has to include generous compensation. An option following the solution implemented after the 1993 Great Flood in the USA is to buy the endangered property with state funds. The enforcement of zoning principles, which have been partly abandoned in the absence of floods over a long time, has to be strengthened.

5. Operational needs

Operational flood forecasting depends on weather conditions, the conditions in the catchment, the river conditions and the potential damage of a flood (i.e., the density of population, vulnerability of industrial facilities, intensity of agriculture, etc.). Longer-term flood forecasting, a week ahead or even more, should be possible in the middle and lower Oder basin, based on telemetry of river discharges in the upper river parts and the weather situation. However, the real-time modelling of flood formation and movement, and now-casting (say six hours ahead), would be needed in fast responding areas ("flash flood catchments"), e.g. in mountainous tributaries. This should be based not only on real-time meteorological observations but also on quantitative precipitation forecasts.

The optimal operation of flood storage reservoirs may require releases of substantial volumes of water before the flood arrival, in order to create a sufficient storage capacity needed to attenuate the flood peak. This may require a long lead-time forecast based on the predicted pattern of movement and development of meteorological conditions, beyond the forecast needed for issuing public warnings.

All flood defense activities have to be strengthened. Depending on the magnitude and severity of the flood, they may include [7]:

- Issuing warnings,
- Monitoring and strengthening flood defense structures,
- Operation of flood storage systems,
- Evacuation of flood plain occupants and closure of transport links, and
- Emergency assistance and rescue.

Improvement of forecasts and upgrade of the whole system of forecast - warning - dissemination is needed and, in particular, improvement of the weakest element in the chain that determines the performance of the whole. In particular, an operational and robust trans-boundary flood warning system, comprising all three countries, and a fast and reliable exchange of information are required. It is necessary to upgrade the logistics and the coordination of actions, to make the division of responsibilities clear and to provide reliable communication systems. Robust instrumentation and data transmission are a must, because the flood unveiled deficiencies of the existing arrangements. There is a need for a trans-boundary data base and registration of hydrological conditions/responses for the whole river and its catchment. This is a long term task that should be initiated and funded by the governments and possibly by the EU. The data and exchange procedures should be harmonized.

In addition, to assist in the translation of pre-flood planning into an effective operational flood management, it may be possible to encapsulate the knowledge from scenario-based contingency plans into a decision support system. Such a decision support system may assist flood forecasting officers to interpret forecast information during events which exceed their previous experience.

In exceptional conditions, it may be necessary to take extreme measures for the overall public good. When considering options for defense of a large town, the sacrificial breaching of an embankment, to provide flood storage within a normally protected area, was considered to mitigate flood levels in highly vulnerable areas elsewhere. This requires an early information of the population on flood risk and the possibility for a strong and active public involvement in the planning of flood defense measures.

6. Research needs

There are several research needs in the context of improving flood protection and flood management systems. In general, a holistic view of floods, from a multidisciplinary and interdisciplinary perspective, with hydrological, meteorological and human components (i. e., social, economic, etc.) is required. In this respect, the following topics have been identified during the meeting:

Meteorological forecasts, databases and climatic conditions:

- Quantitative precipitation forecasts, in particular for small and mountainous catchments. This includes the establishment of a meteorological radar system.
- Creation of hydrometeorological databases with long-term and homogeneous data records. Exchange of data across borders.
- Investigation of possible changes of the climatological conditions in Central Europe affecting flood generation in the Oder basin; is the probability of low pressure weather systems in Central Europe changing?

Hydrological and hydraulic tools

- Design of a hydrological data network with sufficient redundancy to achieve the required accuracy and the security of information for forecasting in the most severe conditions.
- Development of advanced forecasting methods (including real-time approach, now-casting) and related international collaboration.
- Evaluation of the effects of land-surface conditions on flood formation, including reliable models for the quantification of water retention by different types of vegetation, soils, landscape structures, types of agricultural practices, etc. How strong is the effect of deforested areas (or dead forest) on flood generation?
- Development of an integrated hydraulics model (including flood propagation in rivers) and a catchment-wide hydrological model, in conjunction with visualization components for all areas; and, the development of GIS-based software, visualizing flood levels for real-time operation.

Management and assessment of flood risk

- Development of flood hazard maps, digital terrain maps and inundation visualization software.
- Execution of zoning principles, e.g., identifying the areas not to be occupied or zones of high risk.
- Optimization of reservoir control schemes.
- Development of strategies for involving the public in flood management.

7. Human factors

In order to minimize flood losses in any catchment, a great deal of human collaboration and co-ordination is necessary. This is even more critical for an international river, such as the Oder.

The population of the Oder valley has to be made aware of the flood danger. Even if floods had struck quite frequently in the past, people keep envisaging a normal magnitude event and not an unthinkable size of the recent deluge. The range of those who have to improve their knowledge and understanding of floods is broad: from the general public - population of the areas at risk, through civil defense, local municipalities, regional and national authorities to scientists, policy makers and legislators. The precise delegation of responsibilities and a good understanding of "who does what", already before flood arrives, is very important.

In Poland the flood came during the pre-election time. Yet, the nation was not quite united in the face of the common danger. There have been so many political aspects that different players wished to pursue. In fact, the political perspective, that is the time horizon to the next election, does not match the long-term perspective needed for envisaging a flood protection system. "Human" factors typically involve four areas:

- How best to communicate information to the public in terms of identification of the level of risk, action plans to take during floods and the warning of an imminent flood.
- The need for better communication between hydrologists and meteorologists when developing and operating real time forecasting systems.
- The need to communicate research advances from the scientific community to the operatinal authorities and agencies, and
- The need for continued dialogue between researchers in flood risk mitigation (e.g. the RIBAMOD concerted action, e.g. Casale et al [8]).

The Oder flood demonstrates the importance of an international approach to the problem with a trans-boundary cooperation: exchange of information, data, forecasts, and technical assistance. However, the data, models, and forecasts for different countries may differ, with a discontinuity at the interface.

8. Conclusions

Floods are natural phenomena, which will continue to occur. The human impact on floods can be of considerable extent, both in positive and negative ways, but always will be limited. Scientists and water engineers have to contribute to providing answers to the questions related to floods. They have to get actively involved in the public discussion on this topic, in order to transfer their knowledge to the public in an appropriate form.

As no absolute flood safety can be provided, societies have to live with floods rather than trying, in vain, to eliminate them. Flood protection should be recognized as a concern of the whole society and be of the same value as social or economic welfare [2]. The core issue is to improve flood preparedness and to minimize flood losses. In particular, minimization of fatalities must be achieved, following the slogan: "Save lives rather than property".

If the flood return period falls into the range of thousands of years, as was the case of the 1997 Oder flood, there is no way to avoid material losses. One thing is certain, however, the losses could have been significantly reduced. The ultimate cause of the flood, namely intense and long-lasting rainfall on saturated soils was a natural phenomenon beyond anyone's power, but an inadequate flood protection system, deficient organization and information dissemination have clearly contributed to the magnitude of the resulting losses.

Acknowledgements
The authors gratefully acknowledge the support and funding of this workshop by the European Commission (DG XII).

9. References

1. Bronstert, A., Ghazi, A., Hladný, J., Kundzewicz, Z. W. and Menzel, L. (eds.) (1999) Proceedings of the European Expert Meeting on the Oder Flood 1997; Potsdam, 18 May 1998, European Commission, DG XII; Office for Official Publications of the European Communities, Luxembourg, 163pp.
2. Grünewald, U. (1998) Ursachen, Verlauf und Folgen des Sommerhochwassers 1997 an der Oder sowie Aussagen zu bestehenden Risikopotentialen. Eine interdisziplinäre Studie - Kurzfassung, *Deutsche IDNDR-Reihe* **10a**, 46pp.
3. Kron, W. (1998) Insurance aspects of river floods, in: A. Bronstert, et al. (eds), *Proceedings of the European Expert Meeting on the Oder Flood 1997*. Office for Official Publications of the European Communities, Luxembourg, 135-150.
4. Dubicki, A. and Zenon, W. (1993) Streamflow changes of mountainous river in West Sudety after damage of forest ecosystem, *Zeszyty Naukowe Akademii Rolniczej we Wroclawiu, Inzyniera Srodowiska* **III**, Nr. 232.
5. Mareš, K. and Marešová, I. (1998) Overview of the Odra flood from a Czech Perspective, in: A. Bronstert, et al. (eds), *Proceedings of the European Expert Meeting on the Oder Flood 1997*. Office for Official Publications of the European Communities, Luxembourg, 37-42.
6. Kundzewicz, Z.W., Szamalek, K. and Kowalczak, P. (1999) Destructive flood in a holistic perspective: Poland 1997 *Hydrol. Sci. J.* **44** (in press).
7. Kundzewicz, Z. W. and Samuels, P. G. (1998) Conclusions from the Workshop and Expert Meeting. in: R. Casale et al. (eds), *Proceedings of the Workshop/Expert Meeting 25 and 26 September 1997 in Monselice, Padua, Italy*. Office for Official Publications of the European Communities, Luxembourg, 170-177.
8. Casale, R., Havnø, K. and Samuels, P. (Eds.) (1997) RIBAMOD. River basin modelling, management and flood mitigation, Proceedings of the first expert meeting, Copenhagen, 10 and 11 October 1996. Office for Official Publications of the European Communities, Luxembourg, 167pp.

THE 1998 EXTREME FLOOD ON THE TISZA RIVER

ISTVÁN KRÁNICZ

National Water Authority
Budapest, Hungary

1. Introduction

The risk of flooding is especially significant in Hungary, because of unusual hydrometeorological and runoff conditions in the Carpathian Basin. Before the mid 1800s, disastrous floods had occurred in sequence on various rivers. Since the middle of the 19th century, river regulation has been conducted to improve the living conditions by introducing flood management activities and initiating construction of a complex flood protection system. Flood protection agencies, at that time independent from each other, were established almost in parallel with these activities. The flood protection system has been continuously improved and changed several times until now. It can be stated, however, that the existing flood protection system, still in need of some further improvements, has protected the country against flooding, without any breach of embankments and major inundation, during the past almost 30 years.

In the following, the Hungarian approach to flood control will be briefly presented, including such topics as the natural geographical and hydrometeorological conditions, runoff characteristics of rivers, and the organisation of flood defence which has taken almost 100 years to establish. Finally, the extreme flood of November 1998, on the upstream section of the Tisza River in the north-eastern part of Hungary, will be described, with respect to flood formation and propagation. Other issues addressed include flood defence works built in Hungary, the state of these works, interventions made in, and experiences gained with, this system, and the major tasks and expenses in the future.

2. Brief presentation of the flood control in Hungary

Hungary is situated in the deepest part of the Carpathian Basin, in the area confined by the Carpathians and the eastern arm of the Alps. The territory of the country is 93.000 km^2, 52% of which are lowland plains. The highest point of the country barely exceeds 1000 m above sea level.

The headwaters of major Hungarian rivers are formed by the mountains of the near Alps and Carpathians. Ninety-six percent of surface waters and floods originate on foreign territory. The climate of river catchments is equally influenced by West-European oceanic, South-European Mediterranean and East-European continental

J. Marsalek et al. (eds.), Flood Issues in Contemporary Water Management, 85–94.

effects. It is typical for precipitation conditions that a flood generating great rainfall might appear in any river basin any time during the year. However, in the winter, there is a danger of ice related problems.

Fifty-two percent of the Hungarian territory is endangered by floods and excess waters. It represents a unique, incomparable flood risk in Europe, which motivated the government to declare water damage prevention issues as part of the public safety policy.

2.1. THE GEOGRAPHY OF HUNGARY

With respect to annual flows, in the 120 billion m^3/year of surface and subsurface waters which leave Hungary along the southern border, only 6 billion m^3/year originates on its territory. Thus, the water regime in Hungary depends on water management practices outside of its territory, apart from the domestic hydrological and hydrometeorological conditions. Consequently, the international co-operation in this field is extremely important for Hungary.

Seventy to seventy-five percent of Hungarian surface waters are conveyed by the Danube and Drava rivers, and the rest is contributed by the Tisza river and its tributaries. Hydrological characteristics, established from long-term records, indicate occurrences of minor floods every 2-3 years, significant floods every 5-6 years, and severe floods every 10-12 years. The flood duration could be as short as 5-10 days in the upstream sections of rivers, but up to 50-120 days in the downstream sections. The past 10 years confirm the above flood statistics; in spite of long-lasting dry periods in almost every year, significant floods occurred on some rivers in 1991, 1994, and 1995-96, and severe floods occurred in November 1998 and the spring of 1999.

2.2. RIVER REGULATION AND CONVEYANCE OF FLOODS

Major river regulation works were started in the middle of the last century and generally led to the shortening of river courses by cutting off over-developed bends and increasing the river slope. On the downstream Danube section, below Dunaföldvár, the original river length of 494 km was shortened to 417 km and the overall bed slope increased from 5 to 8 cm/km, by 23 cut-offs. After cutting 115 bends in the Hungarian section of the Tisza river, it was shortened from 1211 km to 758 km and its overall bed slope increased from 3,7 to 6 cm/km. On several Tisza's tributaries (Szamos, Bodrog, Korös system and Maros rivers), the total length of 1398 km was shortened to 670 km, after implementing 322 cut-offs.

By 1910, river regulation by cut-offs was practically finished. Since then up to now, the main tasks have been to maintain and improve the so-called uniform river bed, providing flood conveyance and navigation. Recently, water management interventions serving to protect the environment and nature have become important, mainly in the form of restoration of river branches.

2.3. FLOOD PROTECTION

The areal extent of flood-prone areas is 20.700 km^2, divided into 151 sub-catchments. This land includes 1/3 of the national crop lands, 32% of all railway lines, 15% of the whole road network, and almost 2000 industrial plants. Almost 2,5 million inhabitants live on this land in 700 municipalities and produce 30% of the national Gross Domestic Product.

The protection against flood damages has been well developed in Hungary. As a result of the flood protection improvements during the last century, 97% of all the areas endangered by floods is now protected. This protection is assured by about 4200 km of first-class embankments. In this total length, about 2/3 have been built for the critical level of 100-year floods.

The embankments have been built during centuries and strengthened several times, using the building technology of the given period. Therefore, their construction and reliability vary. This fact caused problems, especially in the case of the embankments built of hard soils on the Hungarian Great Plain. The flood protection embankments occasionally cross the primordial river beds in various places and are less safe in these places. According to the explorations done by the most up to date technological means, there are about 1200 locally weak sections, 50-200 m long, where the flood protection safety is inadequate.

2.4. THE ORGANISATION AND MANAGEMENT OF FLOOD CONTROL IN HUNGARY

The first large-scale national Water Act, addressing river regulation, bank protection, and flood and excess water control, was introduced in 1885. This Act charged the owners with the responsibility for flood control and implementation of flood protection activities. This system was in effect until 1948, when the National Government was requested to provide subsidies for the flood protection works and defence. The more recent comprehensive Water Act of 1964 then assigned the full responsibility for flood control to the State.

The Parliament created a new water management law in 1995 (LVII. The Law of 1995), which refers to the protection against damages caused by water, to the defence against such damages and to the related tasks. This law deals with water damage prevention activities, regulation, organisation, direction and defence which surpass local public duties. The law states that in the interest of protection against water damages, the conduct of the necessary tasks is the obligation of the State and/or local self-governments. It is an obligation of those who are concerned with the prevention of damages.

The law also provides direction concerning the tasks of the national management of floods and excess water. It describes who directs the activities at which stage of the protection, and how the competence and power structure are formed. The direction is performed by the minister, up to the highest readiness for protection, recognising that there are three stages depending on the level of danger. At the utmost preparedness stage, direction is provided by the government commissioner [1]. That was the case in November 1998 on the Upper Tisza.

It has to be stressed that beside the National Technical Direction Staff (OMIT) providing operative direction at the national level, local direction is provided by the County Protection Committee, which is headed by the chairman of the general county assembly. Its members are the representatives concerned with flood protection, including the director of the local District Water Authority. The District Water Authority provides technical guidance and organisation of the local protection [2].

International co-operation, with emphasis on co-operation with the neighbouring countries, forms an important part of the improvement, organisation and direction of flood control. This follows from the natural setting of, and situation in, the country. Almost all flood water flow arrives from abroad and passes through Hungary. Generally the co-operation with neighbouring countries is governed by bilateral agreements. Within these, the co-operation is assured by joint flood control regulations and agreements for the exchange of hydrometeorological data.

For the Tisza system, it was the 1998 flood which demonstrated that besides bi-national agreements, it is also necessary to ensure co-operation with all the countries located in the entire river basin. The first meeting of this kind was already held at the beginning of 1999.

3. The extreme flood of November, 1998

3.1 FORMATION OF THE FLOOD

During the 48 hours preceding October 30, 1998, rainfall which fell on the catchment areas of the Upper Tisza and its tributaries in Sub-Carpathia and Transylvania averaged 30-70 mm. The resulting flood wave reached the level of Stage I of preparedness at the municipality of Tiszabecs, on October 30, at 0600, and it resulted in water levels approaching Stage III in the section between Tiszabecs-Záhony, and Stage II between Dombrád-Tokaj and Dombrád-Záhony. Among the tributaries it was necessary to order Stage III preparedness for flood protection on the Túr river and Stage I on the Kraszna river. During the period of this first flood wave, there was no need for special interventions. The area of concern is shown in Figure 1.

After some dry days, a new cyclonic activity began in the Carpathian Basin on November 3, 1998 and resulted in a rainfall of 10-40 mm on the same day, and another 50-80 mm on the following day. At several places, the total measured rainfall depth reached 140-200 mm, on Nov. 4. The rainfall falling on soils saturated by the previous rainfall was almost fully converted into runoff and quickly reached the Tisza river. This quick runoff caused violent rise of water level in the river section upstream of the Hungarian border, where it produced the greatest ever observed flood. Water levels of the Tisza at Rahó and Técső municipalities in Ukraine exceeded the highest ever measured levels and this information led to a forecast of an exceptionally large flood wave in the Hungarian section.

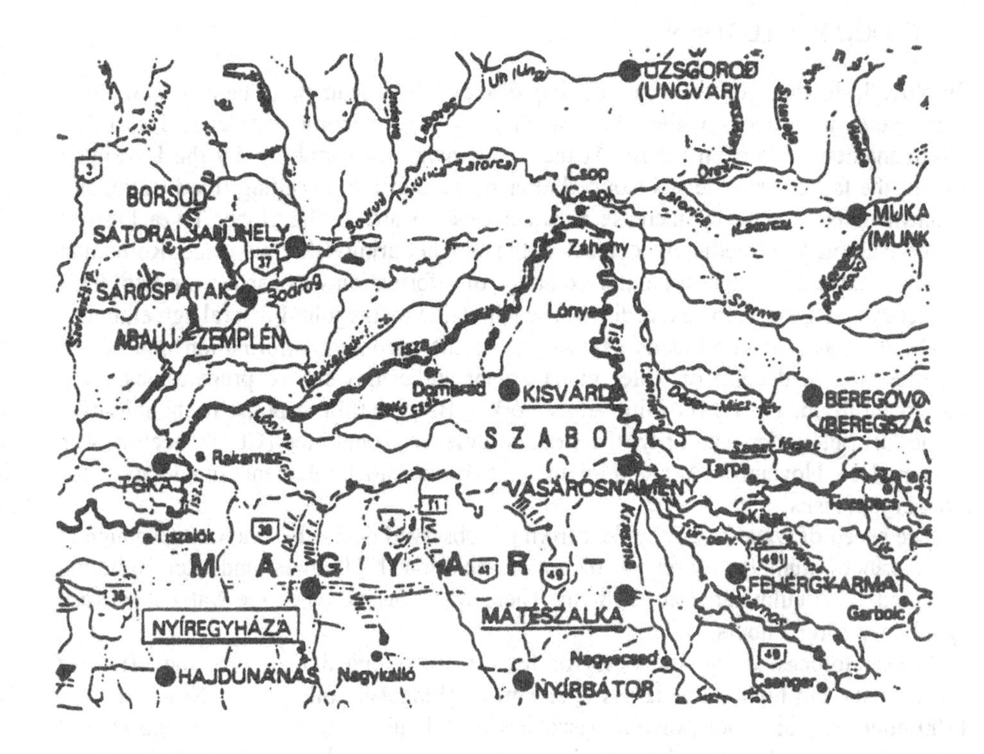

Figure 1. The Upper Tisza River: location map.

Figure 2 summarises annual precipitation in the Tisza region during the past 20 years and in Fig. 3, the 1998 monthly average discharges of the Tisza at the Záhony station are presented.

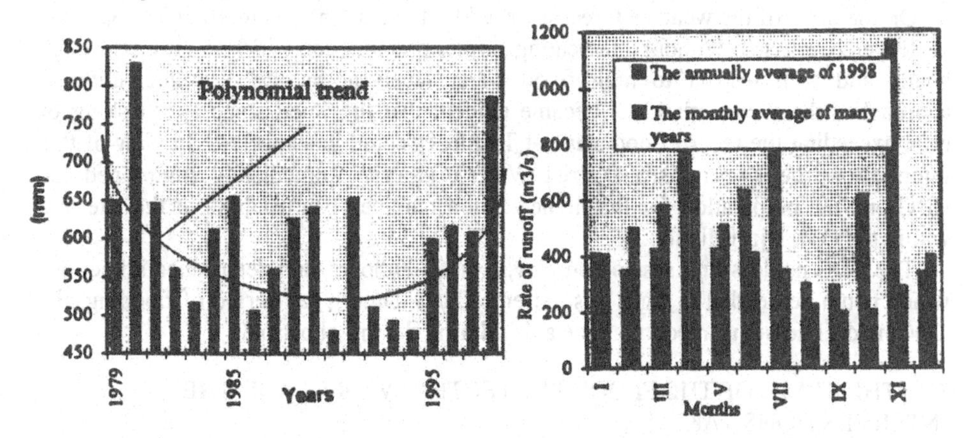

Figure 2. Summary of the annual precipitation in the region during the last 20 years.

Figure 3. Average monthly discharges at the Záhony station.

3.2. FLOOD PROTECTION

On Nov. 3, downstream from the municipality of Vásárosnamény, Stages I-II of flood alert were in effect, while the situation in the section between Tiszabecs-Vásárosnamény was even worse. At the municipality of Tiszabecs, by the Ukrainian border, the fall of water levels stopped after passage of the preceding flood wave, and at dawn of Nov. 4, rapid increase in water levels started. The Upper-Tisza District Water Authority immediately contacted the neighbouring Ukrainian and Romanian water agencies. Since that time, the exchange of information was carried out with high frequency and great scope exceeding those stipulated by the joint bilateral agreements.

On the basis of continuous assessment of the incoming information and actual situation, the leadership of water management authorities started preparations for a significant flood. The OMIT initiated its work in Budapest. In response to the initiative of the District Water Authority, the County Defence Committee (CDC) meeting was convened on November 4, at midnight, and decided to implement immediately the most urgent tasks.

The speed of water rise is evident from the observed data - the water stage gauge at the Tiszabecs municipality read +102 cm on the 4th at 1200 hours, and after 36 hours, the flood wave culminated at +708 cm. There was a period when the water level rose by 68 cm during 2 hours.

It was forecasted, on the basis of hydrometeorological data (i.e. the effective rainfall data and high water levels upstream of Hungary) collected on Nov. 4 before 1300 hours that the flood wave at Tiszabecs would culminate at +526 cm and exceed the previous record. Taking into consideration the minimum lead time, 24-36 hours, on the 4th at 1500 hours, the water-level at Tiszabecs was below that specified in flood alert. The Stage I flood alert was ordered for the embankment section between the municipalities of Tiszabecs and Vásárosnamény. The leaders of the sectional protection and the assigned staff went out to the embankment location immediately and started preparation for protection against a flood wave exceeding the level of the Stage III readiness.

On the basis of the weather forecast issued by the National Meteorological Service, on November 4, at 1600 hours, indicating further precipitation of 30-35 mm in next 20 hours, and with respect to high flood water levels at the Rahó municipality (it approached the record mark), it became evident that the flood stage approaching or even exceeding the record flood stage at Tiszabecs could be expected. In view of this information, the Upper-Tisza District Water Authority immediately summoned the CDC and the preparation for the highest flood readiness started both at the site and within the OMIT jurisdiction.

The CDC held its first session on Nov. 4, at 2400 hours, and determined all actions which could be needed to fight this extreme flood. During the period of flooding, the CDC held 11 sessions, meeting twice a day during the initial period.

3.3. THE STATE OF THE FLOOD PROTECTION WORKS AND THE INTERVENTIONS TAKEN

On the left bank of the Tisza, between Szatmárcseke and Olcsvaapáti, the low height of the protection embankment threatened to cause major problems, which were avoided by a timely build up of the embankment height along a river length of 8 km. At some places, the emergency berms contained flood waters 30-50 cm above the original embankment crest. Analysis is still being conducted with respect to what would have happened in the Hungarian flood defence, if an embankment section would not had broken on the Ukrainian territory. Earlier calculations of flood propagation without the embankment breach indicated water levels 10-15 cm higher in the Tiszabecs municipality (see Fig. 4). In the embankment section, there were two locations where embankment have not been yet fully fortified following the earlier floods. At these sections, the original level of protection had to be restored with the help of building contractors.

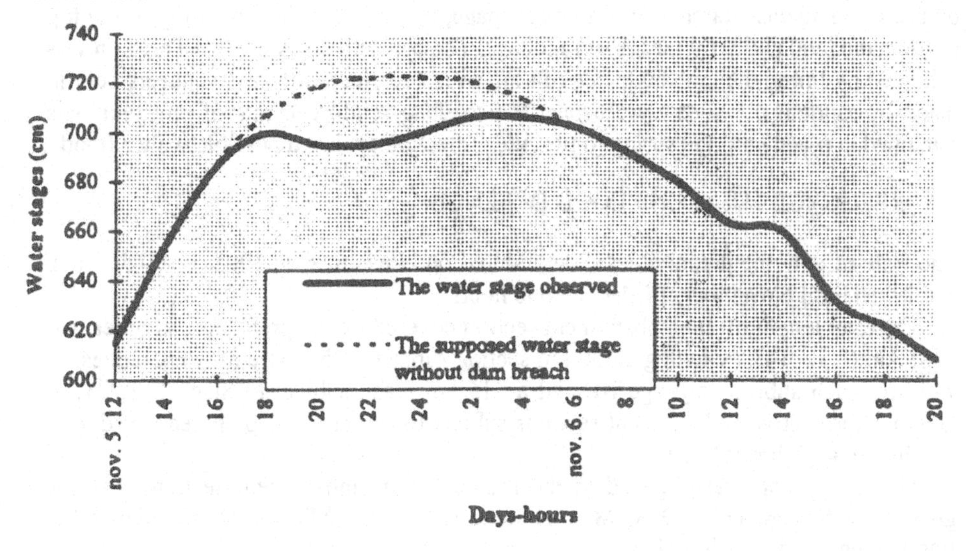

Figure 4. The estimated water stage without an embankment breach.

The flood wave culminated at the municipality of Városnamény on Nov. 7, in the morning. The water level has not reached the embankment crest anywhere, but in certain places, it was within 10-12 cm of the crest of both the left and right embankments of the river reach between Vásárosnamény-Záhony. However, protection activities had to focus on other phenomena threatening the structural integrity of embankments by soaking, softening, water piping, waves, and others.

Similar problems occurred in the reach between Záhony-Tiszabercel and immediately below. However, there was no need for significant interventions with respect to the embankments, just the need to deal with local problems. On November 9, several cases of water piping under the embankment had to be dealt with.

Simultaneously, as the flood wave passed downstream, significant flood fighting activities were conducted in the Lónyay main channel, along the embankment section between the Tisza and Bodrog rivers, and even in the embankment sections of the Bodrog river and in the Taktaköz area. In these areas, embankments were high enough, but the flood which approached the highest ever observed level, called for interventions at several points of the embankment sections with inadequate strength.

As a result of extensive duration of the flood wave, important interventions took place along the embankment section of the Bodrog-köz, on the right bank of the Tisza in the districts of Révleányvár and Ricse municipalities, and also in five other municipalities. At some of these locations, sandbagging operations took place.

It should be mentioned that the emergency storage of flood water in Slovakia mitigated considerably the flood on the Bodrog river, which was forecasted as very high.

As the flood wave passed downstream, the main point of defence was gradually transferred to Taktaköz. Here, Stage III of flood protection readiness had to be kept in force for a long time. In the Taktaköz flood protection system, the embankments had to be protected against water trickling and seepage, and the resulting softening of the top portion and land slides by means of sandbags. The protection against wave erosion was achieved by lining the embankments (on the river bed side) with plastic sheets held in place by sandbags. For safety reasons, the sandbag reinforcements of embankments remained in place until the spring to provide protection against a possible winter flood.

3.4. RESOURCES USED IN FLOOD DEFENCE

The state of extreme preparedness was ordered on November, 5, 1998, at 1145 hours, and lasted until November, 17, 1988, 1600 hours.

The total length of the embankment section covered by the preparedness order was 1605 km (Nov. 14), including a 450 km section subject to the extreme preparedness, a 740 km reach subject to Stage III, and a 415 km reach subject to Stage I. The total length of protective embankment sections subject to any stage of preparedness during the flood duration was 2223 km.

The total manpower engaged in the protection of embankment sections was the greatest on November 7, 1998, 9800 persons. The total number of people involved in flood defence was much higher:

Water services staff	2.300 persons
Civil defence forces	10.961 persons
Task forces organised by mayors and civil defence	20.475 persons
Hungarian Army	2.955 persons
Police Force	1.530 persons
Border Guards	1.227 persons
Fireguards	180 persons
Total	39.628 persons

Number of sandbags: 1,2 million

Vehicles and machinery: 620 public road and water vehicles, 120 pieces of heavy equipment.

3.4.1. *Costs of defence*

The State Government took steps dealing with the costs of flood defence and also approved the allocation of additional 2 billion HUF for the reinforcement of the Upper Tisza embankments in the near future.

The water organisations bore the costs of flood defence during the flood and the winding down of operations and cleanup after the flood (the recovery of materials used temporarily in defence structures, the replacement of materials and tools, the cleanup of banks of the defence structures, the restoration of dike crests and access roads, etc.) and the rehabilitation of the flood protection capacity, in preparation for the future flood events.

3.4.2. *Settling of damages*

In March, 1999, there was another flood on the Tisza and its tributaries, exceeding in some locations the 1998 flood. Consequently, the settling of damages from the two events is done concurrently. Simultaneously with the need to restore the flood defence works, social and political demands for settling of flood damages grew stronger, particularly with respect to those caused by domestic excess waters.

The types of remedial action can be classified as:

* restoration of buildings to make them inhabitable,
* remedy of damages affecting agriculture,
* remedy of damages to structures owned by self-government agencies (buildings, roads, protective structures), and last,
* remedy of damages to the State property (roads, protective structures, pumping stations, canals, embankments).

The State Government further determined that the damages to residential buildings should be remedied without delay, using the following principles:

- providing shelter to the affected persons,
- financing expedient reconstruction with cost sharing among the State, the local self-government, and the affected people (in the last case, advantageous bank loans with 4-5% interest and the distribution of donations by the population were also employed). According to the latest decision, the State Government accepts responsibility for 2/3 of all flood damages.
- the places selected for reconstruction should be safe,
- the assessment of damages should be objective and subject to appropriate controls, and
- the priority should be given to the role of local self-government agencies in distribution of subsidies.

In summary, it can be stated that Hungary was protected effectively against the 1998 extreme flood on the Tisza river, and neither residential building nor agricultural fields were inundated as a result of embankment breaches. This success resulted from combining forces at the national level, good organisation, the precise direction and division of protection tasks, and above all, the dedication of all who participated in flood protection activities.

4. References

1. Tóth, S. (1993) Overview of flood defence problems in Hungary. *Proceedings of the UK-Hungarian Workshop on Flood Defence.* September 6-10, 1993, Budapest, Hungary, pp 33-55.
2. Zorkóczy, Z. (1993) Flood defence organization and management in Hungary. *Proceedings of the UK-Hungarian Workshop on Flood Defence.* September 6-10, 1993. Budapest, Hungary, pp 5-13.

CAUSES AND PECULIARITIES OF RECENT FLOODS ON THE DNIESTER RIVER

A. SHULYARENKO[1], M. YATSYUK[2] and I. SHULYARENKO[3]

[1]*Institute of Hydrobiology, National Academy of Sciences of Ukraine*
[2]*State Committee of Ukraine on Water Resourses, Ukraine*
[3]*Taras Shevchenko Kyiv University, Ukraine*

1. Introduction

The frequency of floods in Ukraine has increased with respect to both spring floods caused by sudden and profound thaws and summer floods caused by rainstorms. In recent years, the frequency of floods especially increased on rivers whose catchment areas are partly or fully located in mountainous regions.

One of the flood-prone regions of Ukraine is the Dniester river basin (Fig. 1). Its catchment area is 72.100 km^2, the river length is 1.362 km, and the average annual flow is 10,2 km^3 (measured since 1885)[1]. The Dniester river basin is unusual with respect to its geography; its upper part with a developed network of tributaries extends into the Ukrainian Carpathians, which are in a zone of considerable rainfall activity (the average annual precipitation in the Carpathian Mountains is 1400-1800 mm [2]). The physico-geographical conditions of the mountain locality are characterised by a high drainage density (1,1-1,5 km/km^2), considerable slope of river beds and valleys (4,0-15,0 m/km and in some cases even 20,0-26,0 m/km [1]). Both these factors play a great role in formation of high floods.

The formation of floods in this region results from interaction of several factors. The hydrometeorological factors (the amount of atmospheric precipitation, its intensity and duration, temperature and moisture of air, areal extent of humidity, discharge capacity of stream channels) and the landscape factors (local relief, soils, and plant cover) are the main ones. They determine the value of hydrological abstractions and their spatial-temporal dynamics, velocity of overland flow in the basin and, consequently, the flood wave velocity.

In the database of the State Committee of Ukraine on Hydrometeorology, a wet period can be discerned in recent years, more or less during the 1990s. It has come after a dry period from 1982 to 1990. In the current decade, high and even catastrophic floods were observed in 1992, 1993, 1998 in the Tisa river basin, in 1993, 1998, 1999 in the Prypyat river basin, in 1995 on German rivers, and in 1997 in Poland, Czech Republic, Slovak Republic and also in the Dniester river basin and on rivers of Crimea.

J. Marsalek et al. (eds.), Flood Issues in Contemporary Water Management, 95-100.

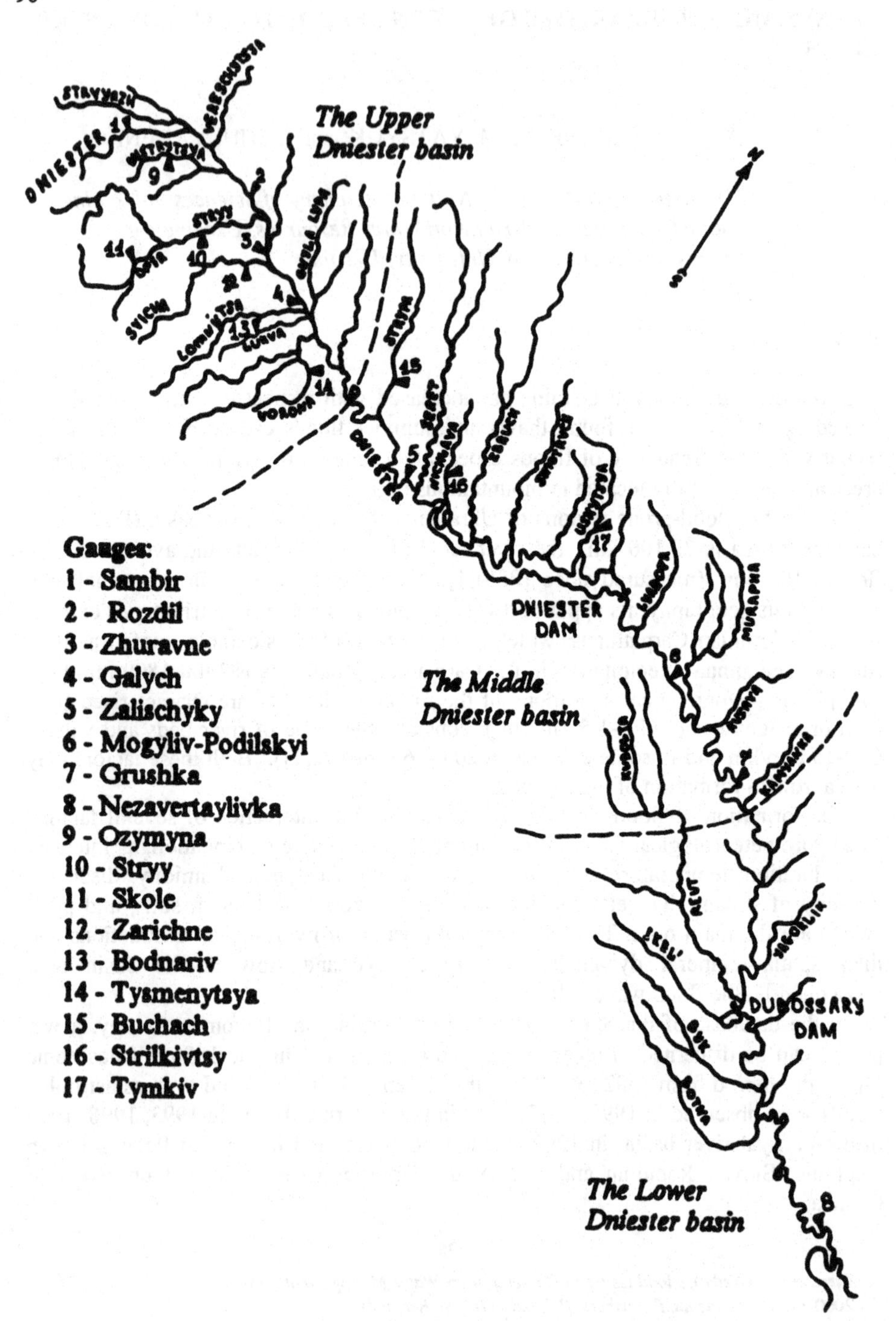
The Upper
Dniester basin
The Middle
Dniester basin
The Lower
Dniester basin
DNIESTER
DAM
DUBOSSARY
DAM
DNIESTER
STRYY
SVICHA
Gauges:
1 - Sambir
2 - Rozdil
3 - Zhuravne
4 - Galych
5 - Zalischyky
6 - Mogyliv-Podilskyi
7 - Grushka
8 - Nezavertaylivka
9 - Ozymyna
10 - Stryy
11 - Skole
12 - Zarichne
13 - Bodnariv
14 - Tysmenytsya
15 - Buchach
16 - Strilkivtsy
17 - Tymkiv

Figure 1. Dniester river basin.

2. Discussion

The most important causes of formation of frequent floods are first of all specific meteorological situations and peculiarities of morphometry of mountain rivers and also anthropogenic activities in river basins. Among these causes, one can name:

- sharp fluctuations of air temperature in winter period and storage of snow in the catchment;
- short-term, intensive, high rainfalls during summer months;
- large slope of river beds (more than 5,0 m/km), considerable surface slopes in basin and insufficient water conveyance capacity of channels;
- non-systematic clearance of forests on mountain slopes that speeds up the velocity of overland flow towards rivers and, consequently, rapid formation of floods with high stages;
- clogging of river channels by wooden debris, rubbish and alluvial deposits of previous floods;
- no control over economic activity in basins and unplanned development on floodplains, up to river channels.

It is necessary to note that the climatic conditions of the Dniester basin are characterised by considerable rainfalls in the summer period that cause a rapid increase of water levels, overflowing of river banks and inundation of adjacent land, especially on the right-bank side of the basin.

The upper mountainous part of the Dniester basin is the main area of formation of rainfall floods. In this part of the basin, the intensity of precipitation amounts to 20 mm per hour, the value of river channel slopes ranges from 20,0 to 26,0 m/km near the headwaters to 4,0-8,0 m/km near the river mouth [1], and water absorption and storage capacity of the surface and underlying rocks is very small, which creates favourable conditions for formation of high floods. About 75-80% of the annual flow comes from upper reaches of the Dniester catchment and its right-bank tributaries Stryy, Lomnytsya, Bystrytsya, and Svicha representing only 30% of the total catchment area of the Dniester river [3].

The most significant floods on rivers of the north-east slopes of the Carpathians occurred in 1911, 1927, 1941, 1955, 1969, 1974 and 1980. All of them were floods caused by high rainfalls. A catastrophic flood, which was observed on the Lower Dniester at the end of July and the beginning of August in 1980, was formed during a period of 6-8 days, under the influence of intense rainfalls occurring in headwaters of the Dniester basin. The amount of rainfall measured during this period by various raingauges varied from 76 to 86 mm.

The territory of the Dniester basin is highly developed, with respect to industry, agriculture, railways, roads, oil- and gas-pipelines, and electrical transmission lines. Frequent floods in the Dniester basin have brought about great losses to the national economy as well as to local inhabitants. There are 224 settlements, including 211 villages, 8 townships and 5 towns in the flood-prone zones, under a permanent threat of flooding. Considerable areas of agricultural fields are also often inundated.

One of the typical extreme floods, which were observed lately in the Dniester basin, is the flood of June 1998 that was classified as a historical flood. Heavy and very heavy rains on June 18, 1998 were caused by a quasi-stationary cold front, and on June 19 and 20, by the influence of a high-altitude cyclone, which was formed over the Carpathian region. The rainfall depths in the Dniester basin during this period were 52-183 mm (41 - 145% of the monthly norm) in the Lviv region, 37 - 165 mm (23 - 102% of the norm) in the Ivano-Frankivsk region, 16 - 151 mm (19 - 176% of the norm) in the Ternopil region, and 23 - 108 mm (29 - 135% of month norm) in the Khmelnytsk region.

Because these rainfalls represent about 80% of the average monthly rainfall, the Dniester river and its tributaries started to flood during the night of July 18/19. Special characteristics of conditions of the flood formation included:

1) Precipitation covered the whole catchment area, but its intensity and amounts varied in different parts of the basin.

2) In the mountainous part of the Lviv province (the Dniester basin upstream of the town of Rozdil, the Stryy river basin, and the Strvyazh river basin), the rainfall depth was about 40 mm on the first day of flood formation and 50-100 mm during the whole period of flood.

3) The highest rainfall was observed in the basins of the Lomnytsya river, the Bystrytsya river, the Lukva river and the Vorona river (the Ivano-Frankivsk province), where it varied from 100 to 200 mm. Such a distribution of rainfall distinguishes this flood from the other ones with the main flood flow formed in the Lviv province.

4) A considerable rainfall also occurred in the zone of formation of lateral inflow into the Dniester reservoir (30-96 mm during the flood).

Saturation and wetness of the catchment caused an intensive rise of water levels in different tributaries of the Dniester river and, consequently, also in different reaches of the Dniester. During the flood, the maximum water levels were observed from June 19 to 21, with exceedance of pre-flood levels by 1,01-2,91 m in the Dniester near the town of Rozdil, 3,23-5,56 m in the reach from the town of Zhuravne to the town of Zalischyky, 0,6-3,7 m in Dniester tributaries in the Lviv and Ivano-Frankivsk provinces, 0,8-1,4 m in the Ternopil province, and 0,2-0,8 m in the Khmelnytsk province. The exceedance probability of this flood on the Dniester was 5 per cent.

Floodplains of the Dniester river and its tributaries within the Lviv, Ivano-Frankivsk, Ternopil and Khmelnytsk provinces were flooded. The Dniester floodplain in the reach from the town of Galych to the town of Zalischyky was inundated by 2,0 to 4,7 m of water during a period of 3-9 days. Flood peaks exceeded the pre-flood discharges 14-21 times in the Dniester upstream from the town of Sambir, 4-11 times in the reach from Rozdil to Zalischyky, 7-52 times in mountain tributaries, 2-5 times in plain tributaries, 127 times in the Lukva river (the gauge in Bodnariv), and 130 times in the Bystrytsya Solotvynska river.

Exceedance of dangerous stages of partial flooding of settlements and economic objects was observed in the Dniester river (the reach from Rozdil to Zalischyky), the Vorona river (the gauge in Tysmenytsya), the Opir river (the gauge in Skole), the Svicha river (the gauge in Zarichne), the Bystrytsya river (the gauge in Ozymyna), the Strypa river (the gauge in Buchach), the Nichlava river (the gauge in Strilkivtsy), the Ushytsya river (the gauge in Tymkiv) and elsewhere.

The combined influence of heavy rains, local flow, increase of groundwater and river water levels caused partial flooding of many settlements and agricultural lands, and damages to many buildings and business objects in the Lviv, Ivano-Frankivsk and Ternopil provinces, in the town of Mogyliv-Podilskyy (Vinnytsya province) and to separate objects in the Odessa province. The harmful consequences of this flood could have been even worse, but owing to the Dniester reservoir, they were prevented.

The Dniester reservoir plays a major role in flow regulation in this river basin. It was built in 1981. In the spring of 1987, the reservoir was filled up to the normal discharge level. Its area is 142 km^2, storage capacity 3,0 km^3, the controlled storage capacity 2,0 km^3, length = 204 km, mean width = 0,8 km, mean depth = 21,0 m, and the largest depth is 55 m [4]. The second reservoir on the Dniester river, the Dubossary reservoir, was built in 1954. It has a capacity of only 0,48 km^3 and the controlled storage capacity of 0,21 km^3 [3]; that does not permit an annual regulation of the flow.

In the history of the Dniester reservoir, the June 1998 flood has been the largest with respect to the maximum inflow to the reservoir. The mean daily inflow into the Dniester reservoir increased from 647 m^3/s on June 18 to 4.190 m^3/s on June 21. The maximum inflow into the reservoir was 4.250 m^3/s at 1200 hours on June 21. The flood volume which passed through the Dniester hydropower station (HPS) from June 19 to 28 was 1,25 km^3. The maximum daily inflow into the reservoir during the flood was estimated to have a 5% probability of occurrence with respect to the discharge and about 10% with respect to the flow volume.

At 0800 hours on June 19, 1998, the water level upstream of the Dniester Dam was 120,96 m and the free reservoir capacity ($7x10^6$ m^3) was insufficient to handle flood waters in the usual way. Considering the forecast of water inflow, the mean daily release was increased on June 19 from 200 m^3/sec to 871 m^3/sec (Table 1). The observed arrival of the expected large flood (the maximum inflow into the reservoir was forecasted on June 21 in the range from 3.500 to 4.000 m^3/s) indicated the necessity to increase the mean daily releases from the reservoir by opening spillway gates. To prevent possible flooding downstream from the Dniester Dam, a decision was made to restrict discharges to the range from 1700 to 1800 m^3/s. A summary of data characterising the operation of the Dniester reservoir is given in Table 1.

TABLE 1. Characteristics of the Dniester Reservoir operation in June 1998

Date	Mean daily values (m^3/s) of			Mean daily	Mean daily
	inflow to reservoir	inflow in line of HPS	discharge in line of HPS	water levels in the reservoir (m) (the Baltic Sea datum)	volume of water in the reservoir (km^3)
18.06.98	647	646	643	120,88	2.982
19.06.98	840	767	871	120,90	2.985
20.06.98	2750	1480	1250	120,88	2.982
21.06.98	4190	3045	1645	121,39	3.057
22.06.98	3280	3260	1860	122,33	3.196
23.06.98	2160	2510	2100	122,76	3.256
24.06.98	1240	1370	1790	122,63	3.238
25.06.98	1090	1250	1620	122,44	3.212
26.06.98	890	959	1150	122,30	3.212
27.06.98	810	820	1200	122,14	3.170
28.06.98	687	725	950	122,09	3.146

The total inflow into the reservoir was 1,54 km^3, and the volume of discharge through the hydropower station was 1,25 km^3. So, in principle, the Dniester reservoir fulfilled its flood-regulation function, by storing some volume of flood flow and reducing maximum releases from the reservoir. The contribution of the Dniester reservoir to the regulation of flood flow is evident from the data in Table 2.

TABLE 2. Hydrological characteristics of large floods on the Dniester river

Water gauge	July, 1980		May, 1989		June, 1998	
	Q max (m^3/s)	Date	Q max (m^3/s)	Date	Q max (m^3/s)	Date
Zalischyky	3910	27.07	2700	11.05	4080	21.06
Dniester Dam	--	--	1650	12.05	2150	23.06
Mogyliv-Podilskyi	3530	28.07	1870	12.05	2400	24.06
Grushka	3420	29.07	1520	13.05	2270	24.06
Dubossary Dam	2630	30.07	1590	13.05	2100	25.06
Nezavertaylivka	1260	1.08	--	--	1500	3.07

In 1980, the Dniester Dam did not exist yet and the flood wave reached the Dubossary Dam without any hindrance. However, in June 1998, thanks to flow regulation by the Dniester reservoir, the maximum flood discharges decreased by almost two times.

3. Conclusions

In view of frequent floods occurring anytime during the year on mountainous rivers of the Carpathian region, including the Dniester basin, and the associated considerable losses for the national economy and population, Ukraine urgently requires to develop effective measures to subdue natural disasters, such as floods. An advanced system of prevention, warning and mitigation of consequences of floods and other extraordinary catastrophical phenomena on rivers must be developed. This system must be based on the contemporary scientific knowledge and technologies, with consideration of all existing achievements in this field.

4. References

1. Resources of surface waters of USSR. *Main hydrological characteristics. Vol.6. Ukraine and Moldavia. Issue 1. West Ukraine and Moldavia* / Editors Gorbatsevych, N.P. & Egorova, E.M. (1976). Gidrometeoizdat Press, Leningrad, 621 p. (In Russian).
2. Marinich, A.M., Paschenko, V.M. and Shyschenko, P.G. (1985) *Nature of the Ukrainian SSR. Landscapes and physico-geographical division into districts.* Naukova Dumka Press, Kyiv, 224 p. (In Russian).
3. Shevtsova, L.V., Aliev, K.A., Kuzko, O.A. et al. (1998) *Ecological state of the Dniester river.* Kyiv, 145 p. (In Russian).
4. Shevtsova, L.V. and Aliev, K.A. (1997) *Methodical recommendations for ecological regime of the Dniester reservoir operation.* Kyiv, 35 p. (In Ukrainian).

MODELLING TOOLS – THE KEY ELEMENTS FOR EVALUATION OF IMPACTS OF ANTHROPOGENIC ACTIVITIES ON FLOOD GENERATION, WITH REFERENCE TO THE 1997 MORAVA RIVER FLOODING

E. ZEMAN[1] and P. BÍZA[2]

[1]*Hydroinform, a.s., Na Vršich 5, 100 00 Praha 10, Czech Republic*
[2]*Povodí Moravy, a.s., Drevarská 1, 601 75 Brno, Czech Republic*

1. Introduction

The International Decade for Natural Disaster Reduction (IDNDR) was announced by the General Assembly of the UN in 1987, to run from 1990 to 2000, with a basic objective to reduce human life losses and damages resulting from flood events. Water can be regarded as the most important element in natural disasters, because it plays the main role in one third of all deaths, injuries and material losses resulting from natural disasters. Events of wide-ranging significance rarely have a single cause. Flooding results from a complex combination of human activities and natural events, where coincidence of conditions, circumstances and factors is an important but complicating consideration. Leavesley et al. [1] published proceedings focused on disastrous floods.

The objective of this paper is to express authors' views on the use of modelling tools for identifying the risk of flood disasters. The paper is focused on a complex modelling tool that can be applied for such purposes as flood mitigation measures, flood risk assessment, disaster reduction measures, parallel design of hydraulic structures, disaster prevention and preparedness, evaluation of the impact of future anthropogenic activities (both structural and non-structural), evaluation of the impacts of restoration activities on flood protection, evaluation of urban planning policy, calibration of flood forecasting (FF) tools, and public awareness and education. Towards this end, the modelling tool's properties needed to fulfill the above requirements are presented. In this process, a distinction is made between anthropogenic activities and natural events in a way that the modeller understands and is able to translate them into boundary and initial conditions for the modelling tool. Some elements of the hydrological cycle are discussed and a schematisation is suggested for mathematical modelling.

The modelling tool of the above described properties can be used for 'if ... then' scenarios, which are essential for flood risk assessment of the current situation, and for the assessment of suggested changes. This modelling tool represents an attempt to replace trial and error based solutions. Finally, an overview of the model set-up is given for the Morava River basin, where disastrous floods occurred in July 1997. The modelling tool aims to provide the best concept for reducing the vulnerability of high-value urban areas, industrial zones and protected natural areas.

J. Marsalek et al. (eds.), Flood Issues in Contemporary Water Management, 101–114.

2. Flood Disaster: Consequence of a Complex Event

A flood event has to be studied as a hydrological phenomenon of great complexity. Hydrological characteristics of the flood reflect the stochastic behaviour of precipitation, interception, infiltration, evapotranspiration, soil moisture, overland and groundwater flows, and river channel hydraulics. This is by no means an exhaustive list of the factors influencing the form, magnitude, timing and volume of a flood.

All these factors may involve some aspects of human activities, some of which are more visible or more important. How can one proceed in evaluating the impacts of progressing urbanisation, forest clearance, and hydraulic structure construction on flood propagation in the basin? In order to describe the 'real world' by means of modelling tools, the underlying factors and interrelations between them have to be simplified by taking into account only the most important ones. It is tempting, but also dangerous, to oversimplify the system by applying a simple modelling tool which will make the problem more tractable. The dividing line between realism and oversimplification is important, when attempting to quantify the level of flooding hazards. Askew [2] gives an introduction to IDNDR, and defines terminology in accordance with the UN Department for Human Affairs recommendations:

- A hazard is a threatening event, or a probability of occurrence of the potential damaging phenomenon within a given time period and space,
- Vulnerability is a degree of loss ranging from 0-100%, resulting from a potentially damaging phenomenon,
- Risk defines expected losses of life, personal injuries, property damage, and economic activity disruptions due to the particular hazard for a given area and reference period. Based on mathematical calculations, risk is the product of hazard and vulnerability.

A hazard is a complex phenomenon. Vulnerability must be referenced to the natural and/or man-made environment that is threatened by flooding. UN conference on Natural Disaster Reduction [3] describes this complexity of vulnerability not only in terms of its composition, but also in terms of its dynamic nature and the stochastic, uncertain, barely predictable outcome of flooding. A natural disaster in the form of flooding, which has such a severe impact on the society, has both natural and anthropogenic causes. It is a typical interaction between the society and the environment, which is continually being dynamically changed by man. What turns a hazard into a disaster is the human factor. Without the vulnerability brought about by human beings, the hazard would not become disaster. Natural occurrences alone, without the risk caused by human beings, are not disastrous.

In vulnerability assessment, it is important to distinguish between a potential loss of life, on the one hand, and damage to assets and nature, on the other. In some countries, due to their natural conditions, the dividing line between the two types of risk is essential. There is high concern about danger to human life in low-lying countries, such as the Netherlands, where the rivers classified as very dangerous are provided with embankments that can be overtopped only once in 1250 years, on average [4].

As societies become more developed, and bring increasingly sophisticated and complex methods and technologies into use, they become more vulnerable to the hazard of flooding. The professional community improves its understanding of the natural processes underlying flooding. Engineers and scientists have developed better modelling tools to understand this phenomenon and to reduce the impact of hazards. Hydroinformatics combines these tools in a single environment, thus assisting engineers in risk analysis.

3. Modelling Tools: Requirements, Properties and Abilities

The use of models for flood protection and flood forecasting has become widespread during the last 20 years. Many models are available from research institutes, software vendors and individuals. Such models were classified in a WMO publication [5]. The WMO guidelines divide the models into stochastic and deterministic. For evaluating anthropogenic consequences, stochastic models are clearly less attractive for description of processes of surface and subsurface flows.

Deterministic models are usually based on physical properties of elements that represent or influence the process under investigation. Three types of deterministic models can be found in the literature:

- Deterministic models based on hydrodynamic laws
- Deterministic conceptual models
- Deterministic black box models

3.1. TIME AND SPACE DEPENDENCE

The basin or catchment has its own more or less effective drainage system (overland flow, stream flow, interflow, groundwater flow). The runoff process and flood occurrence are defined by the properties of the surface cover of, and subsurface layers in, the basin. The flood itself is generated as a consequence of intensive climatic events, including snowmelt, a local rainstorm, or a regional rain event. An unsteady approach to the study of the rainfall/runoff processes on a regional or local scale is unavoidable. A steady flow approach can be used for structure design and for studying the impact of changes that have no time dependent effects.

Space dependency is closely related to the process being described. One-dimensional and two-dimensional unsteady hydrodynamic models are available at both commercial and scientific levels. With proper spatial representation of variables and parameters, three-dimensional (3D) models are still quite expensive tools for quantifying the current situation from a hydrodynamic point of view. One-dimensional (1D) representation is the cheapest and easily accessible, but in some applications it can be rather approximate. Since some parameters are not regularly measured and analysed, it is tricky to base the model set-up on them. However, the cost of the data tends to reduce the enthusiasm of modellers for routine applications of 2D and 3D hydrodynamic models.

3.2. DETERMINISTIC MODELS USED FOR HUMAN IMPACT EVALUATION

Deterministic hydrodynamic models are well suited for stream and reservoir modelling, and to some extent for overland flow. They are very important for evaluating the impact of human activity close to the river, or impacts on drainage. Hydrodynamic models are useful for comparing flood wave propagation under different basin conditions. The comparisons are evaluated for key characteristics (such as discharge, water levels, depths, velocities, and sediment transport rate). The deepening of, and improvements to, the channel are introduced by cross-section geometry and roughness changes. The overall effect of the deepening will be seen clearly in a comparative study (comparing the key characteristics of the current situation and the expected changes). Hydrodynamic models require calibration and verification.

Hydrodynamic and hydrological deterministic models can be used for human impact evaluation. The success of model application depends on the type of the model, accuracy of model set-up, data availability and accuracy of calibration and verification procedures.

The integration of a full hydrodynamic river channel model in a 1D+ form (which enables branching and looping of river channels and flood plains) with lumped models (or semi-distributed lumped models) of subcatchments provides a modelling tool for evaluating the changes in the basin.

3.3. MODEL OBJECTIVES DICTATE SCHEMATISATION AND LEVELS OF DETAILS

Flood protection measures in any basin should start with analysis of the current state, reflecting the combination of human activity and natural development in the basin. Any further changes in the basin, which can change the rainfall/runoff process, should be investigated. It is obvious that the current status of the basin reflects both natural and human development in an integral form. Before making any changes, the new and the existing conditions should be compared.

When dealing with changes, well-conceived flood management, flood protection measures or flood mitigation, it is no longer a question of always building new hydraulic structures in the river network. In some basins, it is necessary to return some parts of an already-trained river channel back to its natural state. In some countries, summer dikes have been breached at selected places, where restoration projects are being developed [4] (the Rhine in the Netherlands, as suggested by Rijskwaterstaat).

3.4. MONITORING AND MEASUREMENTS

Data availability in the basin is the most crucial limitation on wider application of simulation models, since the behaviour of the basin under conditions of flooding reflects all changes in the basin during some antecedent time period. A monitoring network in the basin can provide very important evidence of the extent to which the basin response to rainfall has changed in time. Trends in time series of the investigated variables may provide evidence of the impact of certain changes in the basin, but the measured data are actually more important for model calibration and verification. Calibrated models can often serve better for analysing the basin behaviour, because the model user can test the model with any loading condition and is not limited to historical, recorded events only.

Rating curves at a gauging station provide some of the most important transfer functions for flood evaluation. However, such rating curves are often derived using a primitive methodology, and the corresponding transfer functions then might give completely wrong data for flood evaluation. The upper part of the rating curve is in many cases 'only extrapolated' by the good judgement of an engineer. For this reason, the application of 1D or 2D hydrodynamic models is recommended in order to derive the complete rating curve of any complicated cross-section under unsteady flow conditions. Ingeduldova et al. [6] presented a detailed study of the gauging station at Tyniste nad Orlici on the Orlice River. That study served for evaluation of the existing rating curve, and preceded a proposal for river and flood plain improvement, in order to obtain a rating curve that would be more accurate and suitable even under high flood conditions.

4. Human Activities in the Basin

It is not easy to identify human activities in a basin, since most of the European basins have been influenced considerably by all types of changes, introduced either recently or hundreds years ago. However, one can specify the changes in the basin that have a negative or positive impact on rainfall/runoff processes. The negative impact is reflected in an increasing peak discharge Q max, lower retention capacity of a rural area, greater celerity of flood waves, higher local velocities, higher sediment transport capacity, higher volumes of rainfall excess, higher erodibility, etc. Positive impacts are highly desirable, and recently, approaches have been brought forward that represent a major change in engineering thinking. There is now interest in systems of distributed reservoirs with flood retention volumes, structural changes in agriculture and forestry practices, restoration of wetland functions, manipulation and real time control (RTC) in polders and reservoirs, breaching of parallel dikes, and deepening of river channels and flood plains. A combination of such activities might lead to mitigation of flood hazards in a more natural way. Nevertheless, the importance of retention reservoirs should not be underestimated.

Having in mind the problems of agricultural over-production and increasing social tension in rural areas in both EU and non-EU countries, there is an opportunity for new capacity building in villages and remote settlements. Care for arable land and forests has declined considerably in some countries in recent decades, and it is highly desirable to find a new role for small farmers on their land. The owner of land or forests can be encouraged by grants and/or obliged by legislation to take measures to increase the natural retention of the landscape, and to decrease rapid runoff through structural measures. Such practices include:

- changing arable land into forests or meadows
- applying best agriculture and forestry practices to increase natural retention capacity
- delaying runoff by extending flow paths on higher slopes
- creating new ponds and polders
- creating natural sediment traps
- making improvements to natural creeks and small rivers, and
- avoiding the use of heavy machinery for harvesting.

This approach gives a new value to the activities of landowners in rural areas, provides rural employment, and maintains rural populations.

5. Improved Flood Management Practices Evaluated by Modelling Results

Regional and global flood management strategy deals with structural and conceptual measures that can be combined to achieve the best and most cost-effective flood protection. Several types of structural measures are available. They can be categorised as follows:

A. restoration and environmentally sound measures (rehabilitation of wetlands, restructuring some dikes in order to enlarge the flooded area etc.)
B. local structural measures in the area of high property value (protection of special assets, city centres, industrial complexes, hazardous waste dumps, water supply centres, etc.)
C. major large-scale structures (retention reservoirs, complex flood protection of densely populated areas) and complex schemes for flood protection control in selected reaches with elements of Real-Time Control (RTC), and
D. combined infrastructure schemes (highway development, railway development, city infrastructure development, navigation and river improvements) aiming at multiple benefits including mitigation of the flood hazard.

The proposed methodology is based on the application of deterministic modelling in the area of interest (global or regional view), with the following steps:

- data collection and verification
- schematisation of the basin and model set-up proposal
- calibration and verification of the model on the basis of the existing situation
- conceptual proposal of feasible structural measures in four categories (A-D)
- evaluation of the impacts of the proposed concepts taking into account economic consideration
- best possible combination of structural flood protection measures proved by the model (including the features of the combined measures) for design floods or discharges
- design and implementation of measures (using models for parallel design).

The best flood protection plan has to be carefully compared with the 'do nothing' scenario, which involves only random compensation of owners and flood plain occupants when disastrous floods occur. The most practical flood protection project will often be a combination of structural measures, operational measures and taking no action, in various sections of the basin.

Organisational and legislative measures supplement structural measures. Legislation can help considerably with structural measures, primarily those included in item A.

Early warning systems and effective communication between the flood forecasting service and the integrated rescue services are also essential.

Adequate civil engineering regulations, norms and standards are needed, together with the application of well-proven methodologies that provide good flood protection (design of bridges, culverts, distribution networks, water supply protection, urban drainage facilities, etc.)

Flood mapping and flood plain classification for various hydraulic loading situations form an integral part of city development planning. Flood mapping is based on mathematical modelling applied with a digital model of the terrain. Flood propagation and flood mapping provide essential information for working out evacuation plans for large cities.

6. Flood Modelling in the Morava River Basin

The extreme flood in July 1997 in the Czech Republic opened up a discussion about what should be done in the future to protect people and property against similar floods. The Morava River Basin Administration prepared a basic proposal for flood protection in the basin in May 1998, to provide a platform for discussion. This proposal was based on experience gained from the flood, which indicates that most of the damage arose in sub-basins with insufficient potential for hydraulic structures and storage reservoirs to store, divert or release the necessary quantity of water. It was evident that the storage capacities of reservoirs could play an important regulatory role, and that they can protect downstream areas from the adverse flood impacts.

Several groups of measures have been proposed, which all have equal priority – new storage reservoirs, flood retention basins, and enhancement of the discharge capacities of river reaches by technical works or by embankments. The current task is to select from among these measures the combination that will provide the best flood protection, and it is here that modelling can help.

The Morava River Basin Administration has some experience with using simulation models. In 1996, together with the Danish Hydraulic Institute, a PHARE project 'Improved Water Resources Management in the Morava River Basin' was completed, with an objective of transferring the expertise on applications of MIKE 11 and MIKESHE modelling software [7]. Following this successful co-operation, it was decided after the 1997 flood to prepare a similar project on flood problems and flood modelling. The project, under the title 'Flood Management in the Czech Republic', is now under way. The main task is to develop a flood model of the Morava and Becva rivers using a 1D+ HD model (MIKE 11), and to apply this model to evaluating various flood management measure proposals.

6.1. DATA COLLECTION AND SCHEMATISATION

The Morava river basin was selected for this project, because it was the most affected basin during the 1997 flood [Fig. 1]. First, data had to be collected for preparing the model. There were two main groups of data – survey data and hydrological data. The survey data consisted of measured cross-sections of the rivers and flood plains. Most of the cross-section data were collected from the earlier projects and studies on the Morava and Becva rivers, but some of them were newly measured after the flood. One useful tool was a digital elevation terrain model, prepared for the Morava river basin, covering an area of almost 700 km^2 of flood plain, including the junction of the Morava and Becva rivers. This digital elevation terrain model was created by aerial photography. It consisted of more than 1,5 million points, and was used for preparing cross-sections on the flood plains.

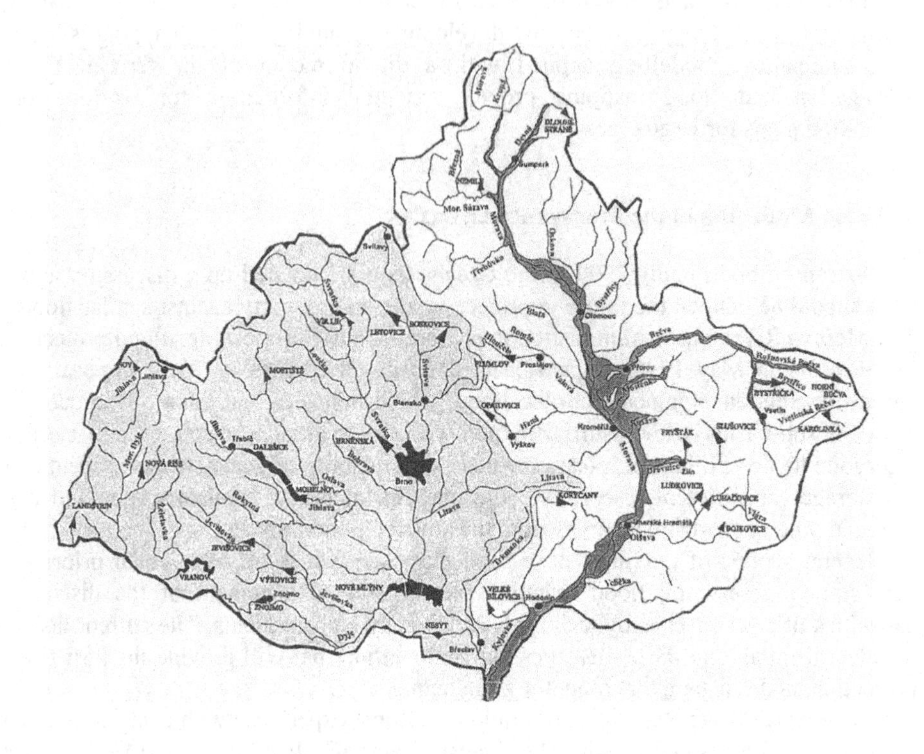

Figure 1. **The domain of the hydrodynamic model.**

Hydrological data were supplied by the Czech Hydrometeorological Institute. Time series of measured values from gauging stations on the Morava and Becva rivers and on the main tributaries provided important information. Daily data from 17 gauging stations in the basin were used for the period 1981–1986 and for the flood period in 1997. Rainfall data from 130 rain gauge stations in the basin were used for the model, and were supplemented by temperature data and potential evaporation data. The largest volume of data was prepared for the 1997 flood situation, which is the best ever documented flood in the Czech Republic. Other supplementary data were also very important for the whole project, e.g., detailed land use maps of the basin, many photos of the 1997 flood, sediment data for the Morava and Becva, etc.

The whole flood model consists of two parts: rainfall/runoff models of the Morava river basin, and a hydrodynamic model of the Morava and Becva.

The total basin, 9700 km^2 in area, was divided into 40 sub-catchments to create rainfall/runoff models for each tributary with a catchment larger than 100 km^2. Runoff in these basins is generated on the basis of information on rainfall, temperature and evaporation. Calculated values of discharges from rainfall runoff models are used as input into the hydrodynamic model. The NAM [Fig. 2] module of MIKE 11 was used as the modelling tool for these rainfall/runoff models.

The hydrodynamic model of the rivers was prepared on the basis of MIKE 11, and described the Morava between the town of Hanusovice in the upper part and the junction with the Dyje River, a total length of 255 river km. The Becva River was included with a 61 km reach from the town of Valasske Mezirici. Also included were some small sections of the Desna, Mostenka and Drevnice rivers, which are tributaries of the Becva. Example of simulation for the Becva is given in Fig. 3.

The hydrodynamic model was prepared as a flood model describing water flowing not only through the main river channels, but also over the flood plains. Flows in flood plains were described by a looped system of branches connected with the main river by link channels and connecting branches. There were about 300 objects on the main river system, mainly bridges and weirs. There were linear structures, such as railways and roads, that crossed the flood plain. The hydraulic structures in the model were described as combinations of weirs and culverts. The actual set-up of the hydrodynamic model consisted of almost 700 branches.

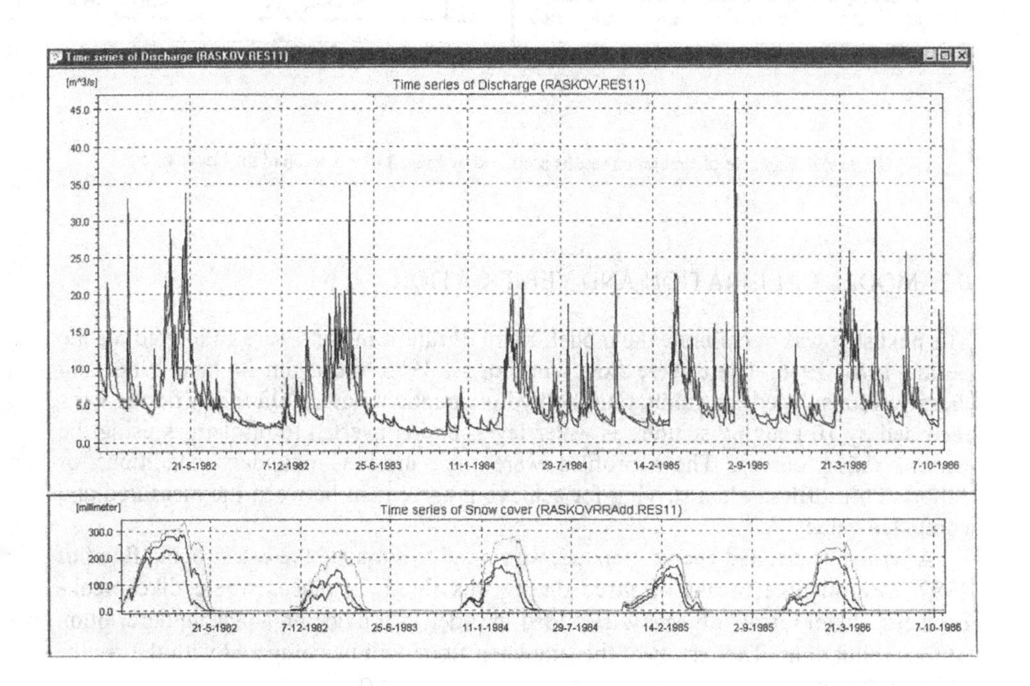

Figure 2. **Discharge calibration of one subcatchment for the period of 1981-1986 with snow cover.**

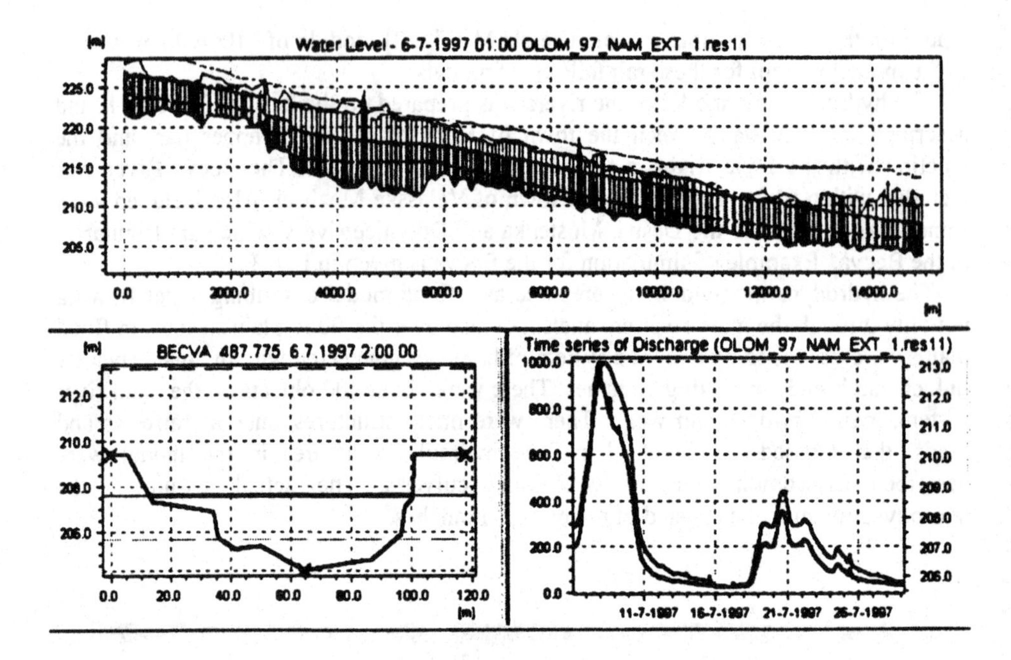

Figure 3. Example of simulation results produced by Mike 11 for a reach of the Becva river.

6.2. MODEL CALIBRATION AND VERIFICATION

The next step was to calibrate the model. Rainfall/runoff models were calibrated for the period 1981–1986. There were floods in August 1985 and again in June 1986, and these were also used for calibrating the hydrodynamic model. Calibration floods were recorded by 10 gauging stations as water levels and converted to discharges using the current rating curves. These profiles were then used as boundary conditions or calibration profiles, when striving for achieving agreement between the measured and computed values.

After the model had been calibrated, we started to simulate the catastrophic flood of 1997. Several dike breaks occurred during this flood. In places where dike breaks occurred, "dam break" objects were added in order to provide an adequate description of those situations. The results of the simulated flood will be compared with the results of other scenarios.

6.3. CONCEPTS FOR FLOOD PROTECTION SCHEMES

Proposals for possible future flood protection measures in the Morava river basin were prepared mainly on the basis of the information collected during the flood of 1997, and from previously processed flood protection materials and studies. Measures were designed in several groups, e.g., new water retention reservoirs, dry polders, increasing river capacity by technical improvements or embankments, and the use of ring dikes to protect towns and villages.

There are also other flood protection proposals. A proposal prepared by environmentalists is based on ideas of better land use, and on increasing native water storage directly in the river basin. The main goal of these ideas is to achieve the maximum water retention effect together with minimum investment in installing technical measures in the river basin. One sub-section is based on the idea of local embankments for towns.

The final category of proposals for improving flood protection in the Morava basin is linked to transportation, by constructing a Danube-Oder-Elbe navigation canal. Parts of the proposed canal and its embankments could be used for water retention and as a safety drain for flood waves. Further simulations have been prepared for each proposal, and the following flood protection measures will be checked by the mathematical model:

Change of land use throughout the catchment. The idea is to increase the area of meadows and forests in order to improve retention in the basin. One needs to know

A. what effects will be achieved if forests are brought into better condition or expanded to cover larger areas.
B. Removing settlements from the flood-prone area. This simulation evaluates the impact of dikes and local embankments on flood discharges in the basin. It is assumed that a smaller flooded area will lead to higher peaks during a flood.
C. Using water reservoirs for flood protection. On both the Morava and Becva, several locations for new reservoirs or dry polders have been proposed, and the simulation can determine the effect of these reservoirs on lower parts of the rivers.
D. Using a navigation canal for flood protection. There is an old idea to construct a Danube-Oder-Elbe navigation canal in the basin. This canal could serve as a diversion channel during flood situations to protect major towns.
E. The results of these simulations will be compared with the simulation of the 1997 flood.

6.4. APPLICATION OF THE MODEL RESULTS IN FURTHER DECISION MAKING

The working team is now involved in preparation of the final set-up of scenarios for comparing the effectiveness of individual proposals in flood protection of the affected area. Based on these comparisons, it will be possible to assess the positive and negative aspects of each scenario and to select the best combination of flood protection measures. It is also possible to demonstrate, for example, the influence of constructing reservoirs or polders on simulated flood flows in the downstream areas.

It is expected that this flood model will form a base for future discussion of flood protection in the Morava river basin. This study serves as an example for other river basins in the Czech Republic with respect to using modelling tools for solving complicated problems in the area of water management.

It is also expected that the calibrated model will be further utilised for other, closely related purposes:

- detailed flood mapping in cities and selected areas of interests [Fig. 4]
- improvements of rating curves in the basin

- training of rescue forces for preparation of evacuation and flood protection plans
- model set-up can be used for deriving new frequency curves for selected profiles
- model will serve for master drainage plans of larger cities providing boundary conditions (which part is flooded and when, where to install pumping stations), and
- parts of the model could be used in the FF model of the Morava basin and for its calibration.

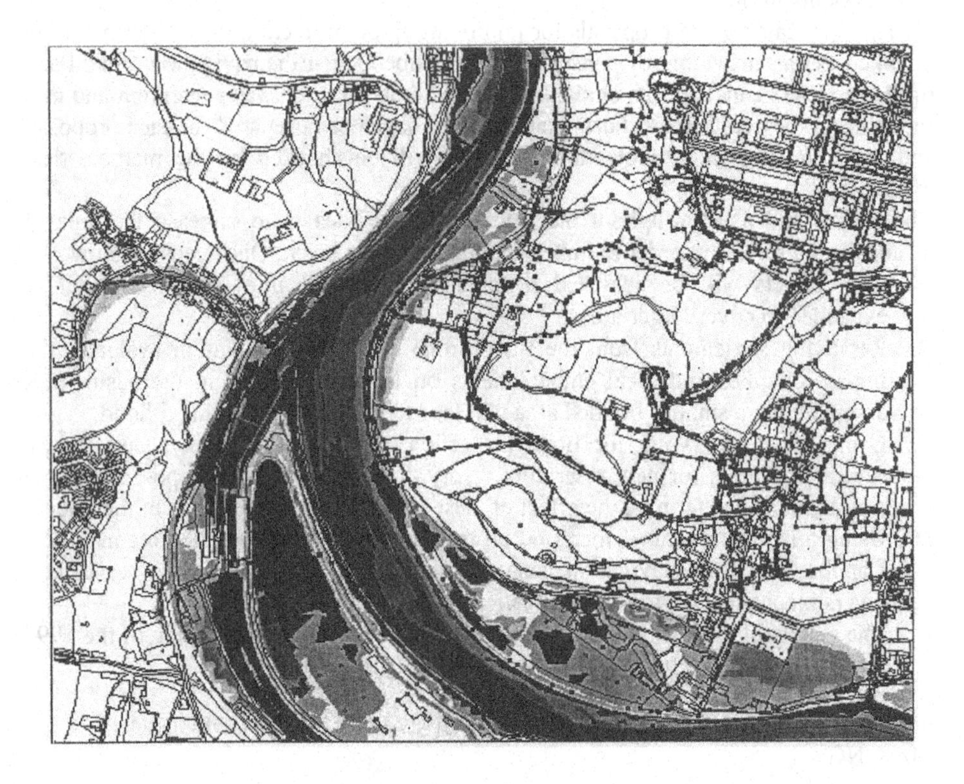

Figure 4. **Example of water levels transferred into GIS.**

7. Recommendations

The paper describes a set of tools applicable to analysis of human activities in the basin and their impacts on flood generation. Towards this end, suggestions and reasons were provided for applying deterministic models with physically based parameters primarily to regional and/or global flood event analysis.

To cope with flood hazards, it is evident that rising or reinforcing river embankments (based on the last disastrous flood) alone will not ensure a long lasting and most suitable solution. Behind the embankments, thousands of inhabitants are living and enormous investments have been concentrated. But in addition to the immediate river improvements after the flood disaster in the river channel and flood plain, other kinds of measures are necessary to reduce the flood hazards. The measures to be taken in the river itself, the adjacent flood plain and in the basin should aim to limit the probability of damage during periods of high water levels and enormous discharges. The river channel and its adjacent active flood plain should be kept free of obstructions for the main functions of this major river system – conveyance of water.

Sustainable river management should focus on structural maintenance and improvement and, where possible, on increasing the channel conveyance and restoring and developing the natural character of the river course.

Application of deterministic models can prove adverse effects of developments on the main river functions. The recognised impacts might be local or global. An execution of selected or recommended measures is a political issue. The political decision can be made easier, when the impact of measures is clearly presented in the form of graphs and maps, and all the participants in the discussion understand the matter.

Basic data for model set-up have to be collected, analysed and verified prior to their use in model preparation. The most modern technologies are recommended for use in data processing, preferably only in the digital form. Application of a digital model of terrain falls in this category. Supervising authority should monitor the status of basic data and plan regular updating for the most important river reaches.

Well designed and operated routine and non-routine monitoring of the key hydrological and meteorological parameters is essential for model calibration. A development plan for the monitoring program and its evaluation should be regularly prepared by the relevant authority. Without calibration and verification, deterministic models are not acceptable for impact evaluations.

Deterministic models have to be connected with GIS and other tools of hydroinformatics in order to disseminate the results in the easiest way. Distributing this knowledge using Internet might be taken as one of the option for reinforcing important communications between politicians and stakeholders.

8. Conclusions

Reassessment of flood protection standards, procedures and mainly experience from recent floods have led to the following conclusions:

- it is a regular pattern that only disastrous situations draw public's attention to the problem of flood hazard and protection; after the event, this awareness fades fast,
- data availability and their accuracy represent the basic condition for setting-up hydrodynamic models of a deterministic nature in an acceptable time period,
- the best evaluation of proposed structural flood control measures, from global or regional view, seems achievable by application of hydrodynamic deterministic models. Simulation of unsteady flow conditions, with a properly calibrated and verified hydrodynamic model, is in most cases the only way for establishing flood maps and evaluating flood control measures,
- it seems most efficient to evaluate different types of control measures – structural changes in the basin - separately and then suggest an overall concept for flood protection afterwards as an optimal combination of the measures investigated,

- the probability of risk of flooding has to be limited by combination of structural, non-structural, legislative and operational measures. Their effects have to be tested by hydrodynamic mathematical models prior to their execution,
- all future measures and activities in basins of large rivers need to be tested as to how they affect the flood generation and propagation in the basin,
- the flood risk could be lowered by a good warning system and suitable emergency measures, which should be based on results of hydrodynamic modelling,
- full protection against high water levels and extreme discharges is actually impossible, because a 100% flood protection is far from being economically acceptable and technically feasible.

9. References

1. Leavesley, G.H. et al. (1997) Destructive water: *Water Caused Natural Disasters, their Abatements and Control*, IAHS publication no. 239, ISBN 1-901502-00-7, Wallingford, UK.
2. Askew, A.J. (1997) Water in International Decade for Natural Disaster Reduction, in: *Water Caused Natural Disasters, their Abatements and Control edited by Leavesley et al.*, IAHS publication no. 239, ISBN 1-901502-00-7, Wallingford, UK, pp. 3-11.
3. DHA (1994) *Yokohama Strategy and Plan of actions for a Safer World*, World Conference on Natural Disaster Reduction, Yokohama.
4. Hallie, F.P. and Jorissen, R.E. (1997) *Protection Against Flooding: a new Delta Plan in the Netherlands, in:* Destructive water: *Water Caused Natural Disasters, their Abatements and Control*, IAHS publication no. 239, ISBN 1-901502-00-7, Wallingford, UK, pp. 361-370.
5. Becker, A. and Serban, P (1990) *Hydrological Models for Water-Resources System Design and Operation,* WMO-No. 740, Geneva , Switzerland.
6. Ingeduldova, Sklenar and Hrncir (1999) *Evaluation of rating curves in Tyniste upon Orlice,* the work ordered by the Labe Water Board, Hradec Kralove.
7. DHI (1999) *Flood Management in the Czech Republic, project supported by DEPA, Denmark,* MIKE11 Rainfall- Runoff Modelling on Morava catchment – phase report, Horsholm, Denmark.

POSTWAR CHANGES IN LAND USE IN FORMER WEST GERMANY AND THE INCREASED NUMBER OF INLAND FLOODS

R.R. VAN DER PLOEG, D. HERMSMEYER and J. BACHMANN

Institute of Soil Science, University of Hannover
Herrenhaeuser Str.2, 30419 Hannover (Germany)

1. Introduction

In recent years, the number of inland floods in Germany has increased markedly. Developments along the River Rhine can be taken as an example. Recordings taken in Cologne since 1895, cited in [1] and [2], show that six of the 12 most extreme stages have occurred since 1983. A number of factors, such as climatic changes and hydraulically adverse development in the Rhine's flood plain and catchment area, have been suggested as explanations for the increased number of extreme stages and floods in recent years. However, these factors alone cannot satisfactorily explain the change in the river discharge regime.

In recent publications, e.g. [2], [3], and [4], it has been argued that part of the inland flood problem must be possibly attributed to changes in land use. It was shown in these papers that surface sealing in urban areas, drainage of agricultural land, and disappearance of meadowland can result in increased amounts of runoff during heavy rainstorms or intensive snowmelt. It is the objective of this paper to examine the extent to which changes in land use in postwar West Germany have taken place. Another objective is to estimate the possible effect of such changes in land use on the formation of surface runoff during periods with extensive precipitation or snowmelt that may lead to excessive river discharge and floods.

2. Changes in Land Use

Regularly since 1957, the West German Department of Agriculture (BMELF) has issued yearbooks [5] with statistics about nutrition, agriculture, and forestry. Among other items, these yearbooks contain information about land use in the country, including agriculture. Table 1 is prepared from these yearbooks, especially from BMELF (1957, table 121) and BMELF (1990, table 102).

J. Marsalek et al. (eds.), Flood Issues in Contemporary Water Management, 115–123.

TABLE 1. Changes in postwar land use in former West Germany

Land use	Total area (Mha)		Change
	1951	1989	(%)
Urban (+ road system)	1.83	3.05	+4.91
Agriculture	14.37	13.36	-4.06
Forest + Woodland	7.05	7.40	+1.41
Uncultivated Land (Marshland, Heath, etc.)	0.25	0.13	-0.48
Water	0.44	0.45	+0.04
Wasteland	0.92	0.47	-1.82
Sum	24.86	24.86	0.00

For comparison, the 1951 data include area estimates for Saarland and West Berlin, even though both areas did not belong to West Germany in 1951. Including Saarland and West Berlin, the total area of former West Germany was 248 620 km^2 (= 24.86 Mha). At a first glance, Table 1 suggests that changes in postwar land use were rather small. It will be shown, however, that these changes are likely to affect the country's surface hydrology considerably. The urban area (including roads, highways, and railways) increased by 4.91 %. Simultaneously, the total agricultural acreage decreased by 4.06 %. Noteworthy are also the changes of the woodland and forest area (+ 1.41 %) and of the wasteland area (- 1.82 %). It appears that much wasteland was converted into forest and woodland in the postwar period.

Table 1 can be used for construction of a hypothetical, but representative (standard), watershed in former West Germany for the years of 1951 and 1989. Table 2 contains data for such a watershed of a 1000 ha size. The table shows that from 1951 to 1989 the percentage of urban area increased from 7.4 % to 12.2 %. The percentage of farmland decreased from 57.8 % to 53.7 %. In comparison, other changes in land use that took place were rather small. The combined area of forest and woodland, uncultivated land, water, and wasteland was 34.8 % in 1951 and 34.1 % in 1989. For brevity, the latter changes are ignored herein.

TABLE 2. Standard watershed (with respect to land use) of a 1000 ha size in former West Germany in 1951 and 1989

Land use	Area (ha)	
	1951	1989
Urban	74	122
Agriculture	578	537
Forest + Woodland	283	298
Uncultivated Land	10	6
Water	18	18
Wasteland	37	19
Sum	1000	1000

In Table 3 the change in land use for agriculture (shown in Table 2) is further detailed. Table 3 was prepared mainly from the data given in BMELF (1957, tables 119, 120, and 122) and BMELF (1990, tables 102, 104, 105, and 110).

Shown are changes in soil use for the major crops cultivated in postwar West German agriculture. For reasons that will become obvious later, soil use has been grouped in different categories: Fallow, Row Crops, Small Grains, ... , and Farm Woodlots, Farmsteads, and Roads. For the last category, the data from BMELF (1957, table 119) and BMELF (1990, table 104) were used that are listed there under horticulture ("Gartenland, Obstanlagen, Baumschulen und Korbweiden"). Neither the urban area shown in Table 2 (74 ha in 1951 and 122 ha in 1989), nor the combined area for forest and woodland, uncultivated land, water, and wasteland (348 ha in 1951 and 341 ha in 1989), are included in Table 3.

Table 3 illustrates that the major changes in soil use in agriculture that took place between 1951 and 1989 pertain to small grains (from 185 ha in 1951 to 223 ha in 1989) and to (permanent) meadows (from 157 ha in 1951 to 108 ha in 1989). In what follows, an attempt will be made to estimate the difference in runoff from the hypothetical, but representative, 1951 and 1989 watersheds, during a heavy rainstorm.

TABLE 3. Changes in soil use in agriculture in a standard watershed of a 1000 ha size in former West Germany, between 1951 and 1989

Soil use	Area (ha)	
	1951	1989
Fallow	2	7
Row Crops	86	91
Small Grains	185	223
Close-seeded Legumes or Rotation Meadow	45	12
Pasture or Range	80	91
Meadow	157	108
Farm woodlots, Farmsteads, Roads and right-of-way	20	5
Sum	578	537

2.1. ESTIMATING SURFACE RUNOFF FROM SMALL AGRICULTURAL WATERSHEDS

The rainfall-runoff model of the US Soil Conservation Service (SCS) [6,7] was used to estimate the amount of surface runoff in our hypothetical watersheds. With this model, runoff from small agricultural watersheds can be estimated for heavy rainstorms of a one-day duration or less. Figure 1 is a graphical representation of this model and displays the so-called runoff curve numbers (CNs), ranging from

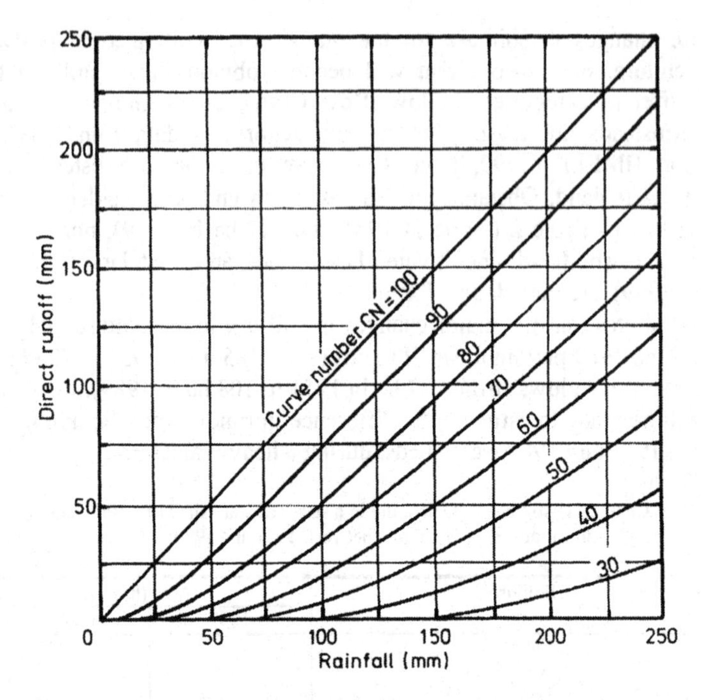

Figure 1. The relationship between rainfall and runoff for small agricultural watersheds, as presented in USDA [6] or Schwab et al. [7].

30 to 100, which allow, for a specific amount of heavy rain, estimation of the corresponding amount of direct runoff. The curve numbers refer to a particular form of land use or soil cover. Table 4 shows, for selected types of land use or soil cover, values of the appropriate curve numbers.

TABLE 4. Runoff Curve Numbers for various Hydrologic Soil-Cover Complexes, for Antecedent Rainfall Condition II, and the Initial Abstraction $I_a = 0.2$ S (after Schwab et al. [7]; abbreviated)

Land or Soil Use	Treatment or Practice	Hydrologic Soil Group			
		A	B	C	D
Fallow		77	86	91	94
Row Crops	Straight row	67	78	85	89
Small Grain	Straight row	63	75	83	87
Close-Seeded Legumes or Rotation Meadow	Straight row	58	72	81	85
Pasture or Range		39	61	74	80
Meadow (permanent)		30	58	71	78
Woods (farm woodlots)		25	55	70	77
Farmsteads		59	74	82	86
Roads and right-of-way		74	84	90	92

These values apply for Antecedent Rainfall Condition II (a 5-day antecedent rainfall of 13-28 mm in the dormant season and of 36-53 mm in the growing season) and an initial abstraction, $I_a = 0.2$ S, consisting of interception losses, surface storage, and water that infiltrates into the soil prior to runoff, where S is the maximum possible watershed retention. It can be seen that the curve number for a particular type of land use or soil cover depends on the so-called Hydrologic Soil Group to which the specific soil belongs. Soil Group A includes deep sands with little silt and clay and also deep permeable loesses. The final infiltration rate for soils belonging to this group after a long period of rainfall is 8 to 12 mm/hr. For Soil Group B (mostly silty soils), this final infiltration rate is 4 to 8 mm/hr, for C (predominantly loamy soils) it is 1 to 4 mm/hr, and for D (clay soils) it is 0 to 1 mm/hr. For details, see [6] or [7].

Not shown in Table 4 is the dependence of the runoff volume on the so-called Hydrologic Condition of the soil cover. For brevity, this information was omitted. Also not shown are the runoff curve numbers for Antecedent Rainfall Conditions I and III. Condition III applies to heavy rainfall (a 5-day antecedent rainfall of more than 53 mm) in the growing season, or light rainfall (a 5-day antecedent rainfall of more than 28 mm and low temperatures) in the dormant season. Factors to convert curve numbers for Condition II to Conditions I or III can be found elsewhere [7]. Thus, the SCS runoff model thus can be used to estimate runoff volumes for a broad range of weather, soil and crop conditions.

2.2. USE OF THE SCS RUNOFF MODEL IN GERMANY

The SCS model may be applied to estimate the runoff volume from the past rainstorms that have caused extensive floods in Germany. Such storms occurred, for example, in the Rhine area on Dec. 19 and 20, 1993. On these two days, large parts of the area received rain that varied in total between 60 and 120 mm [8]. Those storms affected an area of approximately $100,000\ km^2$. The cities like Koblenz and Cologne suffered severe damage from the resulting floods. The SCS runoff model can be used to evaluate hydrologically the change in land use between 1951 and 1989 for such large storms. To this end, the standard watershed of Tables 2 and 3 will be considered again.

Let us assume that the soil in the watershed belongs to Hydrological Soil Group B and that Antecedent Rainfall Condition III prevailed. The latter assumption is reasonable, because the month of December was exceptionally wet in 1993. Let us also assume that the areas used for the growth of row crops and small grains were fallow (bare) at that time of the year. To estimate the total volume of runoff for this watershed, either for the land use conditions of 1951 or 1989, a weighted curve number for the entire watershed must be calculated. A procedure for such calculations is described in [7] and further illustrated in Table 5, with respect to the land use and soil cover conditions of 1951.

TABLE 5. Calculation of a weighted curve number for the standard watershed in 1951, for Antecedent Rainfall Condition III and Hydrologic Soil Group B

Land or Soil Use	Area (ha)	Curve Number CN	Conversion factor for CN	Area × converted CN
Urban	74	86	1.10	7000.4
Fallow	276	86	1.10	26109.6
Row Crops	—			
Small Grains	—			
Close-seeded Legumes or Rotation Meadow	45	72	1.19	3855.6
Pasture	80	61	1.29	6295.2
Meadow	157	58	1.32	12019.9
Woodlots, Farmsteads, Roads	20	71	1.20	1704.0
Forest, etc.	348	55	1.35	25839.0
Total	1000			82823.7

The urban area in the watershed was treated as fallow soil. The weighted curve number, CN, was calculated as 82.8. The runoff volume Q (mm) for a watershed can be calculated for a given weighted curve number and the amount of rainfall I (mm), using the following equation [7]:

$$Q = \frac{(I - 0.2S)^2}{I + 0.8S} \qquad (1)$$

where the maximum possible watershed retention S (mm) is calculated as

$$S = \frac{25400}{CN} - 254 \qquad (2)$$

and CN is the appropriate curve number. It should be noted that Figure 1 is based on these two equations.

As for 1951, a CN-value for 1989 was calculated (CN = 84.1). Subsequently, both CN-values were used, together with Eq.2, to compute the corresponding S-values. For 1951, S = 52.8 mm and, for 1989, S = 48.0 mm, were computed. The values for S, in turn, can be used in Eq.1 to estimate the runoff volume Q for different values of I. In Table 6, the values of Q are presented for (daily) values of I that were recorded on a large scale in the catchment area of the Rhine on Dec. 19 and 20, 1993. The details of rainstorm distribution on these days can be found in [8].

Table 6 shows that for the prevailing rainfall conditions, direct runoff for watershed conditions as of 1989 increased by 8.0 to 16.3 % as compared to the watershed conditions of 1951. The question, whether or not this increased amount of runoff may have affected the stage of the Rhine at Cologne in December 1993, may be answered by looking at Table 6 again. From Table 6 it can be seen that, for a rainstorm of 30 mm, the surface runoff in our standard watershed increased between 1951 and 1989 by 0.9 mm (or 0.9 l/m^2). For the standard watershed, this makes a difference in runoff of 9000 m^3. If it is assumed that the German upstream section of the Rhine catchment that was struck by extreme rainstorms on 19 and 20 Dec. 1993 (100,000 km^2) consisted entirely of similar small watersheds, this makes a difference in the total amount of runoff of 9.0×10^7 m^3/day, or 1042 m^3/s. If instead a rainstorm of 40 mm is assumed, a difference in generation of runoff of 1505 m^3/s is calculated.

TABLE 6. Calculated direct runoff Q in the standard watersheds of 1951 and 1989, as a function of rainfall I

Rainfall	Direct runoff (mm)		Change in runoff	
(mm)	1951	1989	mm	%
I = 30	Q = 5.2	Q = 6.1	0.9	16.3
I = 40	Q = 10.5	Q = 11.8	1.3	11.9
I = 50	Q = 16.9	Q = 18.5	1.6	9.5
I = 60	Q = 23.9	Q = 25.8	1.9	8.0

The calculated increase in runoff (1042 m^3/s for I = 30 mm and 1505 m^3/s for I = 40 mm) for both Dec. 19 and 20 can be compared with the actual discharge values that were measured in the Rhine at Cologne in December 1993. Figure 2 shows the hydrograph of the Rhine, recorded in Cologne in December 1993 [1]. According to Vogt [1] and Engel [8], a maximum water level height of 10.63 m was recorded on Dec. 24, 1993. The corresponding discharge was 10 800 m^3/s. Vogt [1] and Engel [9] indicated that river stages of 10.0 m and more are critical as they cause inundations of large areas in and around Cologne. The Rhine's discharge for a stage of 10.0 m is 9800 m^3. The 10.0 m stage was exceeded between 22 and 25 December (see Fig. 2), which caused an estimated damage in Cologne of 150 million DM. The total damage in Germany caused by the December 1993 flood was 1.3 billion DM [10]. Our runoff, discharge, and stage numbers thus illustrate the possible impacts of postwar changes in land use on river streamflow.

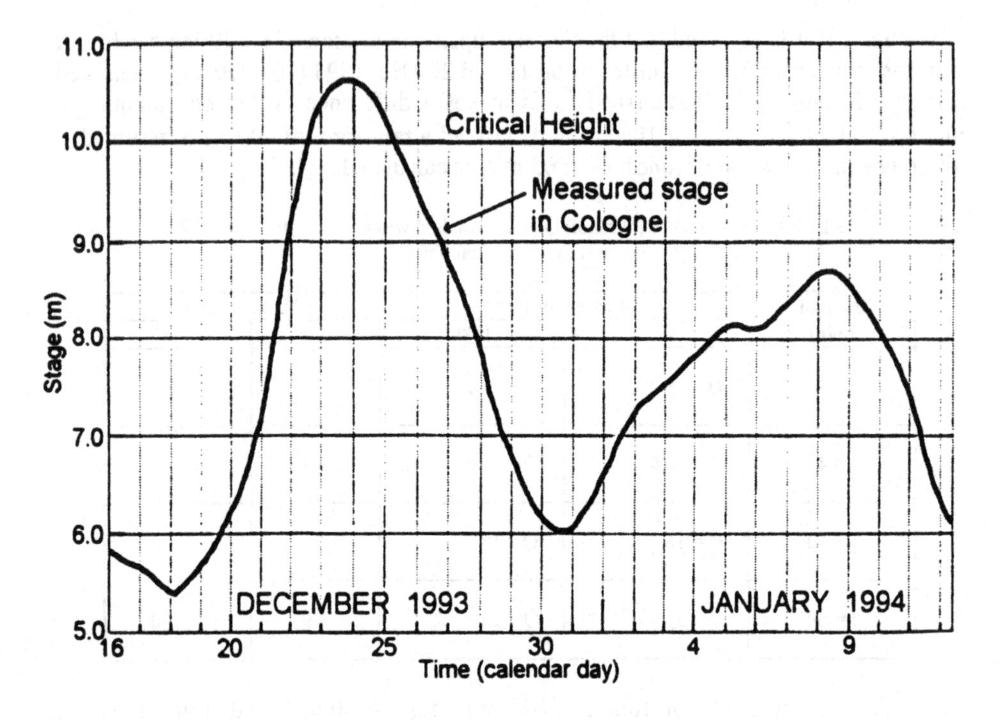

Figure 2. The hydrograph of the Rhine at Cologne in December 1993 (after Engel [8])

3. Other Changes in Postwar Land Use

In addition to postwar changes in the type of land use, the intensity of land use has also changed. This applies in particular to agriculture, where a pronounced mechanization and use of heavy machinery can be observed. This development has caused physical soil degradation on a large scale [2]. A distinct plow pan at the 25 to 30 cm soil depth and aggregate slaking and surface sealing are features that reduce soil infiltration capacity in the cool and wet winter period. Physical soil degradation is another reason for the increased runoff from arable soils in wet periods. A possible way to evaluate the effect of physical soil degradation on the formation of surface runoff during heavy rainstorms is the use of the previously discussed SCS model, in combination with a shift in the Hydrologic Soil Group. For loess soils, as used in the sample calculations for our representative watershed, a postwar shift from Hydrologic Soil Group B to C for all arable soils can be considered. In terms of our previous watershed calculations, such a shift results in an increased value for the watershed's weighted curve number (CN = 87.5) and in additional runoff. For brevity, this topic will not be treated here in more detail.

Another development in postwar agriculture is the large-scale drainage of wetland soils. Van der Ploeg and Sieker [4] estimated that by 1990 in former West Germany an area of 3 Mha had received a subsurface drainage system. This is more than 20 % of the total farmland area. A subsurface drainage system increases, in general, the soil infiltration. Regarding the previous watershed discussion and runoff estimates, this means that a backwards shift from Hydrologic Soil Group C to Group B for 20 % of the farmland area seems reasonable. For this area in our representative watershed the amount of surface runoff during a heavy rainstorm in wintertime would become smaller. However, simultaneously the amount of subsurface runoff would increase [4]. The overall effect would be an increase in runoff and, eventually, an increase in river discharge. Hence, the change in intensity of postwar land use has additionally increased runoff and the risk of floods along our main rivers.

4. Conclusions

Although our reasoning is based on many simplifications and the data used contain uncertainties, the following conclusions seem to be justified. Postwar changes in the type and intensity of land use in former West Germany appear to influence the discharge behaviour of the country's major rivers. Although it is possible that climatic changes and anthropogenic flood plain interferences are partly responsible for the increased number of inland floods observed recently in Germany, it seems that much of the changed discharge behaviour of German rivers can be explained by postwar changes in land use.

Acknowledgement

This research was supported by a grant from the Federal Foundation for the Environment (Deutsche Bundesstiftung Umwelt, DBU).

5. References

1. Vogt, R. (1995) Hochwasser in Köln: Erfahrungen, Maßnahmen, Schlußfolgerungen. Ratgeber "*Mit dem Hochwasser leben*" **1**: 48-55. Stein-Verlag, Baden-Baden.
2. van der Ploeg, R.R., Ehlers W., and Sieker F. (1999) Floods and other possible adverse environmental effects of meadowland area decline in former West Germany. *Naturwissenschaften* (in press).
3. Sieker F. (1996) Dezentrale Regenwasserbewirtschaftung in Siedlungsgebieten: Ein Beitrag zur Dämpfung extremer Hochwasserereignisse ? *Geowissenschaften* **14**(12): 531-538.
4. van der Ploeg, R.R. and Sieker F. (1999) Bodenwasserrückhalt zum Hochwasserschutz durch Extensivierung der Dränung landwirtschaftlich genutzter Flächen. *Wasserwirtschaft* (in press).
5. BMELF (Bundesministerium für Ernährung, Landwirtschaft und Forsten) (1957,... 1990) *Statistisches Jahrbuch über Ernährung, Landwirtschaft und Forsten*. Verlag Paul Parey, Hamburg und Berlin (1957,... 1975), Landwirtschaftsverlag GmbH, Münster-Hiltrup (1976,... 1990).
6. USDA (1972) *SCS National Engineering Handbook*, Section 4, Hydrology. U.S. Government Printing Office, Washington D.C.
7. Schwab, G.O., Fangmeier D.D., Elliot W.J., and Frevert R.K. (1993) *Soil and water conservation engineering*. 4th ed. John Wiley and Sons, Inc., New York.
8. Engel, H. (1994) *Das Hochwasser 1993/94 im Rheingebiet*. Publication BfG-Nr. 0833. Bundesanstalt für Gewässerkunde, Koblenz.
9. Engel, H. (1996) *Das Januarhochwasser 1995 im Rheingebiet*. BfG-Mitteilung Nr. 10. Bundesanstalt für Gewässerkunde, Koblenz.
10. Ebel, U. und Engel, H. (1994) The "Christmas floods" in Germany 1993/94. *Bayerische Rück Special Issue* No. 16. Bayerische Rückversicherung AG, 80526 Munich (Germany).

ANTHROPOGENIC IMPACTS ON THE FORMATION OF FLASH FLOODS AND MEASURES FOR THEIR COMPENSATION

M.W. OSTROWSKI

Darmstadt University of Technology
Institute of Hydraulic and Water Resources Engineering
Petersenstr. 13, D 64297 Darmstadt, Germany

1. Introduction

During the 1990s, frequent occurrences of disastrous floods throughout the world have motivated experts, journalists, politicians, citizens and others to identify reasons for the formation of these floods. Even among experts the reasoning seems fairly subjective at times. Often, under the actual impression of catastrophic damages, basic process understanding gets lost and is replaced by all kinds of irrational assignments of guilt and responsibilities. Some publications and media even suppose that such factors as urbanisation are more important than rainfall for the formation of rare flood events. Quite astonishing is the criticism uttered by individuals and representatives of communities hit by flood damages, even though they have settled, more or less consciously, in the areas endangered by high flood risk, e.g. on flood plains. Yet it has to be clearly remembered that except in few cases, floods are caused by extreme rainfall events. However, anthropogenic impacts related to land use have an influence on the intensity of flood discharge. It must be remembered that each flood event has its own set of causes, which can be categorised as anthropogenic and natural event dependent (meteorology) and event independent. In this paper, results of a model based sensitivity analysis will be presented with respect to quantifying these impacts on floods. The model structure provides an opportunity to consider anthropogenic impacts like land use changes and river training. Some control measures, which appeared suitable to compensate for the anthropogenic impacts, were applied in two case studies.

2. Modelling Approach

The modelling analysis was carried out with a deterministic simulation package, WBM_{TUD}, developed by a research team at the author's institute. The structure and components of the model are given in Figure 1.

The package is flexible with respect to spatial and temporal scales, and has special modules for adequate consideration of urbanisation, flood routing and reservoir regulation. The structure of the model is determined by a basic module, which allows to use quasi non-linear relationships between storage and flows, with a numerically fast

J. Marsalek et al. (eds.), Flood Issues in Contemporary Water Management, 125–133.

analytical solution. The model can be efficiently applied in both event and continuous modes.

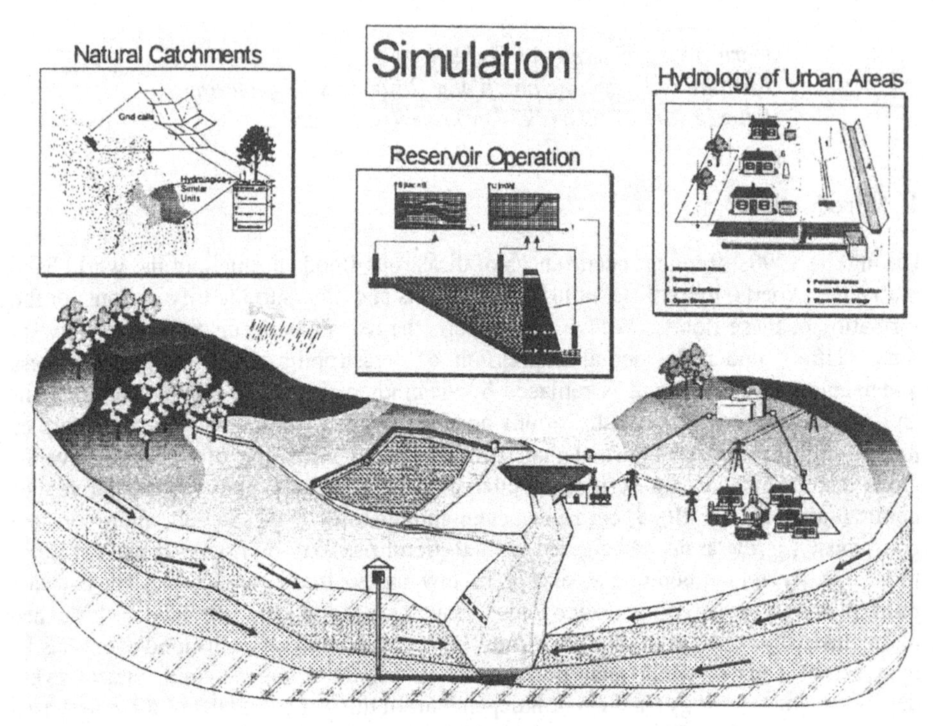

Figure 1. The structure and components of the WBM_{TUD} model.

3. Sensitivity Analysis

It must be mentioned that in addition to rainfall, there are few other reasons for flood formation, which can be, e.g., the operation of reservoirs serving for production of peak hydropower or the failure of hydraulic structures, like dams and dikes, which might be caused by flash floods. Both of these cases are caused by human activities, but will not be considered in this paper.

In principle, the effects of different factors on the formation on floods caused by precipitation is known. This section summarises and quantifies the effects of different influences.

3.1. EVENT-INDEPENDENT SYSTEM PARAMETERS

Shape of the catchment - seven different types of catchment geometry were compared. The variation of the largest and smallest peak flow of the response function was ±20 % of the mean value.

Slope - an average hill slope was varied from 1 to 10%; the related variation of peak flows was ±15%.
Soil types - in this case, the sensitivity is extremely large. Peak values ($m^3/s\ km^2$) of the response functions for different soils are given in Table 1. The related rainfall has a probability of 0.1.

TABLE 1 Peak runoff for different soil classes ($m^3/s\ km^2$)

Soil type	Sand	Silt	Loam	Clay
Unit Discharge	0	.03	.25	.31

3.2. EVENT-DEPENDENT SYSTEM PARAMETERS

Initial soil moisture conditions depend on the antecedent weather conditions and on soil types prevailing in the catchment. In Figure 2 the response of a test catchment with sandy loam soils to different initial soil moisture conditions is shown. In this case, the effect on overland flow was relatively small for moisture contents between 50 % and 100 % of field capacity.

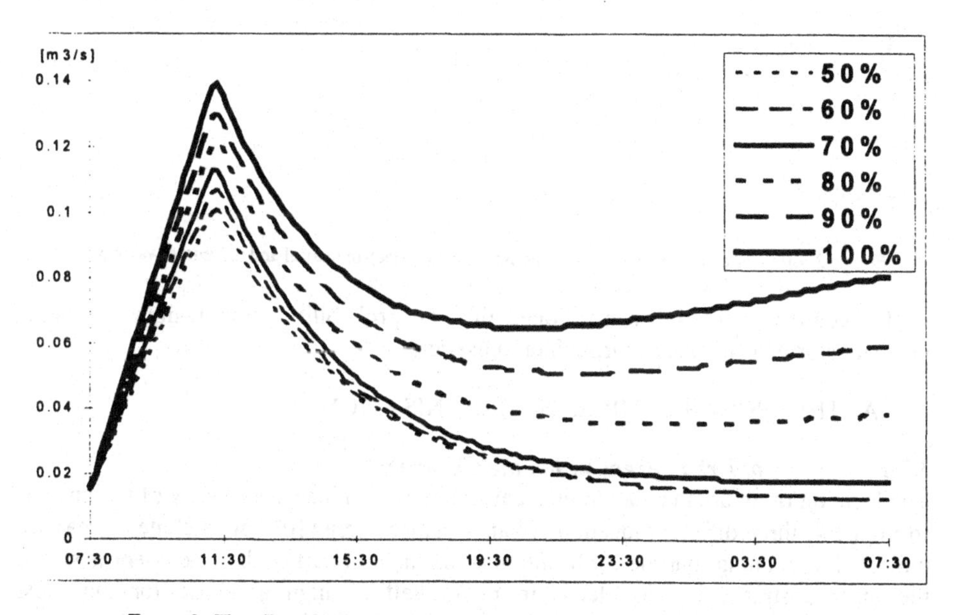

Figure 2. **The effect of initial soil moisture conditions on the formation of flood waves.**

Precipitation duration and intensity - with an increasing initial moisture, the faster lateral subsurface runoff produces increasing discharge volumes, which can become significantly effective in larger catchments and contribute to the overall flood peaks through superposition of runoff from sub-catchments. In Fig. 3, the effect of varying storm duration is demonstrated for a natural (undeveloped) catchment and for a slightly urbanised catchment. Here again it can be seen that each flood has its own characteristics. Because of a unique mixture of temporal and spatial scales and catchment characteristics, it is most difficult to define reliable formulae for their interrelationship. This is confirmed by the relatively large uncertainty in attempts of statistical multivariate regionalisation. Although many characteristics look similar on average, a process might be predominantly governed by small-scale singularities which are considered unimportant on larger scales.

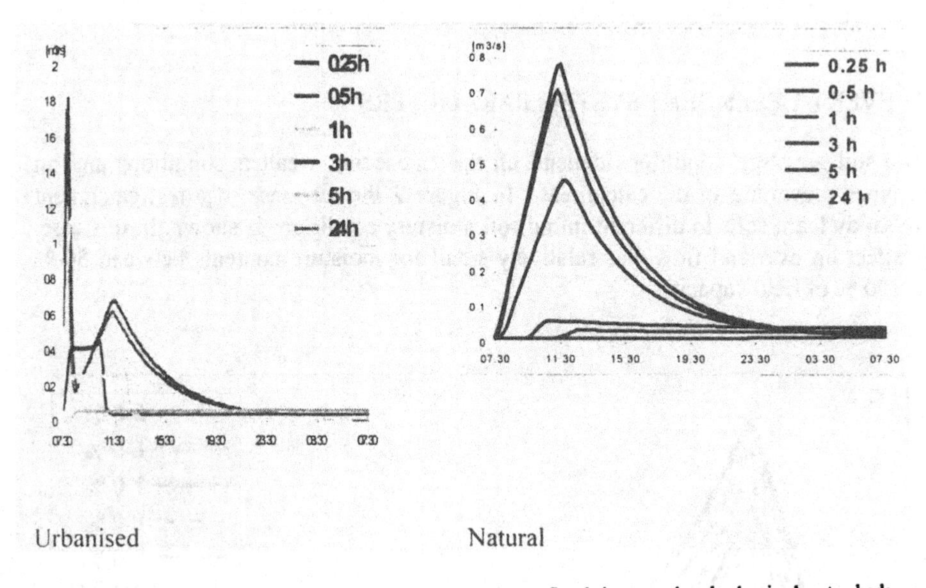

Figure 3. Effect of precipitation duration and intensity on floods in natural and urbanised watersheds.

In this context it should be mentioned that the probability of short-duration, high-intensity storms is increased through urbanisation.

3.3. ANTHROPOGENIC IMPACTS - URBANISATION

3.3.1 *Location of Urban Areas within the Catchment*

The location of an urban area within a catchment determines the severity of the impact. In Figure 4, three different locations of an urbanised area (10% of catchment area) are indicated. Assuming that the contribution of the natural part of the catchment is small, the most upstream location leads to peaks half as high as those for the most downstream location. The general statement "everybody lives downstream" seems to be somewhat misleading in this context, because it matters, how far downstream one lives from the source of an impact.

3.3.2 *Degree of Imperviousness*

When discussing the effect of increasing urbanisation, a frequent opinion is that it generally increases flood peaks, which is true in many cases. However, this is not always the case. The magnitude of this increase depends on the degree of imperviousness, which itself depends on the catchment size and on the location of the downstream control section.

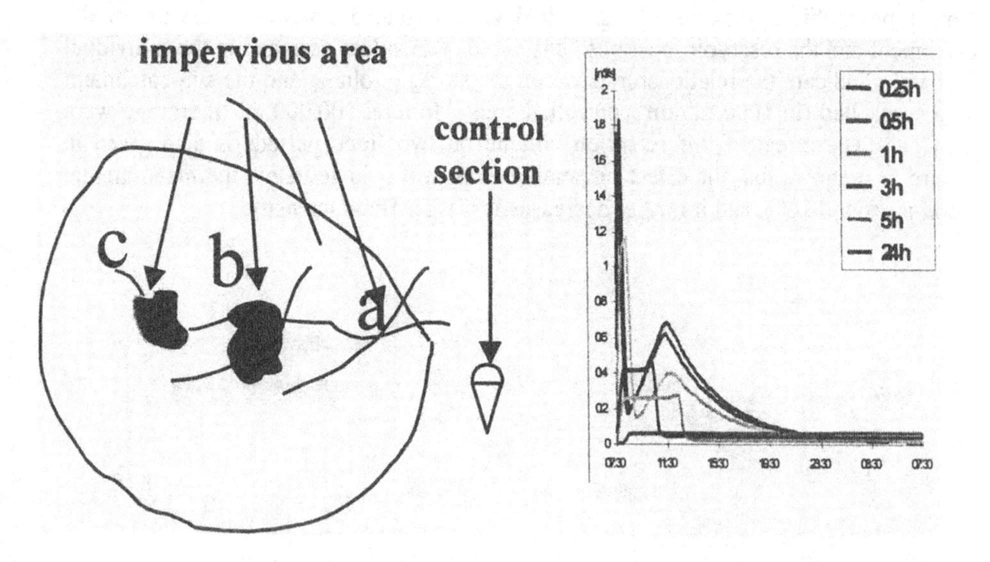

Figure 4. Effect of the location of urbanisation in a catchment.

At a control section right downstream of an urban area, and for a 15% imperviousness of the total catchment, the 1-year flood peak might even decrease as the urban runoff might have passed through this section before the main flood, from the reduced remaining area (85% of the total) arrives. Concerning the degree of imperviousness, small densely populated catchments may be up to 80% impervious, while very large catchments, like those of the major European streams, are characterised by the effective imperviousness of 4 to 5 %. If impervious areas are uniformly distributed throughout the catchment, this also represents the upper limit of their impact. However, communities often develop on bottom of downstream valleys.

4. Case Studies on Impact Compensation

The investigations described herein were carried out for the state agency for water resources of Rhine Palatinate, for the 580 km^2 Glan catchment in the south-west of Germany [1]. For the total catchment, measures were investigated to activate or restore spatially distributed retention volumes. Within the Glan catchment, the sub-watershed of the Bledesbach with an area of 7 km^2 was investigated in more detail to examine the effect of land use changes and source control measures. Model parameters were first estimated both in long-term simulation and in event-based calibration runs.

4.1. ACTIVATION OF RETENTION IN THE RIVER CORRIDOR

4.1.1 *Spatially Distributed Flood Control Reservoirs*

Ten possible locations for small distributed flood control reservoirs were identified from topographic maps and during a field visit. Figure 5 gives an overview of the catchment and the reservoir locations considered. The values assigned to the individual reservoirs indicate the relationship between the storage volume and the sub-catchment area controlled (in 1000 m^3/km^2 controlled area) . In total 500,000 m^3 of storage were provided. The effect of this retention volume on two flood periods is also given in Figure 5. It shows that the effect on small events in the range below the mean annual flood is around 10%, and it further decreases for larger flood intensity.

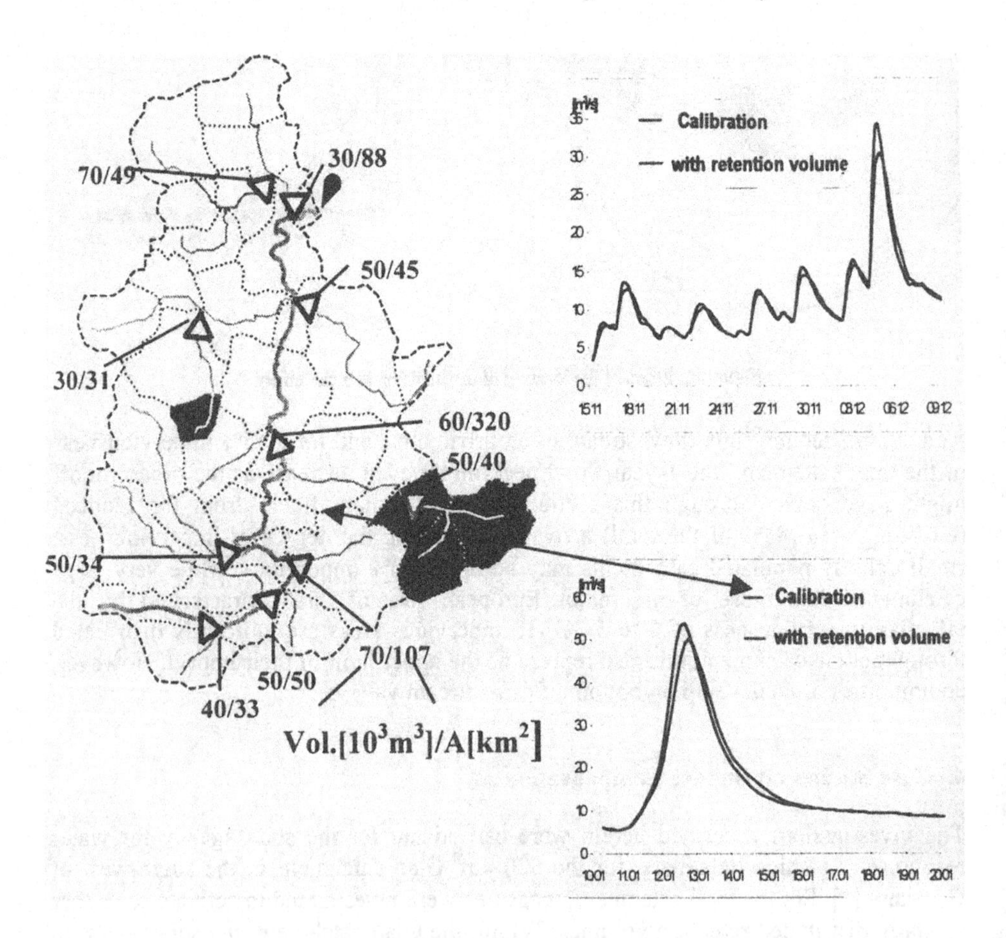

Figure 5. **Effect of distributed flood control reservoirs.**

4.1.2 *Restoration of Flood Plains*

The next step was to analyse the removal of dykes where technically and socio-economically feasible. Thus, the mean river slope would decrease and additional storage would be created. Figure 6 shows the effect of river corridor restoration. The effect is compatible with that of the distributed reservoir activation.

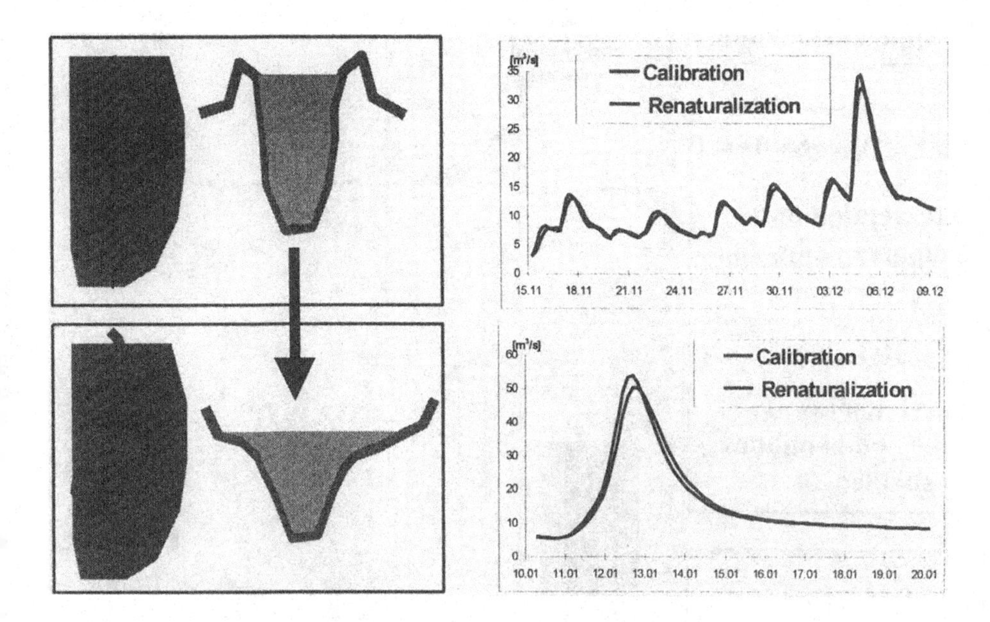

Figure 6. Effect of flood plain restoration through removal of dykes and reduction of the mean bed slope.

4.2. LAND USE MANAGEMENT AND SOURCE CONTROL

The analysis was carried out on the Bledesbach sub watershed with a mean slope of 10% and about 5% of imperviousness. Twenty-five % of the soils are highly permeable, the rest is intermediately permeable. Of the 95% permeable soils, 5% are forested, 26% are grassland and 64% are used for agricultural cultivation. Four scenarios were analysed, including (1) complete afforestation, (2) complete use as grassland, (3) application of improved agricultural practices (reduce compaction of soils, increase surface roughness and reduce slopes) and (4) infiltration of surface runoff from impervious areas. The results of this analysis are given in Figure 7.

The effects can be seen for an annual flood hydrograph as well as the components of the annual water balance. The results can be summarised as follows: (a) frequent (small) floods are mainly caused by urbanisation, thus infiltration of urban storm water (source control) has the best effect for these floods, and (b) afforestation has the biggest effect on flood peaks, but improved agricultural practices come closest.

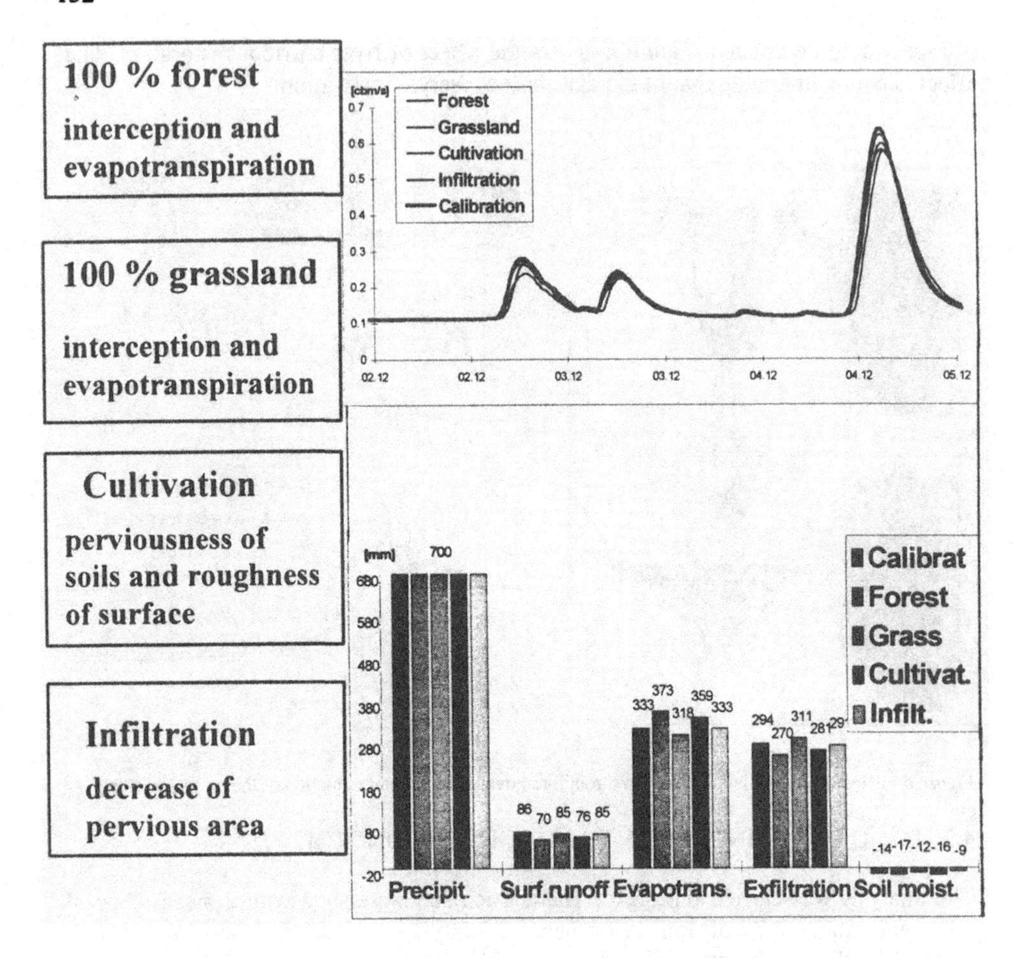

Figure 7. **Effect of land management and source control.**

5. Interpretation and Summary of Results

Before interpreting the results presented herein, some remarks are necessary concerning the uncertainty involved in the simulation work presented. It seems to be agreed that the models applicable to this type of investigations are available in principle. An important question, however, is the accuracy of such computations. Many uncertainties are involved in model development and application [2,3] which are related to data and model accuracy. The performance of the model is related to adequate spatial and temporal resolution of information as well as the use of reliable algorithms. If no measured runoff data are available to optimally estimate model parameters, large additional uncertainty is involved. This point is of special interest and shall be briefly discussed.

As in the statistical analysis, it is not clear how far a model application can be extrapolated into the range of very rare events that have never been observed. Many models use linear approximations of non-linear physical processes, such as the ordinary 1st order nonhomogeneous differential equation used in the linear reservoir approach. The extrapolation of such linear models probably, but not necessarily, involves higher uncertainty than a non-linear, more physically-based approach.

In total, the error range possible in such computations in the worst case easily reaches much more than 10 % even with reliable data and an adequate model structure. In this investigation it was assumed that the errors and uncertainties involved do not influence the relative comparison of different scenarios. Still the above discussion should be borne in mind, because in an absolute quantification of impacts, the inherent errors can exceed by far the changes caused by the impact.

From the analysis of modelling results, the following principles can be concluded:

Natural factors like the catchment shape, soil classes and slopes produce considerable variability in flood peaks. Activation of distributed, quasi-natural retention storage can compensate for downstream impacts.

Urbanisation impacts highly on the formation of more frequent and extreme floods, at a local and, to some extent, at the regional scale. As land use is intense, immense damages can occur. In addition dense urbanisation leads to more frequent convective storms. With an increasing catchment size, the mean rate of effective imperviousness decreases to less than 5 %, which can be considered as the upper limit of intensification for stream catchments. Compensation can be achieved through modified stormwater treatment (by source control techniques).

Agriculture with the application of heavy machinery leads to soil compaction, characterised by reduced pore volumes and hydraulic conductivity. Both these factors increase the so-called net storm rainfall. The negative impacts can be compensated through deep ploughing and adequate cultivation techniques.

Channelisation of rivers causes higher peaks and faster propagation of hydrographs. Compensation is possible through restoration of stream corridors (flood plains).

However, it must be stressed that all negative anthropogenic impacts and related compensation measures become less and less effective with the increasing flood return period. It must be kept in mind that flood damage risk is hardly increased by flood formation, but rather through growing damage potential, i.e. the use of flood-prone river corridors for intensive land use including urbanisation and industrial development.

6. References

1. Ostrowski, M.W., Leichtfuss,A. and Kivumbi, D. (1995) Quantifizierung von Vermeidungs- und Ausgleichsmaßnahmen (am Beispiel des Glan), Untersuchungsbericht für das Landesamt fuer Wasserwirtschaft des Landes Rheinland-Pfalz, unpublished.
2. Beck, M.B. (1983 a) A procedure for modeling, p.11-41, in Mathematical Modeling of Water Quality: Streams, Lakes, and Reservoirs, edited by G.T. Orlob, International Series on Applied Systems Analysis, John Wiley & Sons.
3. Beck, M.B. (1983 b) Sensitivity Analysis, Calibration, and Validation, p.425-467, in Mathematical Modeling of Water Quality: Streams, Lakes, and Reservoirs, edited by G.T. Orlob, International Series on Applied Systems Analysis, John Wiley & Sons.

1998 FLOOD DAMAGES AND REMEDIAL ACTION

VACLAV JIRASEK

Povodí Labe, a.s.
Víta Nejedlého 951
50003 Hradec Králové 3
Czech Republic

1. Introduction

In 1967, five special agencies, river boards, were established to implement the goals of the General Master Plan within the natural boundaries of main river basins in the Czech Republic. One of these boards, Povodí Labe, a.s., is in charge of water use in the hydrological watershed of the Elbe river and is responsible for operation and maintenance of certain types of rivers. These rivers include those that are (a) used for drinking water supply, (b) important from the water management point of view (i.e. with watershed greater than 5 km^2), or (c) border rivers. The Board's principal policy is directed towards administration and protection of water resources and creation of conditions for their sustainable use, while respecting ecological interests of the society. Besides such tasks, as for example the operation of dams, weirs and navigation facilities, the Board looks after the operation and channel maintenance of important rivers. Consequently, it is responsible for recovery actions directed to repair flood damages in river courses.

The July 23, 1998 disastrous flood was caused by an extreme downpour that exceeded the highest historical rainfall, recorded in 1956, 1.5 times and which was assessed as a torrential rain with a return period greater than 500 years. Within 8 hours, nearly 200 mm of rain fell on an area of 500 km^2, and the peak burst lasted tens of minutes. Flood discharges in important rivers, the Bcla river and the Dedina river, exceeded their 100-year peaks, and discharges in their tributaries were even more severe.

The purpose of this paper is to describe flood damage, and the system of flood damage evaluation and remediation. This system was adopted by Povodí Labe, a.s. just after the 1997 devastating flood and further tested under the conditions of the 1998 flood.

J. Marsalek et al. (eds.), Flood Issues in Contemporary Water Management, 135–138.

2. Flood Damages

2.1. OVERALL FLOOD DAMAGES

Flood damages in the region affected, principally in a single district, Rychnov n/Kn, reached 2.0 bill. Czech crowns. More than 26 thousand hectares of agricultural land and 31 villages were flooded in few hours, 23 houses were totally destroyed, 1322 houses were damaged, tens of footbridges and bridges collapsed and more than 200 km of river and creek courses were devastated. The worst was the loss of six human lives.

2.2. RIVERS CONCERNED

Among the rivers, which Povodí Labe, a.s. is in charge of, four rivers, the Divoká Orlice, the Dedina, the Belá and the Olešenka, and their tributaries, were affected by this flood event. While the flood discharge of the Divoká Orlice river was safely retained in the Pastviny Dam reservoir, the other rivers had no similar facilities and were severely damaged. In the Belá river, which is known for its narrow valley without storage space and for a steep longitudinal slope, the flood wave reached the confluence with the Divoká Orlice at the town Castolovice (about 32 km) in 9 hours.

The Dedina river has similar characteristics in its upper reach, from the spring to the town of Dobruška (about 25 km). The flood wave travelled through this river reach in two hours. The lower part of the Dedina, downstream from Dobruška to the confluence with the Orlice river at the town of Trebechovice (about 24 km), is characterised by a broad valley with a mild longitudinal slope and beautiful river meanders. The flood wave was significantly transformed here and took 31 hours to travel through this reach.

2.3. DAMAGE CHARACTERISTICS

Destructive effects of flood waves varied according to the character of the river reaches. In upper parts of both rivers, the Dedina and the Belá, high flow velocities (reaching in some places 5 m/s) heavily eroded some sections and filled others with gravel and boulders, some weighing several tonnes. Under these circumstances, both rivers frequently changed their course. Fallen trees were lodged under bridges and footbridges, forming dams across streams. The forces exerted by these dams eventually tore down bridges and the resulting sudden releases of water caused secondary flood waves devastating everything in their way – buildings, roads, railways, wells or developed sites. In the lower part of the Dedina river, the flood caused a gradually enlarging inundation with high water-mark covering a large territory that has never been known to flood before.

3. Remedial Activities

3.1. RESCUE AND RECOVERY OPERATIONS

Rescue works were conducted from the beginning of the flood by the District Authority in co-operation with firemen, local government agencies and organizations in charge of flooded rivers, such as Povodí Labe, a.s. Beside participating in rescue operations, Povodí Labe, a.s. staff were updating The Operation Plan on the basis of the most recent reconnaissance. One of the main goals of this plan was to restore unobstructed flow in river channels. Above all, the fallen trees lodged under bridges or across the river channels had to be removed as soon as possible. Simultaneously, breaches in embankments or dikes were repaired. Subsequent work focused on dredging river reaches obstructed by large sediment deposits. For this work, Povodí Labe, a.s. used its own crews and machinery, and simultaneously reported information on the actual situation to the District Authority. This information was then used to co-ordinate activities of the Army and the units of Civil Defense. This action represented the so- called 1st stage of recovery activities, to be followed by the 2nd stage, consisting in restoration of the affected area by repair of flood damages.

3.2. EVALUATION OF COSTS OF DAMAGES

Immediately after the flood, during the 1st stage of recovery operations, a special team started to evaluate the extent of significant damages in the affected river courses. In one week, all the necessary field work has been completed and its findings were presented in the Situation Report. For each of the 105 sites checked, damages were characterised by compiling the name of the river and location (in km along the course), name of the place, character of damage with measurements obtained by a field survey, the cause of damage, possible effects on the stream, the type of remedial works and their urgency, the planned year of repair, costs, responsibilities, photos and a location map in 1: 50 000. To expedite this work, blank forms were prepared ahead of time. The Situation Report formed a basis for next step – the bidding on, and contracting out, remedial/repair works.

3.3. REMEDIAL WORKS

First, the 105 sites were grouped into four repair and four reconstruction projects along the rivers concerned. Using the cost estimates produced, financial plans were prepared for 1998 and 1999.

While the state of emergency was still in effect, it was acceptable, according to the law, to exempt the recovery projects from general bidding and ask selected licensed contractors for their bids. In this process, the main requirements were not to exceed the cost estimates in the financial plan, and to prepare structural projects by registered engineers for construction permit application. Individual site remediation projects were prioritised.

Recovery contracts were signed on September 7, 1998 and 85 % of remedial works had to be finished, including project negotiation and permit process, by Dec. 15, 1998. Unfortunately, another flood event had occurred, even though not so severe, and immediately after, the freezing weather arrived in November. Because of such weather conditions, the financial plan as well as the ongoing contracts had to be changed in such a way that the percentage of activities deferred to 1999 had to be increased from 15% (originally planned) to 40 %. The new deadline for all recovery work was May 31, 1999.

4. Conclusions and Recommendations

On the positive side, the recovery system, comprising the Operation Plan and Situation Report, worked really well. The Operation Plan was prepared before, and updated during, the flood and was directed towards restoration of unobstructed flow in river channels. The Situation Report presented the extent of all significant damages in the affected river courses. This process saved time and facilitated correct and timely decisions in both stages of remedial activities.

Several shortcomings were also encountered during the recovery process. Even though repair works started as soon as possible, work involving pouring concrete (repairs of weirs, embankment walls and so on) is weather dependent and its timing has to be carefully assessed. Even though the contracts were awarded to well qualified licensed contractors, there were few cases of problems with the quality of work of subcontractors, who were not subject to approval under the contract award process. Finally, the Czech Water Law (No. 138/73 Sb.) and the law on The Protection of the Nature (No. 114/92 Sb.) are inconsistent and require different approaches, especially with respect to the so-called "natural" rivers.

In view of the above experiences, it is recommended to: (a) retain the procedure comprising the Operation Plan striving to restore unobstructed flow in river channels and the Situation Report evaluating damages, and (b) include recovery activities after severe floods in the exceptions to the bidding process and contracting law.

A DATABASE OF HISTORICAL FLOOD EVENTS IN THE NETHERLANDS

MARTINE JAK[1] and MATTHIJS KOK[2]

[1] *Ministry of Transport, Public works and Water Management*
Road and Hydraulic Engineering Division
P.O. Box 5044
2600 GA Delft
The Netherlands

[2] *HKV Consultants*
P.O. Box 2120
8203 AC Lelystad
The Netherlands

1. Introduction

The Netherlands is situated on the delta of three of Europe's main rivers: the Rhine, the Meuse and the Scheldt. As a result of this, the country has been able to develop into an important, densely populated nation. But living in the Netherlands is not without risks. Large parts of the Netherlands are below mean sea level and water levels which may occur on the rivers Rhine and Meuse. High water levels caused by storm surges on the North Sea or by high discharges of the rivers are a serious threat to the low-lying parts of the Netherlands. Construction, management and maintenance of flood defences are essential conditions for the well-being of the population and further development of the country.

Without flood defences much of the Netherlands would be regularly flooded. The influence of the sea would be felt principally in the west. The influence of the waters of the major rivers is of a more limited geographic impact. Along the coast, the protection against flooding is principally provided by dunes. Where the dunes are absent or too narrow or where the sea arms have been closed off, flood defences in the form of sea dikes or storm surge barriers have been constructed. Along the full length of the Rhine and along the parts of the Meuse, protection against flooding in provided by dikes.

Flood protection in the Netherlands is organised within dike rings. A dike ring is an area which will be flooded if there is an extreme hydraulic load from the sea, the lake IJsselmeer or one of the big rivers. Figure 1 presents an overview of the 53 dike rings.

J. Marsalek et al. (eds.), Flood Issues in Contemporary Water Management, 139–146.

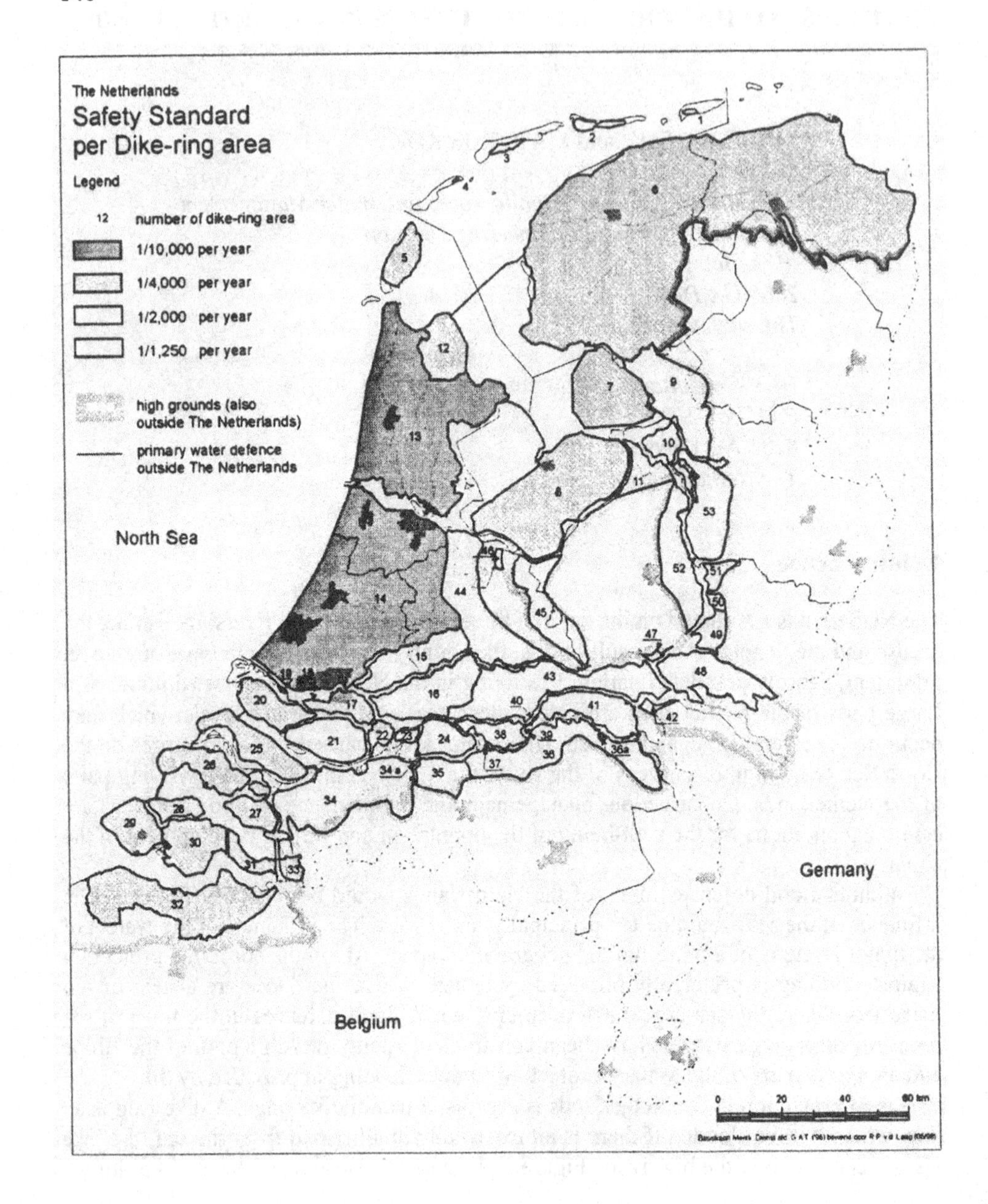

Figure 1. **Dike rings in the Netherlands.**

The goal for the near future is to adopt a safety approach which is based on the risk (including damage assessment) of flooding for each dike ring area. To achieve this goal, a tool is being developed for calculating damages and loss of life for each dike ring area in a uniform and practical manner. Good estimates of flood damage functions (which relate for example flood damage to inundation depths) can be obtained from historical flood data.

In this paper, first results are presented with respect to the design of a central database containing all relevant damage estimates of flood events in the Netherlands. In the past 50 years, five large flood events have occurred. Because of the wide diversity of data, a flexible database has been built. A major effort was spent on finding the relevant data in reports and local databases, and to include them in the database. Towards the end of this paper, a case is made for including flood events from outside the Netherlands in this database.

2. Flood Management in the Netherlands

For determining the required height of dikes, the traditional method used in the Netherlands until well into this century was to take the highest known water level, plus a margin of 0.5 to 1 metre. After the catastrophic 1953 flood disaster in the Netherlands, the design method for flood protection was improved considerably by the Delta Commission [1,2]. The starting point as proposed by the Delta Commission was to establish a desired level of safety for each dike ring area or polder [3,4]. This safety level would need to be based on the costs of construction of dikes and on the potential damages caused by flooding. This economic analysis led to an 'optimum' safety level expressed as the probability of failure for the coastal dikes. In practice, however, the safety level was expressed as the return period of the water level, which provides a better description of the most dominant hydraulic load. One of the main reasons for simplifying the description of the safety standard was the lack of knowledge for describing the dike failure process with sufficient accuracy. The Delta Commission approach is shown schematically in Figure 2.

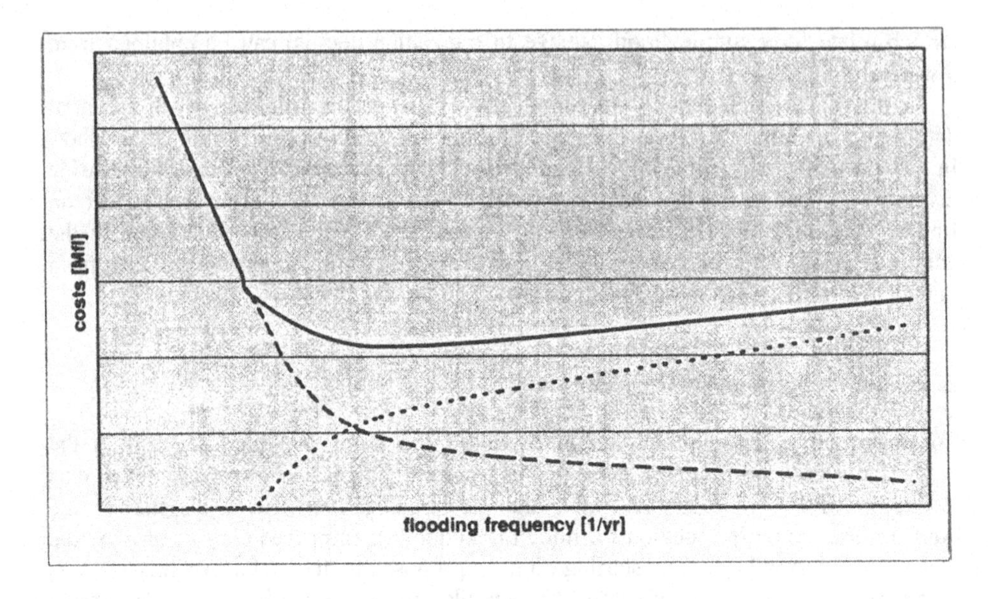

Figure 2. The optimal safety approach in economical terms. The total costs (straight line) is the sum of the flood damage (dashed line) and costs of dike construction (dotted line).

The economic analysis has been used to vary the safety standard according to the expected damages in various polders. Thus, a safety standard has been established for each dike ring area. This standard is expressed as the mean yearly frequency of exceedance of the prescribed flood level. The standards vary from 1/10000 to 1/1250 per year (see Figure 1), depending on the economic activities in and population of the protected area, and the nature of the flood threat (river or sea). In 1996 these standards were included in legislation when the Flood Protection Act came into effect (Ministry of Transport, Public Works and Water Management 1996).

In the near future, a risk–based flood protection policy will be completed. At present the tools required for such a policy are being developed. In the risk-based concept the safety assessment of flood protection structures is replaced by a total risk assessment. It is evident that the probability of flooding is not the same as the probability of the design load being exceeded.

Flooding of a dike ring area causes enormous damage. The extent of this damage depends on the nature of the flood (a sea surge or river flood), and the properties of the area, for example the potential depth of inundation and land use. Escape possibilities and early evacuation can play a mitigating role here.

The concept of flooding risks also offers the possibility of introducing other measures than building dikes. For example, it will be possible to decide whether it would be more useful to invest in spatial planning of an area (for example constructing vulnerable elements on relatively high grounds) or in strengthening the dikes. In calculating the consequences of flooding, the research in the Netherlands is concentrating on developing a tool for calculating damage and loss of life in each dike ring area, in a uniform and practical manner. One of the projects deals with setting up a flood-damage database with historical data. These data can be used to produce better estimates of flood damage functions and uncertainty distributions of flood damages [5].

3. Set-up of a Flood-Damage Database

Five extreme flood events have occurred in the Netherlands during the past 50 years. The most catastrophic was the 1953 flood, with almost 2000 fatalities. These five events include river floods and resulting damage, caused by heavy rainfall in the Netherlands. Because the Netherlands territory is partly below the sea level, heavy rainfall, exceeding limited pump capacity, can cause a lot of damage. Locations of the five extreme flood events are shown in Figure 3.

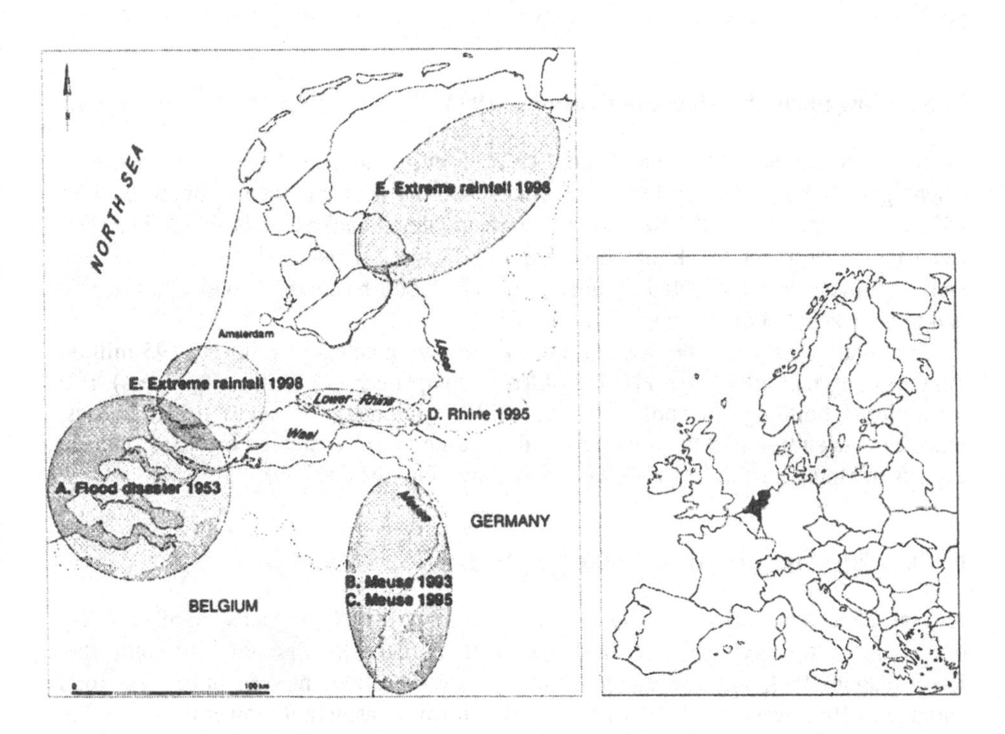

Figure 3. **Flood events in the Netherlands.**

The database is linked with a Geographic Information System, so that the geographical distribution of damages can be readily shown (e.g., the hot spots can be immediately identified).

4. Flood Disaster in South West Holland in Winter 1953

The vulnerability of the Netherlands to flooding became clear in the winter of 1953. A weather depression, combined with an exceptionally high spring tide, caused a storm surge on the North Sea and brought water levels to record heights. Dikes failed in several places, especially in the south-western part of the country. In almost all cases, the dikes were too low and collapsed. Overflow and wave overtopping led to large scale failures of the inner slope. The direct result of the disaster was 1835 fatalities and economic damage of NLG 1.5 billion (the 1956 price index). The indirect economic damage is estimated as a multiple of this cost. Flooding of Central Holland was barely avoided. The concentration of economic activities in the central part of the country would have meant much greater damage.

The database contains the number of fatalities and economic damage per polder, as estimated by the National Office of Statistics, and some other characteristics of these polders.

5. Flooding along the Meuse in December 1993

Around Christmas 1993, the Dutch river Meuse flooded because of extreme discharges, greater than 3100 m^3/s, with a probability of occurrence of about once in 150 years. In this part of the Netherlands, there are no embankments or dikes. The river has a length of about 150 kilometres. The flood caused damage of about 250 million Dutch guilders, and inundated an area of about 18,000 hectares. About 8,000 people had to be evacuated during several weeks [6].

The total damage can be divided into the following categories: houses (95 million guilders), agriculture (20 million guilders), companies (75 million guilders) and institutional buildings (schools, etc., 60 million guilders). Damages to all houses, industry, agriculture and the governmental buildings were estimated by flood damage experts of Dutch insurance companies and entered into the database.

6. Flooding along the Meuse in January 1995

A year after the Christmas flood in the South of the Netherlands, another event occurred in the same area. The discharge at the border with Belgium was somewhat lower than in 1993, but the duration of the January 1995 flooding was longer. The total damage in 1995 was 165 million guilders, which was considerably lower than in 1993. These lower damages can be explained by the fact that in this region 1995, the people were much better prepared for floods than in 1993 [7].

The total damage can be divided into the following categories: houses (40 million guilders), agriculture (20 million guilders), companies (60 million guilders) and institutional buildings (schools, etc., 45 million guilders). Again, all damages were estimated by insurance companies and entered into the database. The main difference between the 1993 and 1995 damages was noted for houses, a reduction of 55 million guilders.

7. Evacuation along the Rhine in January 1995

In 1995 there were also problems along the Rhine river. In Germany, much damage occurred, and in the Netherlands, there was a lot of tension concerning the river dikes. Their safety could not be guaranteed, and therefore some 200,000 people had to be evacuated for more than one week. Fortunately, there were no dike failures nor flood damage caused by this event. However, the evacuation was expensive, and it would have been very helpful to have even a rough estimate of these costs.

The expenses incurred by all companies were estimated by Dutch insurance companies, and partly paid by the government. However, it was not possible to retrieve these data (as of May, 1999), because it was not known if and where they had been stored for future use. The costs of evacuation of 200,000 people is very difficult to estimate, because about 98% of all evacuees did not go to the evacuation centres, but preferred to stay with friends and relatives all over the country. The government paid to each family 500 guilders (in total about 50 million guilders).

8. Heavy Rainfall in Autumn 1998

In September and October 1998, heavy rainfalls caused a lot of water damage in the Netherlands. Due to constraints in the drainage of the polders and other water systems, much agricultural damage occurred. The average return period of this rainfall event was about 125 years. In the Netherlands, there is a law which states that the government can compensate the damages in the case of catastrophic events, such as heavy floods, earthquakes and others. When this law is invoked, all damages are estimated by experts.

The first incomplete estimates of the total damage are now available; the total damage, about 900 million guilders, can be divided into the following categories: housing - 10% (90 million guilders), agricultural operations - 85% (765 million guilders), and government agencies - 5% (45 million guilders). The total number of objects in the database is 14,700 (10,660 agricultural companies; 2,470 houses; 1,220 other companies; 350 government agencies). Damages to all houses, industry, agriculture and the government buildings were estimated by insurance companies.

9. Discussion

The database described in this paper includes the most important events causing flood/water damage in the Netherlands. The data are directly linked with a geographic information system, so that a spatial distribution of the damage is directly known. In this way the flood damage data can be linked with other data [8]. In authors' opinion, it would be useful to extend this database by adding more (European) flood damage data, from all types of floods in other countries. This extension would be even more fruitful, if the damage data could be linked to hydraulic and land use data. These data can be used to produce better estimates of flood damage functions and uncertainty distributions of flood damages [5].

10. References

1. Jorissen, R. (1996) *Safety, Risk and Flood Protection Policy.* Ministry of Transport, Public Works and Water Management. Internal report.
2. Jorissen, R. and Kok, M. (1997) 'A physically-based approach to predicting the frequency of extreme river stages', a review from the property owner and engineering perspectives. In: Roger Cooke, Max Mendel and Han Vrijling, *Engineering Probabilistic Design and Maintenance for Flood Protection.* Kluwer Academic Publishers, 1997.
3. Delta Commission (1960) *Report of the Delta Commission*, The Hague, The Netherlands.
4. Van Danzig, D. (1956) Economic decision problems for flood protection. *Econometrics*, 276-287, New Haven.
5. Van Noortwijk, Jan M., Kok, M. and Cooke, R.M. (1996) Optimal decisions that reduce flood damage along the Meuse: an uncertainty analysis. In: S. French and J.Q. Smith (eds). *The practice of Bayesian analysis,* Arnold, London, 151-172..
6. Delft Hydraulics (1994) *Onderzoek Watersnood Maas. Deelrapport 1: wateroverlast 1993.* Delft/De Voorst, Dec. 12, 1994.
7. Nierop, M. (1997) *Schade in Kaart. Schadeanalyse van de Maasoverstromingen 1993 en 1995 in Limburg.* MSc thesis, Universiteit Twente, 1997.
8. Hogeweg, M., Jonge, J. de, and Kok, M. (1996) Flood damage model of the river Meuse based on GIS. *Proceedings of the fourth conference on GIS*, Vienna, 1996.

FLOOD MANAGEMENT IN POLAND

A. LEWANDOWSKI

Geomor Co., Ltd.
Koscierska 5; 80-328 Gdansk; Poland

1. Background

Several floods have occurred in Poland over the years, mainly along the Odra and Vistula rivers. The devastating flood that hit Poland in July 1997 was particularly severe along the Odra river. More than 140,000 people were evacuated as 86 cities and 845 villages were inundated. It is estimated that the damages caused amounted to about 3 billion US$.

A clear need for provision of new flood management technology in Poland has been identified. For operational purposes there is a demand for timely and accurate forecasts of river flows and water levels within the flood-prone areas. For planning and development purposes there is a need for a flood management system, which can be used for flood control studies, flood mapping, flood risk analysis, selection of strategy for flood protection, etc.

Based on a request from the Ministry of Environmental Protection, Natural Resources and Forestry in Warsaw, the Danish Environmental Protection Agency (DEPA) has decided to finance the transfer of Danish flood management technology to Poland. DEPA has also decided to support a similar flood management project for the authorities in the Czech Republic. By this dedicated effort in the two neighbouring countries, DEPA has been aiming both to provide the authorities with the state-of-the-art flood management technology, but also to emphasise the need for co-ordination of flood prevention activities in the two countries with trans-boundary rivers.

Following the devastating floods in 1997 the World Bank has launched a 200 million US$ Emergency Flood Recovery Project in Poland with a duration of 3 years. The proposed project is of an emergency nature, aiming to restore basic infrastructure in communities and rural areas affected by the flood, to make urgent repairs to the flood management system, and to improve the policy framework and institutional capacity for better flood management and mitigation, concentrating on the most heavily affected region of Poland.

The two principal objectives of the proposed project are to assist the Government of Poland in:

- restoring basic infrastructure in municipalities and rural areas affected by the flood by repairing structures and facilities of economic and social importance, and by

J. Marsalek et al. (eds.), Flood Issues in Contemporary Water Management, 147–153.

restoring lifeline facilities and services for currently displaced people – component A,

- reducing vulnerability and risk of future floods – component B.

Thus, the aim of component B is to create an efficient flood management system covering co-ordinated actions from legislation to implementation. The project covers modernisation of hydrotechnical infrastructure, modernisation of flood management systems, as well as improvement of forecasting and planning.

Component B consists of 4 sub-components:

- B.1 – Basin flood protection planning
- B.2 – Monitoring, forecasting & warning
- B.3 – Investment in flood protection infrastructure (not financed by the World Bank)
- B.4 – Flood risk reduction and prevention

The Danish technology transfer project fits well into the World Bank project and efforts have been made to co-ordinate both projects.

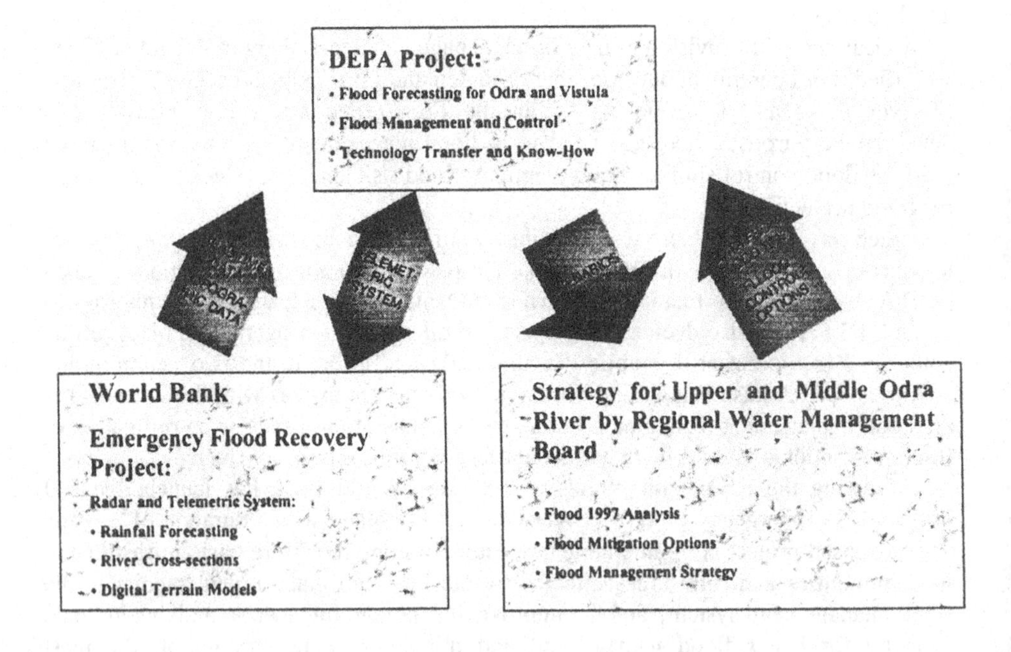

Figure 1. Description of the Polish-Danish project (DEPA project).

2. DEPA Project Description

The project focuses on strengthening the capabilities of the Polish authorities in the use of modern flood management technology. The participating Polish agencies are the Institute of Meteorology and Water Management (IMGW), with branch offices in Wroclaw and Krakow, and the Regional Water Board (RZGW), with branch offices in Wroclaw and Krakow. The consulting company GEOMOR is acting as a local consultant throughout the project. DEPA has entered into a contract with the Danish Hydraulic Institute (DHI) for provision of a flood management system and consulting services including training.

The project started in May 1998 and will be completed in December 2000. The core of the flood management system, which right now is installed in branch offices of the participating institutions, is the advanced river modelling system MIKE 11 for flood management, flood forecasting and flood control studies (Fig. 2). The model simulates water levels and flows in the river system including rainfall-runoff processes for predicting river flows and reservoir inflows generated by rainfall and snowmelt floods. The hydrodynamic formulation allows branched and looped networks and quasi two-dimensional flow simulations in polders and floodplains, as well as flow through a variety of structures. Facilities are also provided for preparing maps of the simulated floods in GIS for comparison with other spatial data to assess, for example, possible flood damage.

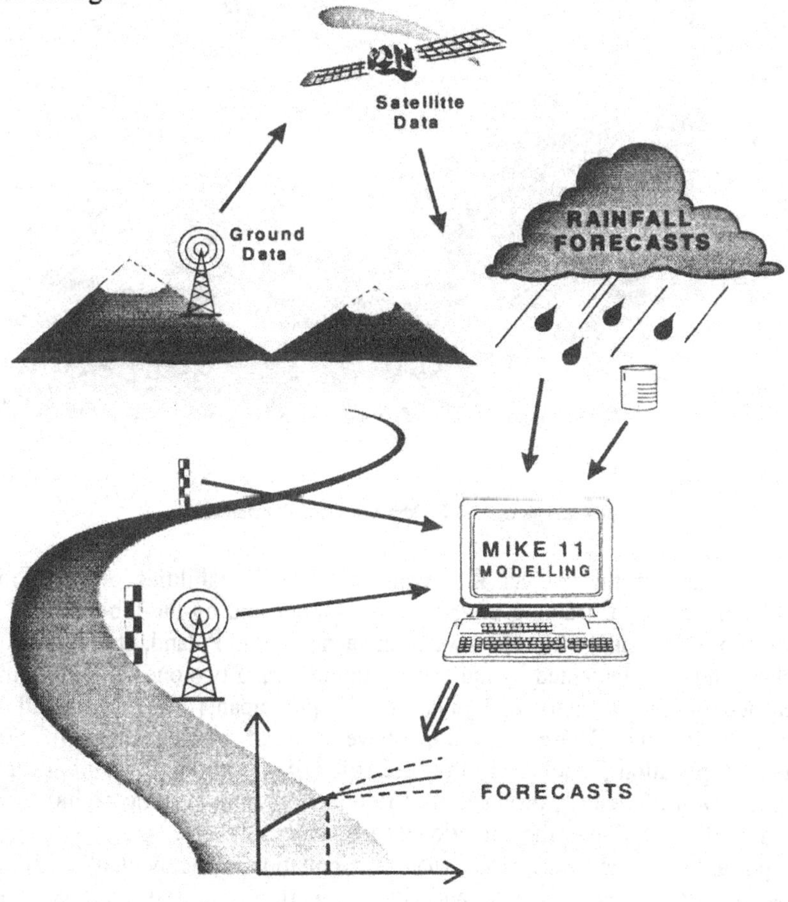

Figure 2. MIKE 11 - flood management.

The main output of the project is the implementation of operational forecasting systems for the Upper and Middle Odra and the Upper Vistula river basins (Fig. 3). The operational forecast system using MIKE 11 and FLOOD WATCH includes flexible data acquisition and database management as well as GIS features.

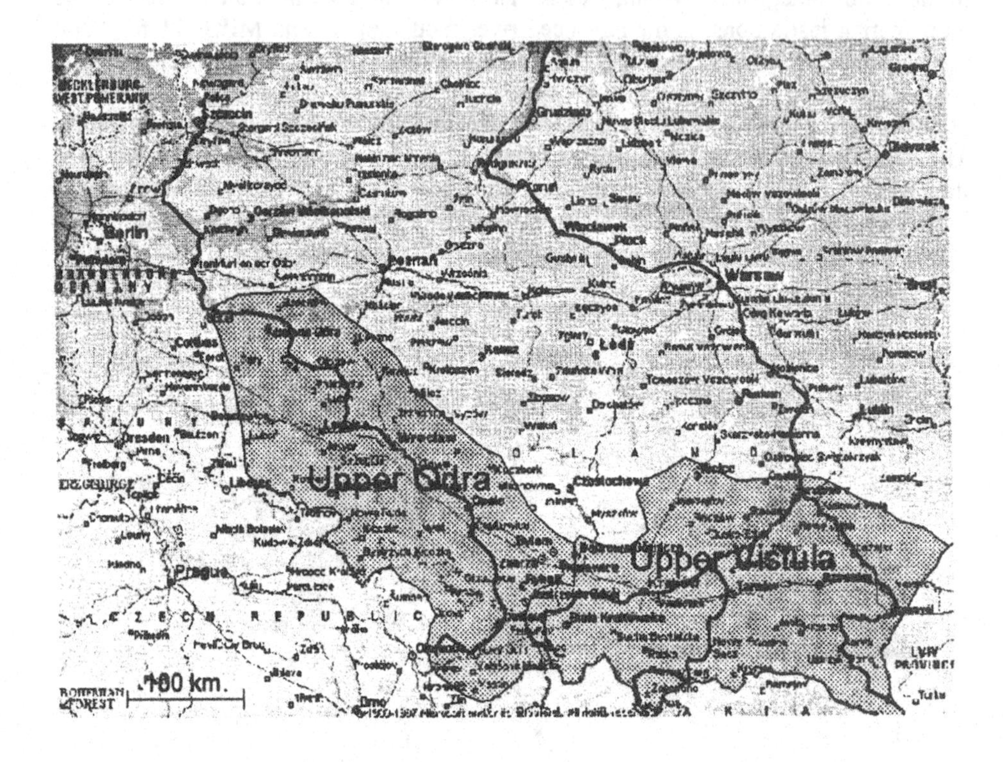

Figure 3. **Upper, Middle Odra and Vistula catchments.**

An integral part of the project is strengthening the capabilities of the Polish authorities by provision of training programs, workshops and on-the-job-training. To date several shorter training programs have been carried out in Poland, at both Krakow and Wroclaw, and complemented by intensive training during two one-month training courses carried out in Denmark and involving 12 participants representing all the participating institutions. These workshops have covered a wide range of flood management applications, including the use of GIS technology, rainfall-runoff modelling, reservoir modelling, production of flood maps using existing digital terrain models, and implementation of the operational forecasting software.

An important outcome of the project has been collation of the available historical meteorological and hydrological data, and river network and cross-section geometry data required for flood control modelling and flood forecasting.

3. Database

The DEPA project has collected and established a comprehensive database with relevant flood management data. The database has been organised in a user friendly GIS-environment and forms a very good basis for future analysis. Historical time series data from floods during the last 30 years have been collected on a daily basis, including the data from the 1997 flood event. Water level and discharge data have been collected from 92 stations in the upper and middle Odra catchment (between Chalupki and Polecko), and rainfall, temperature and evaporation data have been collected from 77 meteorological stations. Similar water level and discharge data have been collected from 55 stations in the upper Vistula catchment and meteorological data have been collected from 108 stations.

A number of GIS-themes have also been selected and established in a common GIS-project. The GIS-themes include among others location of rivers, cities, hydrological stations, structures, reservoirs, polders and cross-sections. The 1997 flood lines have been digitised and can now be used in future flood management analyses. The GIS-project has finally been prepared for real time applications and a database has been established in the GIS-project with access to all historical data. GIS-based maps will form the basis for flood zoning or development regulation on the flood plains. Integration of the information on current and proposed development with flood inundation extent, flood hazard, or flood depth mapping derived from flood modelling tools will provide a powerful tool for appropriate and sustainable floodplain management (Fig. 4).

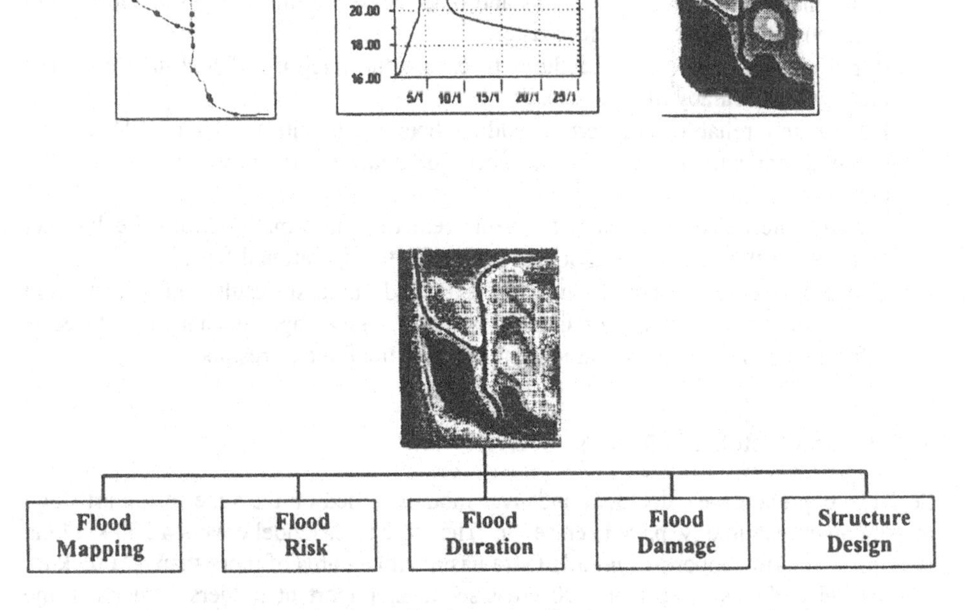

Figure 4. **Flood control studies.**

4. Database Management

The future flows and water levels in a river basin depend on both the current state and water storage in the basin, and the current and future meteorological events. Further requirements are:

- Monitoring must reflect the hydrological regime and the time scale of flood events. Thus rapidly responding mountain catchments require rainfall data as frequently as in 15-minute intervals, while it may be sufficient to monitor water levels in a large river basin twice a day.
- The monitoring network shall be reviewed in relation to flood forecasting. Many established data networks are designed for collection of data for long-term water resource planning rather than for flood forecasting.
- Both the data collection equipment and transmission systems must be robust and reliable even during devastating floods.
- Incorporation of modern rainfall measurement and forecasting technology, such as satellite and radar imagery, as well as meteorological modelling for precipitation forecasting.

An effective real-time flood forecasting database shall include the following features:

- Be capable of rapid storage and retrieval of a range of real-time monitoring data such as temperature, rainfall, flow and water levels.
- Provide a visual and intuitive presentation of the status of the catchment and of the real-time reporting stations.
- Effective links to other real time databases and existing historical databases.
- Flexible database structure to include access and possible storage of satellite and radar images, real-time flood maps and other real-time information to support the decisions of the operational forecaster.
- Rapid quality assurance procedures to ensure the integrity of real-time data and thereby the accuracy of forecasts.
- Robust and reliable hardware including back-up facilities such as CD-writers, mirror systems to ensure continued operation under failure, power system backup, etc.
- Include Internet links to access on-line real time information from the Internet sources of satellite and forecast data for use by the operational forecaster.
- Use the Internet/intranet to provide data and forecast results for private and institutional users of the forecast information, allow rapid exchange of forecast information and feedback and interpretation of the forecast results.

5. Odra and Vistula HD Models

So far, comprehensive catchment and river models, based on the state-of-the-art flood management technology, have been set up. The Odra river model covers a 500 km long river section in the upper and middle Odra basin with an area of more than 49,000 km^2. The model includes more than 500 cross-sections, important polders, structures and floodplains. Similarly, a 300 km river section model has been established for the Vistula and includes more than 200 cross-sections and important structures in the 50,000 km^2 upper Vistula basin. Both models use updated databases including all the available topographical and hydrometeorological data provided by the participating local branches of IMGW and RZGW. When new topographical data become available, the models can be easily updated.

6. Pilot Models

The pilot models, developed for the main river systems in the Upper and Middle Odra, and the Upper Vistula, have been refined throughout the project. In addition, pilot catchment models are being developed for two key basins, the Sola River in the Upper Vistula and the Nysa Klodzka in the Upper Odra basins, where incorporation of reservoir operation is important for accurate prediction of river stages and flows, and for optimising reservoir flood control strategies. These key basin studies form a basis for extending the modelling to the entire Upper and Middle Odra, and the Upper Vistula basins.

7. Project Output

The main outputs of the DEPA project are the models developed and calibrated for flood forecasting in the two river basins, the Upper Vistula and the Upper and Middle Odra. To obtain maximum benefits from these models (river network and catchment definition, numerical delineation and model calibration), the development and testing of flood control scenarios will re-use, and where necessary refine, these scenarios to describe specific flood measures. Flood maps showing the extent and depth of flooding, based either on design or observed discharges/water levels, will increase the public awareness of the flood-prone areas for a range of flood levels. The same flood maps will be used together with the data on the location and nature of flood-prone properties and stage damage curves to estimate total flood damage for a given water depth. The resulting damage analysis can then be used to carry out cost-benefit analyses of different flood mitigation measures. These maps will also form the basis of the proposed flood insurance analyses.

TWENTY YEARS OF FLOOD RISK MAPPING UNDER THE CANADIAN NATIONAL FLOOD DAMAGE REDUCTION PROGRAM

W. EDGAR WATT

Queen's University
Kingston, Ontario, Canada K7L 3N6

1. Introduction

In the early 1970s, Canada faced a problem similar to the United States and other countries that resulted from a reactive strategy towards floods and flooding: ever increasing expenditures of flood control structures together with increasing expenditure on flood disaster assistance without any reduction in flood damage potential. Canada's cities were growing and there was pressure for development in the nation's floodplains and increasing demands on governments to provide flood control. Canada's response to this problem, the National Flood Damage Reduction (FDR) Program, is described by Bruce [1]. "Past governmental contributions to flood relief and flood control structures have not curbed floodplain investment processes nor the concomitant increase in damage potential. The new program is intended to coordinate federal and provincial strategies by clearly defining flood-risk areas, by discouraging continuing investment in those areas, and by following up with appropriate measures to limit damage to existing development."

The cornerstone of the FDR program was a program to produce high quality flood risk maps for all of the nation's urban floodplains. The flood risk areas delineated on these maps were to be designated (as flood prone lands) and land use in these areas would be restricted to undertakings that were not susceptible to flood damage under conditions of the design flood event.

Over the period 1976-1998, mapping was completed for almost all of the urban floodplains identified at the beginning of the program. Land use restrictions were enacted for the majority of these areas and the FDR program is generally viewed as having been extremely successful in identifying urban flood risk areas across Canada and in redirecting damage-prone development away from flood risk areas. Many lessons have been learned and the job is not done. The objective of this paper is to describe the flood risk mapping component of the program in the hope that some of the lessons may be applicable to other countries.

The mapping program is best described in the context of being a key component of a broader program to reduce flood hazard (i.e. the FDR program). Development and implementation of a national program such as this posed significant political and technical challenges because of the Canadian setting in terms of political jurisdiction over water resources and the characteristics of floods in various parts of the country.

J. Marsalek et al. (eds.), Flood Issues in Contemporary Water Management, 155–165.

Accordingly, the following sections provide brief reviews of (i) the Canadian setting and (ii) the Flood Damage Reduction Program.

2. The Canadian Setting

2.1. GEOGRAPHY, CLIMATE AND POPULATION

Canada, the second largest country in the world, has an area of 10 million km^2 with fresh water covering 7.6 percent of this area. It stretches over almost 90 degrees of longitude and 40 degrees of latitude. Approximately 6,400 km separate St. John's, Newfoundland and Victoria, British Columbia. Canada shares a border of 8,800 km with the United States of which some 3,500 km traverse water bodies. Figure 1 shows domestic and international boundaries and selected cities.

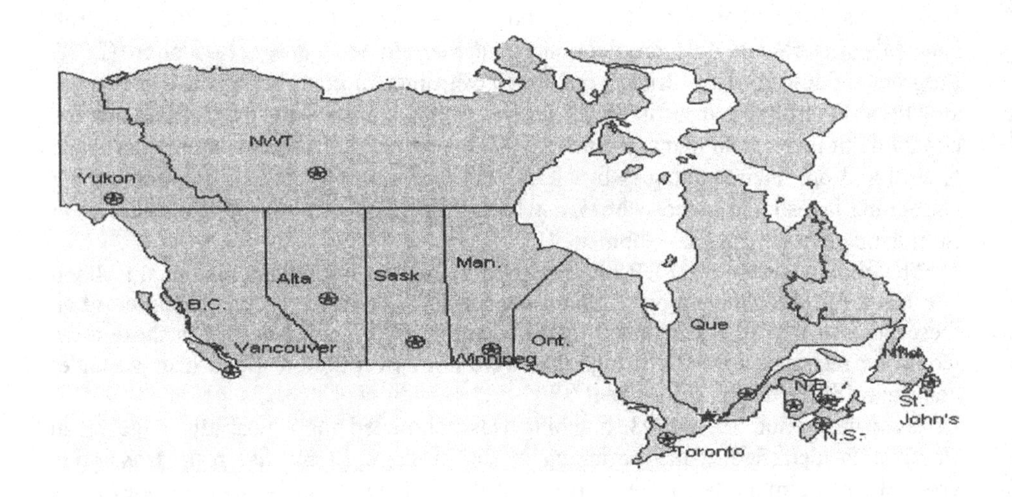

Figure 1. Canada: provincial boundaries and principal river basins.

The climate is influenced by the movement of weather systems, the distribution of land and water, the western mountains of the Cordilleran Region, and the broad plains of the interior. These factors combine with highly differentiated seasons to produce certain characteristic weather features: relatively mild rainy winters on the Pacific coast, cold winters and hot summers of the Prairies, summer intrusions of humid air masses in the Great Lakes area and maritime storms in the Atlantic Provinces. In most regions, snow covers the ground during winter, and rivers and lakes are ice-covered for several months of the year. Permanent glaciers exist in the mountains in the Western Cordilleran Region.

Coupled with these geographic and climatic factors is the distribution of the country's population, the bulk of which is concentrated within 100 to 200 km of the Canada-United States Boundary. About 60 percent of the population and industrial activity is found in the Great Lakes - St. Lawrence basin, which contains a number of large urban industrial complexes including Montreal, Ottawa and Toronto. Major metropolitan centres in western Canada include Vancouver, Calgary, Edmonton and Winnipeg. Other urban centres are more modest in size. Surrounding such centres or lying between them are towns and small cities. Communities which lie outside this general descriptive framework usually exist for historic reasons - such as fishing or native settlements, or are a consequence of mineral, forest and energy development or the transportation and utilization of such products.

2.2. CHARACTERISTICS OF FLOODS IN VARIOUS REGIONS

Urban flood prone areas exist in communities on the shores of the Atlantic and Pacific Oceans and the Great Lakes, and on rivers ranging in size from the St. Lawrence to small urban creeks. Natural causes of riverine flooding include rainfall, snowmelt, rainfall plus snowmelt, ice jams, etc. Many of the larger rivers are regulated but the degree of flood control for larger floods is limited. With the exception of the Great Lakes System, none of the drainage basins is large enough to have over-year memory of high flood potential. Andrews [2] provides a comprehensive overview of flooding in Canada including causes of floods, major flood events, response to floods, flood control and floodplain management. As might be expected, flood characteristics vary considerably across Canada. Watt *et al.*[3] provide a comparison of selected extreme flood discharges in Canada with extreme discharges on U.S. and foreign rivers, summary statistics by region, and a brief description of the characteristics of floods in various regions. Knowledge of such characteristics is relevant to the flood risk mapping program - framing of criteria for delineation of flood risk areas, development of specifications for hydrologic and hydraulic analyses and for mapping - and in other programs such as flood forecasting and warning, flood proofing and structural measures.

2.3. GOVERNMENT JURISDICTION AND INVOLVEMENT

Canada is organized into ten provinces and three territories. Within the provinces, there are Indian lands or Reserves, federal properties, which vary in size from building sites to major properties such as defence installations, and National Parks.

Ownership of resources is primarily provincial, with the exceptions noted above. However, a distinction is made in the Canadian constitution between legislative jurisdiction and proprietary rights. The provinces' ability to legislate in water resources matters is derived from their exclusive authority to legislate over property and civil rights, over matters of a local and private nature, and over local works. The federal government has legislative powers over navigation, fisheries, and interprovincial and international undertakings and may legislate concerning works, which, although situated in one province are declared to be to the general advantage of Canada or two or more provinces. It can also legislate for "peace, order and good government" for emergency situations.

In addition to its involvement in areas of federal jurisdiction, the federal government was involved in areas of provincial jurisdiction prior to the FDR program in two ways: federal assistance for structural flood control works and cost shared disaster assistance. Before 1975, federal assistance to the provinces on structural works was considerable but limited to capital costs and tended to be confined to flood protection of major urban centres. Examples include diversion of a portion of the flow of the Red River around Winnipeg via the Red River Floodway, dyking along the Fraser River to protect Vancouver and the Lower Mainland, and flood control dams in the Metro Toronto area. Disaster assistance was also considerable but, in general, was restricted to large flood events because of a cost-sharing formula whereby the federal government portion depends on the total damage and the population of the province so that as per capita damages increase, the federal share increases.

Flood risk mapping was ongoing in three provinces prior to the FDR program. Doughty-Davies [4] describes the initiation of floodplain management in British Columbia which was stimulated by severe flooding in 1972. In Alberta, legislation was introduced for the establishment of flood control zones and terms and conditions regulating the use of lands within these zones. Through time, a program of floodplain mapping developed to adequately evaluate proposed land use and development. The most comprehensive program, in Ontario, preceded the FDR program by over a decade and is described by Giles [5]. It arose as a result of the Hurricane Hazel flood in Toronto in 1954 and was implemented through the province's conservation authorities. Mapping and hydrotechnical work was done by private contractors to province-wide specifications. Land use restrictions were in the form of provincial regulations.

3. The Flood Damage Reduction Program

3.1. STIMULUS FOR CHANGE

The need for a change from a reactive strategy of structural flood control projects and disaster assistance became evident during the early 1970s. In the author's opinion, the fact that a change took place was the result of a fortuitous joint occurrence of a number of factors: the three official factors set out at the beginning of the FDR program: (i) recognition of increasing pressure on floodplain land, (ii) increasing demands on federal and provincial budgets, (iii) the notion of fairness, and two additional factors: (iv) the human factor, and (v) the flooding factor.

(i) Canada's population was changing from a mixed urban-rural base to one that would be predominantly urban. With the development of these expanding metropolitan areas, pressures would increase to occupy flood prone land for housing, industrial use and infrastructural services. With the greater mobility of the Canadian population, increasing urbanization, and a high premium on land for development, major developments could happen in flood prone areas through ignorance of the risk or unscrupulous actions by speculators.

(ii) In the 1970s, at least three budgetary factors combined to stimulate thought for a change in strategy: an increasing demand on governments to provide flood control combined with limited budgets for these works; a 'run' on the disaster assistance 'pots' as a result of large 'per-capita damages' floods; and a growing environmental movement with demands on governments to commit funds to clean up the environment, maintain open space, preserve flora and fauna in wetlands, and protect the aquatic ecology in lakes and rivers.

(iii) Another factor pointing out the need for a re-assessment of strategies was the 'income transfer' aspect whereby funds were transferred from the general taxpayer to floodplain property owners. Structural measures tended to not only increase the values of developed properties but, in many instances, permitted the development of other flood prone lands thereby allowing windfall gains which accrued to speculators. Development on flood prone lands could also be promoted on the grounds that the federal and provincial governments will eventually "have to do something". Similarly, disaster assistance, in the event of actual floods could be regarded as "premium-free" insurance thereby encouraging existing development to remain and new development to take place in flood prone areas.

(iv) Change, especially in large bureaucracies, required 'the right people' at this 'right time'. The establishment of the FDR program is due in large part to two human factors. First, a dedicated team of professionals from many fields worked for the federal government to achieve a common goal under a leader who was respected nationally and internationally. Secondly, senior water resource engineers and managers in the provinces, who were experienced in program implementation, supported this new approach because it was in line with the co-operative approach which characterized water resource management in Canada, made use of existing programs and government agency activities, and held promise to reduce potential flood damages.

(v) Thomson [6],in a comprehensive review of the development of the FDR program, suggests that the occurrence of several large flood events across the country in 1973 and 1974 (e.g. the Saint John R. in New Brunswick, the Gatineau R. in Quebec, the Grand R. in Ontario, and the Red R. in Manitoba) made it easier for federal and provincial officials to 'sell' this new program to elected representatives.

3.2. PRINCIPLES AND POLICIES

After extensive review and federal-provincial consultation, the federal Minister of the Environment announced a new approach to flooding problems, the National Flood Damage Reduction Program, in April 1975. The "new federal strategy" was described by Bruce [1], the key official responsible for its implementation, and was to be based on the following principles.

(a) Programs of federal agencies concerned with flooding will be coordinated, both at the federal level and with related programs at the provincial level.

(b) The cornerstone of a coordinated program will be flood risk maps, as a basis for joint agreement on the definition of flood prone lands.

(c) Information on floods, on federal policies and programs and on the susceptibility of specific areas to flooding, through flood risk maps, must be provided to the public, the municipalities and all others concerned.

(d) Federal agencies will not develop or support development in areas identified by the mapping as high-risk areas.

(e) Federal disaster assistance will be refused, with respect to new or further developments within defined high flood risk areas.

(f) The Provinces will be asked to restrict their investments in agreed high flood risk areas and would be asked to encourage appropriate zoning regulation in such areas.

These principles were put into practice through federal-provincial agreements under which the two levels of government agreed to jointly undertake a program of flood risk mapping, based on mutually agreed priorities with costs shared on a 50-50 basis. Preliminary (1975) estimates were that over 200 areas in Canada would require mapping at an estimated cost of $20 million and an estimated time to completion of 5-10 years. The program initiation date also varied - from 1976 for most provinces to 1989 for Alberta and British Columbia - largely because of differing needs for federal funds, as cited in [6], and differing attitudes towards federal involvement in provincial matters.

3.3. PROGRAMS AND PROGRAM DELIVERY

The cornerstone of the FDR Program was the Flood Risk Mapping Program which is the focus of this paper. The primary impact of this program and associated policies was to reduce flood damage and risk of loss of life as a result of future developments in floodplains.

Other programs, also cost-shared, were directed towards reducing flood hazard resulting from existing developments. These programs included (i) non-structural measures such as flood forecasting and warning, acquisition of floodprone lands, (ii) structural measures such as dyking, channel improvement, flood proofing and (iii) studies, research or surveys related to flood problems.

In general, all programs were delivered using the same basic model. The province administered the program subject to co-ordination by a federal-provincial Steering Committee and technical control by a federal-provincial Technical Committee. One exception to this model was the mapping program on Indian lands which was administered by federal officials alone.

4. Flood Risk Mapping Program: 1976-1998

4.1. FEDERAL - PROVINCIAL AGREEMENTS

As indicated above, the mapping program was carried out under federal-provincial FDR agreements. Such agreements were signed for all provinces and territories except PEI and the Yukon. The number and form of the agreements vary but there are common elements and similar sections and subsections. Common elements relevant to this paper are as follows:

(a) definitions of terms used in the agreement such as design flood, floodproofing, floodway, floodway fringe, flood risk area, zoning authority, undertaking;
(b) the process for official "designation" of a flood risk area;
(c) policies (and exclusions) that apply upon designation;
(d) a list of each province's or territory's potential flood risk areas;
(e) hydrologic and hydraulic specifications for flood risk mapping; and

(f) surveying and mapping specifications for flood risk mapping.

4.2. HYDROLOGIC CRITERIA AND SPECIFICATIONS

All agreements include the definition of the flood risk area as the area inundated under design flood conditions. Some include subdivision of the flood risk area into two zones (floodway and flood fringe). Because of respect for provincial jurisdiction and pre-FDR provincial programs, design flood criteria vary from agreement to agreement and even within a province under one agreement. However, there are certain common conditions.

- The magnitude of the design flood must equal or exceed the 100-year flood (federal minimum). In fact, in most provinces, the design flood is the 100-year flood.
- In the western provinces, the design flood is a frequency-defined event with return period sometimes exceeding 100 years (e.g. BC:200, Sask. cities: 500, Man. - one case, Winnipeg:160).
- In Ontario (except eastern Ontario), the design flood (Regional Flood) is based on historical rainfall events provided that it exceeds the 100-year flood.
- Floodways in the western four provinces and some floodways in Ontario are defined in terms of hydraulic specifications (increase in depth, or a function of depth and velocity).
- Floodways in Ontario are the areas inundated by the 100-year flood (where it is less than the Regional Flood).
- Floodways in Quebec and the Atlantic provinces are the areas inundated by the 20-year flood.

Maps were prepared according to published procedures, "Survey and Mapping Procedures for Flood Plain Delineation", prepared by the federal government, or provincial procedures where they met these federal minima. Procedures were specified for inspection, aerial photography and targeting, vertical ground control, horizontal ground control, aerial triangulation and numerical adjustment, mapping, cartography, and printing.

Hydrologic and hydraulic analyses to determine design flood elevations were conducted according to published procedures, "Hydrologic and Hydraulic Procedures for Flood Plain Delineation", prepared by the federal government, or provincial procedures where they met these federal minima. Procedures were specified for single station and regional frequency analyses, deterministic hydrologic modelling, steady-state hydraulic modelling, and the contents of the technical report. When a few of the early mapping projects revealed physical situations which were not explicitly covered in the original procedures, an addendum was prepared. It attempted to resolve uncertainty regarding appropriate criteria for five situations: areas subject to ice jam flooding, tidal areas, lakeshore areas, areas at the confluence of large tributaries, and small drainage basins for which the regional flood frequency analysis cannot be applied.

4.3. STATUS OF MAPPING, DESIGNATION AND LAND USE REGULATIONS

The objective of joint federal-provincial mapping and designation of the nation's flood risk areas is to minimize flood damage to future developments in these areas by guiding and restricting types of land use in them. Federal and provincial investment (either direct or through financial assistance guarantees) in unauthorized undertakings vulnerable to flood damage is forbidden in designated flood risk areas. The plan is that all other investment would be controlled by regional and municipal governments who, after designation, are to be encouraged by the two senior governments to enact zoning restrictions that reflect flood risk in the designated areas.

Therefore, the progress of the mapping program cannot be assessed solely in terms of mapping completed. A better indicator is areas mapped and designated, in that all government investment would be restricted in such areas. An even better indicator is areas mapped and designated and for which land use regulations are is in place to restrict private (and non-government assisted) development.

The status and progress of the mapping, designation and regulation phases of the FDR program are summarized by Watt [7]. Information reported therein by province is summarized for Canada as a whole in Table 1. The number of flood risk areas/ communities for which mapping has been completed or designation or regulation has taken place will have increased since 1995 but not to a large extent because the mapping program was essentially complete by 1995. The difference between the areas thought to be floodprone and those mapped can be attributed to a revision of priorities based on a review of flooding problems, development trends and fiscal restraints. The smaller differences between areas mapped and designated (20%) and areas designated and regulated (13%) can be largely attributed to bureaucratic delays in the designation process in the case of the former, and both bureaucratic delays and political interference in the land use regulation process in the case of the latter.

TABLE 1. Status of FDR program after 20 years

Number of Flood Risk Areas/Communities[1]			
Listed in Agreement[2]	Mapping Completed[3]	Designated	Regulations in Place[4]
1298	712	565	493

1. In some provinces, particularly British Columbia, a flood risk area may include more than one community. Includes Great Lakes shoreline communities in Ontario.
2. Including extensions, revisions, amendments etc. of original agreement; not all of these areas/communities will be mapped.
3. Completed and approved.
4. Regulations by municipal or regional or provincial governments.

4.4. OTHER FLOOD DAMAGE REDUCTION ACTIVITIES

Although the Flood Risk Mapping Program consumed most of the budget and human resources devoted to the FDR Program on a national basis, other FDR activities of note were carried out. These included flood forecasting studies and programs in Ontario, New Brunswick (Saint John River) and Manitoba (Red River), regional flood frequency analyses in Ontario, New Brunswick, Nova Scotia and Newfoundland, general application studies in a number of provinces (e.g. development of flood depth-damage relations), and site-specific flood damage reduction planning studies.

5. Evaluation of the FDR Program

5.1. FEDERAL PERSPECTIVE

After fifteen years, a comprehensive review of the Program was initiated. An early element of this review took the form of a workshop in Burlington, Ontario of federal officials involved in the Program. The principal conclusions of this workshop were reported by Watt [8].

(a) The FDR Program has been singularly effective in redirecting damage-prone development away from flood risk areas, and is a model of federal-provincial cooperation in developing public support and initiating a long-term policy of discouraging such development.
(b) The policies of the FDR Program need to be maintained into perpetuity.
(c) This policy maintenance should be accomplished through long-term, federal-provincial commitment through FDR continuance agreements.

5.2. FEDERAL-PROVINCIAL PERSPECTIVE

The perspective of provincial government officials responsible for implementing the mapping program and encouraging municipal land use restrictions was obtained in a second national workshop in 1992. The conclusions of that workshop reported by Environment Canada and the Canadian Water Resources Association [9] are given below.

(a) The FDR Program has been extremely successful in identifying urban flood risk areas across Canada and in redirecting damage-prone development away from flood risk areas. It is a model of federal-provincial co-operation in developing public support and promoting long-term policy for sustainable development.
(b) The current investment in flood risk maps must be protected and maintained, and the policies of the FDR Program need to be maintained into perpetuity. Otherwise, floodplain development will lead to the old spiral of increasing damage and disaster assistance.
(c) Over most of Canada, flood risk maps have been used effectively to serve their intended purposes of information and as a tool in preventing floodprone development and redevelopment. In addition, there are many examples of other uses: planning redevelopment, emergency preparedness planning, ecosystem analysis, identification and protection of source areas for groundwater supplies, etc.
(d) There is a need to apply FDR principles in a more comprehensive manner to aboriginal lands. Implementation of the FDR Program on aboriginal lands has varied across the country but overall has been quite limited. Although the FDR principles should apply to such lands, the mapping should be commensurate with historical circumstances and cultural attitudes of the persons involved.
(e) Provincial representatives agreed that by initiating the national FDR Program, the federal government took on a leadership role in developing a national program which was to prove extremely successful but was in the process of abandoning the FDR Program.

5.3. OTHER PERSPECTIVES

Millerd et al. [10] evaluated four Canada-Ontario Flood Damage Reduction Projects. They determined that benefits exceeded costs for three of the projects; the fourth would likely become viable when the local economy expands. Non-quantifiable benefits included improvement of local land use zoning, identification of the flood status of land and important contributions to environmental preservation. They concluded that "The Program, with its emphasis on planning and environmental benefits, is a much more rational approach to flood damage reduction than flood proofing, compensation for flood damage, or many structural programs." Finally, they noted that although this was a national program to reduce flood damages, it provided an impetus for other environmental mapping (e.g. wetlands) and planning activities (e.g. subwatershed planning) by provincial agencies and municipalities.

5.4. EFFECTIVENESS OF CANADIAN APPROACH

Brown et al. [11] provided a comparison of flooding and flood damages on 14 watersheds in southern Ontario and Michigan resulting from four large summer rainfall events in 1986. They reported that "The climatological and watershed responses of Michigan and Ontario basins were similar. The flood damages in the two jurisdictions were, however, substantially different". Estimated total non-agricultural damages were $203,000,000 for Michigan and $480,000 for Ontario. The authors conclude that "The large differences in non-agricultural flood damages cannot be accounted for by differences in the intensity or volume of rainfall; the volume or frequency of the runoff; differences in physiography; or differences in population densities and amount of urban development. A major reason for the large differences in damages is the large differences in the amount of flood plain development in the two jurisdictions as a result of two different long-term approaches to flood plain management." As indicated earlier, Ontario began mapping in the mid 1950s and accelerated in the mid 1970s under the FDR program. The authors reported that the Michigan approach has been more reactive and less comprehensive in the application of land use and development controls. The focus of floodplain regulations has been to flood proof or elevate new development rather than keeping new development out of the floodplain.

5.5. A PERSONAL PERSPECTIVE

The following points are given from the perspective of a hydrologist/water resources engineer who has been involved in the FDR program since its inception.

(a) The system is far from perfect but is infinitely superior to earlier reactive strategies and is an early Canadian model of sustainable development.

(b) The availability of flood disaster assistance for flood events more frequent than the design flood event (i.e. zero premium flood insurance) is a powerful disincentive to the wise use of urban floodplains. However, it is "good news" politically and therefore inevitable.

(c) In the Canadian context, municipal regulations are crucial. In general not enough work has been done to involve urban and regional planners at on early stage.

(d) The criterion for the design flood event is important; it can be arbitrary (e.g. the 100-year event), but it must be consistent and reasonable.
(e) Guidelines for hydrologic and hydraulic analyses to cover all conceivable situations are essential.
(f) In Canada, the future of urban floodplains is largely in the hands of municipalities and provinces/territories because the federal mapping program is over.

6. Summary

- The FDR Program was developed and initiated in the mid-1970s because of the joint occurrence of a number of factors: recognition of increasing pressure on floodplain lands; increasing demands on federal and provincial budgets for flood control structures, disaster assistance and 'cleanup' projects; the notice that the existing system resulted in a transfer of funds to floodplain dwellers and was unfair; the dedication and talent of federal and provincial officials; and the occurrence of a number of damaging floods across the country.
- The Flood Risk Mapping Program was the cornerstone of the FDR Programs. The designated flood risk areas on high-quality maps are the basis for land use restrictions enforced by federal, provincial and municipal governments.
- Over the period 1976-1998, the FDR Program, based on flood risk maps, has been extremely successful in identifying urban flood risk areas across Canada and in redirecting flood prone development away from flood risk areas.
- The FDR program is now in a transition phase. The challenges are to maintain and update the mapping and to adapt to the "downloading" of responsibilities from federal and provincial governments to municipal governments.

7. References

1. Bruce, J.P. (1976) The national flood damage reduction program, *Canadian Water Resources Journal*, **1**, 5-14.
2. Andrews, J. (Ed.) (1993) *Flooding – Canada water book*, Environment Canada, Ottawa.
3. Watt, W.E., Lathem, K.W., Neill C,R., Richards, T.L. and Rousselle, J. (1989) *The hydrology of floods in Canada: a guide to planning and design*, National Research Council, Canada.
4. Doughty-Davies, J.H. 1976. Floodplain management in British Columbia, *Canadian Water Resources Journal* **1**, 67-74.
5. Giles, W. (1976) Floodplain policies in Ontario, *Canadian Water Resources Journal*, **1**, 15-20.
6. Thomson, K. (1984) *An evaluation of selected Canadian flood plain management policies*, Ph.D. Thesis, Department of Geography, York University, Toronto.
7. Watt, W.E. (1995) The national flood damage reduction program: 1976-1995, *Canadian Water Resources Journal*, **20**, 237-247.
8. Watt, W.E. (1990) FDR program Burlington workshop, conclusions. A report prepared for the Inland Waters Directorate, Environment Canada, Hull, Quebec.
9. Environment Canada and the Canadian Water Resources Association. (1993) National workshop on the flood damage reduction program continuance phase, conclusions and recommendations. A report submitted to the Ecosystems Sciences and Evaluation Directorate, Hull, Quebec.
10. Millerd, F.W., Dufournand, C.M. and Schaefer, K. (1994) Canada-Ontario flood damage reduction program – case studies, *Canadian Water Resources Journal*, **19**, 17-26.
11. Brown, D.W., Moin, S.M.A. and Nicholson, M.L. (1997). A comparison of flooding in Michigan and Ontario: 'soft' data to support 'soft' water management approaches, *Canadian Water Resources Journal*, **22**, 125-139.

THE IMPORTANCE OF MAPS FOR FLOODPLAIN MANAGEMENT AND FLOOD INSURANCE

MARK A. RIEBAU

Technical Mapping Advisory Council to
The Federal Emergency Management Agency
Short Elliott Hendrickson, Inc.
421 Frenette Drive
Chippewa Falls, Wisconsin, 54729, USA

1. Introduction

Floodplain management is defined, for the purposes of this paper, as the planning for, and regulation of, the location and construction of residential, institutional, and industrial buildings. It is implemented through land-use control ordinances. In the United States, floodplain management is delegated to local government through enabling statutes from state government.

Flood insurance has, historically, not been available through private insurance companies. People, in a market-driven economy, tend to purchase casualty insurance only when they perceive a need. Floods, unlike fire or wind, can be predicted. Consequently, only people who recognize their property is likely to be flooded are willing to purchase flood insurance. Thus, in the United States, insurance companies have not historically provided coverage against flood damages, except through the National Flood Insurance Program (NFIP). Under the NFIP, the federal government underwrites all losses.

2. History of Floodplain Management in the United States

The United States, like many countries, was dependent on rivers and oceans for the transportation of goods and services. As the country grew, ocean-going vessels transported people to the North American continent and river-going vessels aided the migration, and the development of the country, from the East Coast to the West Coast. Communities grew around both coastal and river ports.

Rivers also became a source of power, initially for mills and later to generate electricity. These locations also developed into communities. Today, water front property along rivers, lakes and the ocean coasts is sought for its recreational value in addition to its continuing commercial purposes. Communities that have grown adjacent to the ocean coasts, lakes, and rivers suffer flood damages. Flooding is a natural phenomenon. It is only after permanent structures are built in the floodplains that flooding becomes a problem.

J. Marsalek et al. (eds.), Flood Issues in Contemporary Water Management, 167–173.

For decades the only federal response to flooding was to construct levees, floodwalls and flood control dams, yet the damages continued to escalate. In 1966 the total damages caused by flooding was estimated to be $1 billion. A special task force was created by President Lyndon B. Johnson to recommend ways of reducing the trend of increasing flood damages. Their report, A Unified National Program for Managing Flood Losses [1], became the framework for the National Flood Insurance Program (NFIP) passed by congress and signed into law in 1968.

Both the "Report" and the NFIP anticipate local government regulation of new development, and reconstruction of existing development, so that it is not subject to future flooding. The minimum federal criteria do not prohibit development in flood hazard areas, but rather require new development and redevelopment of existing structures be elevated to, or above, the Base Flood Elevation. The Base Flood Elevation (BFE) is the elevation of the 100-year, or 1% chance flood.

Communities that adopt land use regulations that comply with the federal minimum standards are considered "participating communities." Flood insurance, underwritten by the federal government, is available for all structures in participating communities, regardless of when, or where, they are built. There is a difference in the amount of coverage available, and whether flood insurance is mandatory, depending on whether a structure is within or outside the floodplain of the 1% chance flood. Communities may adopt regulations more restrictive than the minimum federal standards, and many have. For example, in some communities new development is simply prohibited within the 1% chance floodplain.

Today, there are nearly 20,000 local units of government "participating" in the National Flood Insurance Program. Of these, most are implementing at least the minimum federal standards for floodplain management through building codes, land-use control ordinances, or combinations of both.

3. The History and Role of Maps for Floodplain Management and Flood Insurance

Both the local regulation of new development and the sale of flood insurance at correct rates are dependent on the availability of maps that accurately define the limits of the 1% chance (100-year) flood. When the National Flood Insurance Program was instituted there was no comprehensive program to develop and maintain maps of flood prone areas. Several federal agencies and some state and local government agencies were providing some flood maps but the NFIP was the first federal program to have a direct mandate to produce maps for communities specifically for the purpose of implementing floodplain management.

There was very little data upon which to base the maps. Consequently, the NFIP was to be implemented in phases. The initial phase, or "emergency phase," included an attempt to map all of the known flood hazard areas in the entire country. An intense effort was made to provide maps for every community in the country known, or thought to have a flooding problem. The maps produced under the emergency phase of the program were called Flood Hazard Boundary Maps (FHBM). Under the "emergency phase" of the program, flood insurance was available, but little or no regulation of land use was required of the local government.

The federal government produced thousands of FHBMs for communities in a very short period of time based on approximate study methods. Few of these early maps were based on detailed hydrologic and hydraulic analyses, and many lacked even the most basic information on land elevations. The quality and accuracy of these maps ranged from adequate to poor. Some of the maps were so inaccurate that neither landowners nor community officials believed they were valid.

The intent was to provide maps, based on detailed hydrologic and hydraulic analyses and accurate topographic maps, subsequent to completion of the emergency phase. The "Regular Phase" began in the early 1970s to replace all of the approximate study-based Flood Hazard Boundary Maps (FHBMs) with Flood Insurance Rate Maps (FIRMs). FIRMs were to be based on detailed hydrologic and hydraulic analyses and sound topographic mapping. Under the "regular phase" of the program, local government was expected to adopt land-use controls that met the federal minimum standards. While the "emergency phase" of the National Flood Insurance Program ended in the mid-1980s, the conversion of all maps based on approximate study methods to detailed study methods has still not been completed.

Local regulation of new development and reconstruction of existing development is dependent on the legal principle of "due public notice." Local government has the legal authority to place restrictions on new development, such as requiring buildings be elevated, or prohibit development altogether in the most hazardous areas, but only if the regulations are adopted subsequent to "due public notice" of the proposed restrictions. This provides an opportunity for landowners to raise objections, question the validity of the rules or ordinances, and evaluate the basis of the proposed regulations. Because written descriptions of land or property can be very confusing, and not easily understood, courts have ruled that maps are necessary to describe where local land use regulations are to be imposed.

Regulation of land use by local government takes many forms. In some communities, the only restriction is that new construction be elevated to, or above, the elevation of the 1% chance flood. In other communities, local regulations completely prohibit any new development in a mapped floodplain. In yet others, some portions of the floodplain can be developed, providing structures are properly elevated, and other portions can not. Typically, the areas of the floodplain that are restricted to open-space are the most hazardous area, where flood waters are the deepest or flow the fastest.

Under the NFIP, the Base Flood Elevation, or BFE, is the elevation of the 1% chance flood. The BFE is provided on most FIRMs, but not all and often not for every flooding source in a community. Detailed studies based on hydrologic and hydraulic analyses are necessary to determine the BFE. Detailed studies are also necessary to determine the limits of a riverine floodway so it can be distinguished from the remainder of the floodplain, or flood fringe. The floodway is typically that part of the floodplain where an obstruction to flood flows would result in an increase in flood height.

The lack of flood elevation information presents a serious challenge for proper administration of local ordinances even if the floodplain delineation is perceived to be adequate. It has been estimated that approximately 60% of the mapped riverine floodplains in the United States are still based on approximate methods and where the Base Flood Elevation has not yet been determined.

Over the course of the past 30 years, the rules and regulations regarding the purchase of flood insurance have changed from being strictly voluntary to being mandatory in some instances. From 1968, when Congress passed the program, until 1973 the purchase of federally subsidized flood insurance was purely voluntary. Flood damages continued to rise.

In 1973, changes in the NFIP required that property purchased with loans from lending institutions that were federally insured must also be insured against flood damages. This change in the law also placed greater emphasis on local administration of land-use controls. There had been few penalties for banks if they failed to require that loans they made were insured against floods, and there were very few communities where the original maps, based on approximate study methods, had been replaced by maps based on detailed studies. Consequently, neither the flood insurance program or floodplain management was working as expected. Flood damages continued to rise.

In the late 1970s, funding for the mapping program was increased in an effort to provide detailed studies and Flood Insurance Rate Maps for the communities which had been provided a map based on approximate study methods. This resulted in many communities receiving detailed studies and that had the effect of improving local administration of the land-use controls. At the same time, more and more communities were experiencing flood damages and the number of communities that needed maps grew.

In the early 1980s, a new initiative to increase the sale of flood insurance encouraged private insurance companies to sell flood insurance policies underwritten by the federal government. The Write Your Own (WYO) program was designed to provide an incentive to private property insurance companies to use their marketing skills and resources to promote the sale of flood insurance. The companies wanted maps for their agents that were easier to read.

The maps were still undergoing the transition from maps based approximate study methods to maps based on detailed hydrologic and hydraulic analyses. And, the maps were still being produced on a "community" basis. Cities or villages would be shown on maps, but the surrounding area that had not yet been annexed would not be shown at all. Similarly, maps for counties would show the floodplain area for the entire county but the cities and villages within the county would be left blank. This policy began when the program was started due to the relationship between the availability of federally subsidized insurance and the participation of the community in the program. The rationale for this policy was that if maps were created on a broad geographic basis, and more than one community was included, there would be confusion as to whether flood insurance could be sold, particularly if one community was "participating" but the other wasn't.

Few other types of maps are produced in this fashion, however, and insurance agents, lenders, and other users, had difficulty interpreting the maps to determine where property was located. The maps did not resemble other maps that users were familiar with.

The conversion of the maps based on approximate study methods to maps based on detailed studies resulted in two similar, but different, maps. One map, the Flood Insurance Rate Map, or FIRM, was intended to be used by lenders and insurance agents. The other, the Flood Boundary/Floodway Map (FB/FM) was intended to be used by local government for regulatory purposes. The Flood Insurance Rate Map depicted the floodplain limits, but not the floodway. Flood Boundary/Floodway Maps depicted both the floodplain and the floodway limits for riverine flood-prone areas. This caused confusion and resulted in problems for both floodplain management and flood insurance.

While the rates for insurance were based on depth of flooding there was no difference in either the requirement for the purchase or the rates between the floodway and flood fringe. New development in the floodway, however, was as to be regulated differently by local government than new development in the flood fringe. The policy of creating two maps for one community-one for insurance purposes and another for regulatory purposes-resulted in many problems and complaints.

In the mid-1980s, due to complaints by the users of FIRMs (insurance agents, lenders, and realtors) and the users of FB/FMs (state and local floodplain management agencies) FEMA embarked on a study to evaluate the format for maps they produced for the NFIP. Several significant policy changes resulted from this study.

First, the problems inherent in producing maps on a "community" basis prompted a major change in policy–maps were to be produced on a "county-wide" basis. All incorporated cities and villages within each county would be included on the maps produced for the county where they were located. There would be no more "blank" spaces on county maps, and city and village maps would no longer be published as if they were an "island" surrounded by nothing. This was a significant change in format which improved both floodplain management programs and the flood insurance program.

Next, the two-map concept was discarded and FEMA began to produce one map for both insurance and floodplain management purposes. This change in policy reduced errors in interpretation. Having two maps being used in a community that attempted to serve similar purposes but with different features and a different appearance caused confusion for all users.

Another major change came in the mid-1980s. FEMA was under pressure to end the "emergency phase" of the NFIP. There were thousands of communities that had received Flood Hazard Boundary Maps–the maps based on approximate studies-that had not received a Flood Insurance Rate Map based on a detailed study. Congress pressed FEMA to conclude the emergency phase of the NFIP but did not provide sufficient funding to do the necessary detailed studies to produce Flood Insurance Rate Maps. FEMA decided to do a whole scale conversion of the approximate study based maps. The Flood Hazard Boundary Maps were, in most cases, simply republished as Flood Insurance Rate Maps by simply changing the title of the map without the benefit of a detailed study.

For the past decade, FEMA has been in what could be described as a maintenance mode. They have attempted to update maps as it is brought to their attention that the maps are out of date or inaccurate. They have also processed map revisions submitted by local government or the private sector when new information is provided. But, due to funding limitations they have not been able to undertake any comprehensive program of improving the maps used in the program. There are now nearly 100,000 map panels for almost 20,000 communities that have been produced by FEMA for the NFIP.

In 1994, Congress directed FEMA to form the Technical Mapping Advisory Council. The directive resulted from continuing and growing dissatisfaction with the state of the maps used for the NFIP. The charge to the Technical Mapping Advisory Council was to evaluate the production, distribution, and use of Flood Insurance Rate Maps and provide recommendations for the improvement in the accuracy, quality, utility and distribution. The Council was also charged with recommending improvements to the standards and guidelines used in preparing Flood Insurance Rate Maps.

The Technical Mapping Advisory Council comprises representatives of ten professional organizations and federal agencies and FEMA. The organizations and agencies that comprise the Council are the American Society of Civil Engineers, the American Congress on Surveying and Mapping, the Association of State Floodplain Managers, the Association of American State Geologists, the National Flood Determination Association, NationsBanc Insurance Services, Inc., the Federal Home Loan Mortgage Corporation, the Federal National Mortgage Association, the National Oceanic and Atmospheric Administration, the U.S. Geological Survey, and FEMA.

The Technical Mapping Advisory Council began its work in May 1996. Since then, three reports have been submitted to FEMA by the Council with specific recommendations for improving Flood Insurance Rate Maps and the mapping processes. These recommendations have been endorsed by FEMA and have been incorporated into a progress report entitled Modernizing FEMA's Flood Hazard Mapping Program. The report, referred to as the Map Modernization Plan, outlines extensive yet important improvements to FEMA's flood hazard mapping program. The Map Modernization Plan has also formed the background for a budget request to provide some of the funding necessary to begin to implement the recommended improvements starting next year.

Among the many recommendations the Council has made, the following are the most critical: (1) An archival system be established for maintaining, in perpetuity for historic and legal purposes, all Flood Insurance Rate Maps and supporting technical data (1996); (2) Base maps used in the creation of Flood Insurance Rate Maps should comply with minimum specified standards (1997); (3) All new map products resulting from flood insurance studies, restudies and revisions should be prepared, produced, and made available digitally (1997); (4) FEMA should be proactive in involving local and state government in the mapping process (1998); and (5) FEMA should document the extent of flooding immediately following a disaster and use the documentation to improve the maps for the communities affected.

4. Conclusions

Floodplain management reduces the potential for future flood damages where it is implemented. One of the keys to implementation is the availability of reliable, accurate, and believable maps of the floodplain. Where the mapping is done well, and people believe the maps are accurate, regulations are acceptable. Where the maps are not believable, it is very difficult to impose land-use controls.

Flood insurance can reduce future flood damages, and support floodplain management, if it is done in a wise manner. A flood insurance program can be implemented where people who are required to purchase the insurance believe their property could be subject to flooding. Accurate mapping is instrumental in the administration of a flood insurance program.

Accurate risk assessment is the basis for both floodplain management and flood insurance and both depend on accurate, reliable, and believable maps.

5. References

1. Goddard, J.E., Hand, I., Hertzler, R.A., Krutilla, J.V., Langbein, W.B., Schussheim, M.J., Steele, H.A., and White, G.F. (1966) *A Unified National Program for Managing Flood Losses,* (A Report by the Task Force on Federal Flood Control Policy), U.S. Government Printing Office, Washington, DC.

URBAN WATERSHED FLOOD MITIGATION PLANNING AND FLOODPLAIN MANAGEMENT

BEN R. URBONAS, P.E. and WILLIAM G. DEGROOT, P.E.

Urban Drainage and Flood Control District
2480 w. 26th Avenue, Suite 156B; Denver, CO, 80209, U.S.A.

1. Introduction

The Urban Drainage and Flood Control District (*District*) is a regional government agency serving the Denver metropolitan area. It was created by the State of Colorado in 1969. The District currently covers an area of 4320 km^2 that includes 39 cities and counties and 2.2 million people. There are 2560 km of rivers, streams and gulches, which are referred to in this paper as *major drainageways,* serving at least 400 hectares of tributary area.

The District was created in response to a disastrous flood in 1965 (see Figure 1) that destroyed or damaged 52 bridges, and flooded a large number of residential, industrial, and commercial structures. The resultant damages were estimated at $110 million. The District grew from a staff of two and revenues of less than $100,000 in 1970, to a full-time staff of 19 and revenues of $12 million in 1999. It accomplishes most of its work by using the private sector. While the planning and floodplain mapping studies, and final designs are performed by consultants, the District's staff closely controls the scope of work and the technical content of the projects.

The first major activity of the District was to inventory drainage basins and sub-basins to determine the extent of problems and to develop a plan to attack those problems. The inventory indicated that approximately 25% of the major drainageway miles within the District were developed, with the remaining 75% undeveloped and amenable to preventive approaches. The District Board therefore made a commitment to correct the problems existing along the developed drainageways while developing a comprehensive floodplain management program to prevent new problems from being created by new development along the remaining undeveloped drainageways. As a result of this policy, even though the District's population has increased by approximately 800,000 people in the last 20 years, along with all of the structures that accompany those people, the total number of structures located in identified 100-year floodplains has decreased by about 4000 units.

The District is organized in five programs: Master Planning, Floodplain Mangement, Design and Construction, Maintenance, and South Platte River. Briefly, Master Planning prepares plans for both mitigation of existing flood hazards and guidance of new development; Floodplain Management works to insure safe new development; Design and Construction implements the mitigation master plans; and

J. Marsalek et al. (eds.), Flood Issues in Contemporary Water Management, 175–184.

Figure 1. **Aftermath of the 1965 flood in Denver. Note the debris blocking the bridge. Photograph by permission of Urban Drainage & Flood Control District.**

Maintenance maintains flood control facilities. The South Platte River Program performs all of the above functions for the South Platte River.

The activities of the first two programs will be described in more detail here, demonstrating how they have implemented the goal of preventing and reducing drainage and flood control problems.

2. Master Planning Program

The Master Planning Program has four basic areas of activity: (1) preparation of master plans, (2) maintenance of the *Urban Storm Drainage Criteria Manual* (USDCM), (3) special projects in technology development, and (4) assisting local governments with stormwater quality issues.

2.1. MASTER PLANNING

The first major drainageway master plan was completed in 1971. It provided an inventory of the major drainageway and flood control system in the Denver region. Subsequently, more than 100 master plans were completed covering more than one-half of all drainageway kilometres within the District. Each master plan is based on a consistent planning methodology that identifies drainage and flooding problems, identifies potential solutions, and selects the most appropriate solution. The final master plan shows a detailed conceptual preliminary design of the selected plan.

Figure 2 is an example of a portion of a plan and profile sheet from one of the master plans.

Each master planning project requires the affected local governments to contribute 50% of the mapping and engineering costs and the District pays the balance. The District contracts for the preparation of the plan with a consulting engineer that is jointly selected by all project's sponsors. The finished plan has to be acceptable to all affected local governments. The completed master plans serve as a basis for new capital improvements along major drainageways and help in the review of land development proposals.

2.2. SPECIAL PROJECTS

One of this program's special projects is the collection and analysis of rainfall and runoff data. This was started in 1970 and has provided the data needed to improve the District's hydrologic methods. This resulted in a calibrated Colorado Urban Hydrograph Procedure (CUHP) and the Rational Method. These two hydrologic methods represent two of only a few calibrated regional urban runoff simulation procedures in the United States. The District is continuing to collect stormwater data which will be used to further refine runoff simulation methods in the future.

The District has conducted many technical investigations. These led to the publication of reports or criteria on the design of hydraulic structures in sandy soils, benefit-cost analysis, wetland issues, gravel mining review guidelines, storm sewer pipe material selection, stability of high gradient rock lined streams, the development of storm drainage analysis and design software. It has also contributed funds to the United States Geological Survey to study foothills hydrology.

Other special projects of the District include continued development of hydraulic and hydrologic analysis tools, development of stormwater quality treatment technology, and water quality data collection.

3. Floodplain Management Program

The Floodplain Management Program was established in 1974 to prevent new development and redevelopment in 100-year floodplains. The program encourages the use of non-structural methods whenever possible, but will consider all options that are in accordance with the USDCM. Its major activities are (1) coordination with the National Flood Insurance Program (NFIP), (2) floodplain regulation, (3) flood hazard area delineation, (4) flood warning, (5) flood damage surveys, (6) land development review and determination of maintenance eligibility, and (7) dissemination of public information on flood hazard areas.

3.1. NATIONAL FLOOD INSURANCE PROGRAM

The U.S. Federal government established the NFIP to make affordable flood insurance available to all floodplain occupants. The District works with local governments to encourage them to remain in the NFIP to keep flood insurance available to their citizens. The District also works with local governments and the Federal Emergency

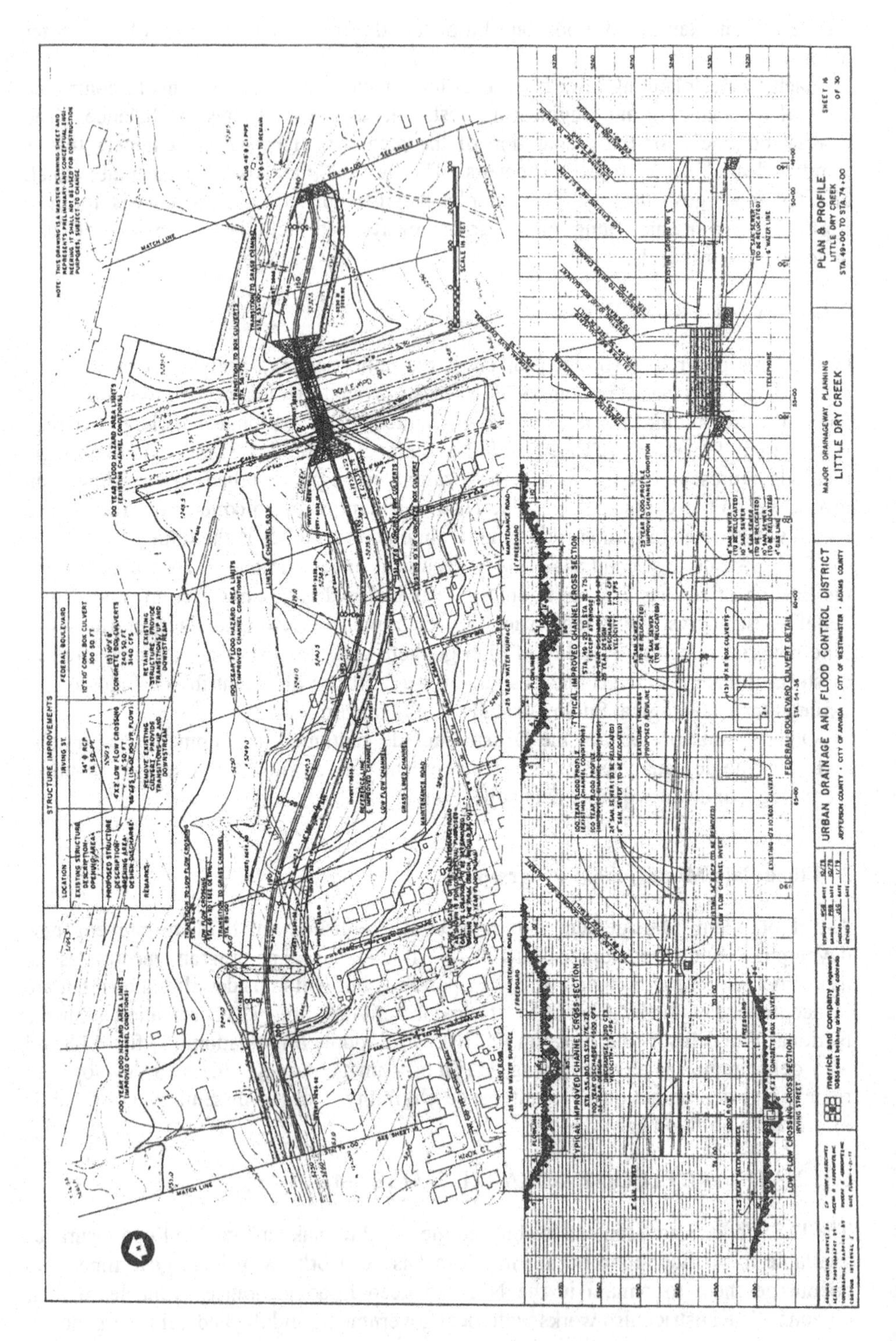

Figure 2. **A plan and profile sheet from a master plan.**
Copy by permission of Urban Drainage & Flood Control District.

Management Agency (FEMA), the agency administering the NFIP, to maintain consistency between District flood maps and the maps published by FEMA.

3.2. FLOODPLAIN REGULATION

The District has the authority to regulate floodplains, but has chosen not to do so provided the local governments implement their own regulations. Instead, the District assists local governments with their floodplain regulations, including meeting the requirements of the NFIP. If, however, a local government does not implement its own floodplain regulation, the District can step in and regulate the 100-year floodplains in accordance with its own floodplain regulations. Figure 3 is a photograph of a floodplain preserved through floodplain management.

3.3. FLOOD HAZARD AREA DELINEATION

The District has published Flood Hazard Area Delineation (FHAD) reports for the majority of the major drainageways within the District. These form the basis for floodplain regulation when land use changes are taking place. Most of the FHADs were developed and published as an add-on scope of service to either major

Figure 3. **A floodplain reserved during land development.**
Copy by permission of Urban Drainage & Flood Control District.

drainageway planning studies or Flood Insurance Studies, thus eliminating duplications and realizing considerable savings over doing these studies separately.

3.4. FLASH FLOOD WARNING

The District maintains an extensive real time rainfall and flood flow monitoring network that is used to predict and confirm the occurrence of floods. A private meteorological service provides daily forecasts of flood potential and when flood potential exists, provides timely updates to this information. These forecasts are made available to all local governments. In addition, the District assists local governments in the development and maintenance of flood warning plans, which provide instructions on how emergency services need to mobilize and react when floods are predicted.

3.5. FLOOD DAMAGE SURVEYS

After any significant flood event, the District mobilizes its staff and private consultants to assess flood damages. This information is valuable in developing emergency relief requests and helps the District to estimate flood damage potential during planning studies.

3.6. DEVELOPMENT REVIEW AND MAINTENANCE ELIGIBILITY

The District reviews and comments on proposed new land development in or near floodplains at the request of local governments. Any drainage and flood control facility constructed pursuant to the District's maintenance eligibility review and approval process is eligible for assistance by the District's Maintenance Program. The determination of this eligibility is provided by the Floodplain Management Program.

3.7. DISSEMINATION OF PUBLIC INFORMATION

The District annually mails over 23,000 floodplain information brochures to all addresses identified as being within or next to an identified 100-year floodplain. The brochure advises the occupants of the flood potential and provides information on floodplain boundaries and on what each individual can do to protect themselves and their property from flooding.

4. Ancillary Issues to Flood Mitigation Activities

4.1. NEED FOR CONSISTENCY IN HYDROLOGY

Whenever two or more hydrology studies are performed independently of the same area, results differ. Our experience shows that the variability in results has not diminished with the use of even the latest, physically based, models. This is because each investigator segments a given watershed into a number of sub-watersheds in his or her unique fashion, makes different assumptions about rainfall losses and surface flow roughness, and applies different principles in the use of design storms, or recorded storms. Differences in simulated peak flow of -50% to +150% have been observed.

To avoid these disparities between different investigators, the District developed and calibrated a simple lumped-parameter model called the Colorado Urban Hydrograph Procedure (CUHP). Specific guidance is given in the selection of parameters and the use of design storms for CUHP. Other models may be used if an investigator so desires, but it has to be calibrated against the peak flow and volume results generated using the CUHP. This approach significantly reduces disparity in design flows estimated by different investigators and has introduced consistency in design for new development and for other facilities.

4.2. RECREATION, GREENBELTS, WILDLIFE HABITAT AND PARKS

Over the years the District has supported the concept of multiple use of publicly owned right-of-way. Many major drainageways offer opportunities for multiple uses. For example, a maintenance access road can also serve as a hiking and biking trail; or a flood plain or a detention basin can be used as a passive recreation area such as wildlife reserve, or have active recreation areas such as play fields, parks, tennis courts or golf courses.

The success or failure of multiple use facilities depends to a large degree on the design of the facilities. It is important to recognize that actively used recreation facilities cannot be flooded frequently. Thus, a park, tennis courts, picnic areas, golf courses, etc, need to be located above the frequent flood level. Experience indicates that if these facilities are located above the 2- or 5-year flood level and are not subjected to high velocities when floods do occur, very little loss in their use is experienced. It is important to realize that when flooding does take place, the flood control agency needs to assist the recreation interests in cleaning up and restoring the facilities. Figure 4 contains photographs of an example of one active recreation activity along a major drainageway while Figure 5 is an example of a preserved riparian zone.

Obviously, this is not always possible since urban right-of-way can be very expensive, especially when retrofitting flood mitigation facilities into previously developed urban areas. Nevertheless, areas of new development often provide opportunity for multiple use of major drainageway facilities.

Figure 4. Recreational hiker-biker trail along the South Platte River.
Photograph by permission of Urban Drainage & Flood Control District

Figure 5. Major drainageway riparian wildlife habitat preserve within a floodplain
Photograph by permission of Urban Drainage & Flood Control District.

4.3. SYSTEM STABILITY VS. DEGRADATION AND AGGRADATION

In addition to conveyance or flood routing of stormwater, drainage and flood control facilities need to provide for a stable system. Properly designed, concrete lined channels and storm sewers are not susceptible to erosion. However, if their grade is too flat they can loose capacity due to excessive sediment buildup. This especially is a problem if soil erosion is not controlled during construction within the tributary watershed.

Soil deposition due to lack of erosion control can be a much more serious problem in grass-line channels, which have to be cleaned at public's expense. In addition, sediment deposition is a major environmental concern since it buries and chokes aquatic habitat. Recent Federal government laws and regulations mandate erosion control during construction activities in United States.

On the other hand, as a watershed urbanizes, runoff volumes for a 2-year storm can increase by a factor of 2 to 100, depending on local meteorology, topography, and soils. Increases in frequent runoff events destabilize unprotected channels, and erosion has been observed to occur in practically every drainageway in the Denver area. To arrest this degradation, grade control structures (i.e., drop structures) are installed along both new grass-lined channels and natural waterways. The photograph in Figure 6 shows channel degradation along the South Platte River where the channel's thalweg dropped three to four meters within a three year period. Figure 7 is of a grade control structure. This grade control structure is a modified baffle chute that fits nicely into a park setting, showing that these types of structures do not have to be unattractive.

5. Summary

The activities of the Urban Drainage and Flood Control District in addressing drainage and flood control problems in the Denver area were described. Since its creation in 1969, these problems were addressed through master planning, floodplain delineation and floodplain management, design and construction of flood control projects, and the maintenance of major drainageways. As a result, significant reduction of drainage and flooding problems has resulted, while floodplain management has significantly reduced the building of structures within 100-year flood prone areas. Where flooding was a common occurrence in the past, recent heavy rainstorms have produced little or no flood damage. Although much was accomplished since 1969, much still remains to be done to mitigate flood damage potential in remaining urban flood prone areas of the District. Some of the technical developments that are instrumental in solving regional flooding problems in the Denver metropolitan area were also discussed.

Figure 6. Streambed degradation along the South Platte River (Note exposed foundation piles under bridge piers).
Photograph by permission of Urban Drainage & Flood Control District

Figure 7. A grade control structure in a city park.
Photograph by permission of Urban Drainage & Flood Control District

THE NATIONAL FLOOD INSURANCE PROGRAM: A U.S. APPROACH TO FLOOD LOSS REDUCTION

EDWARD T. PASTERICK

Federal Emergency Management Agency
Washington, D.C.

1. Introduction

The National Flood Insurance Program (NFIP) is the United States Government's primary vehicle for nationwide reduction in damages due to flooding and for financial recovery by the victims of flooding. The purpose of this paper is to describe the components of the NFIP, how they are integrated with one another, and the specific challenges associated with maximizing their effectiveness.

1.1. BACKGROUND

The NFIP was authorized by the National Flood Insurance Act of 1968. It was the result of recommendations made by the 1966 report from the U.S. Secretary of Housing and Urban Development (HUD) entitled "Insurance and Other Programs for Financial Assistance to Flood Victims." This report followed years of experience in flood control that raised doubts among many as to the effectiveness of sole reliance on structural floodworks for reducing damages due to flood disasters and the ever increasing costs of providing financial relief to flood victims through general tax revenues. The report proposed that a flood insurance program to replace reliance on public assistance to repair flood-damaged property would be feasible if it contained the following essential elements:

- accurate estimates of risk,
- compensation to the risk bearer,
- the possibility of some level of premium subsidy, if publicly desirable,
- incentives to policyholders to reduce risks,
- incentives to states and local governments for wise management of flood-prone areas, and
- continuous reappraisal.

J. Marsalek et al. (eds.), Flood Issues in Contemporary Water Management, 185–196.

2. Structure of the National Flood Insurance Program

The NFIP structure includes three essential components: risk identification, hazard mitigation (i.e., actions taken to protect people and property from the flood peril), and insurance. Effective integration of these three components requires cooperation among the federal government, state and local governments, and the private property insurance industry. The authorizing legislation defined a role for each of these, which, in some cases has been altered over the program's history.

2.1. RISK IDENTIFICATION

One of the major obstacles preventing the private insurance industry from providing flood insurance was the inability to adequately identify all the areas throughout the country that were vulnerable to flood hazards and then to effectively define the nature and extent of the risk. Because of the costs associated with conducting the hydrological studies needed to secure this information and the nationwide scope of the effort, the legislation assigned this task of risk identification to the federal government. This risk identification process takes the form of a flood insurance rate study that results in a Flood Insurance Rate Map (FIRM). The FIRM serves both as the guide to the community for purposes of floodplain regulation and overall land-use decision making, and as the source of risk information for property insurance agents to accurately rate policies. A property owner is eligible to purchase flood insurance only if the community in which his property is located enacts the necessary ordinances to regulate the construction of new buildings in the high risk flood zones delineated on the FIRM.

Until 1974, both community participation in the NFIP and individual purchase of insurance was completely voluntary, and the studies conducted and maps produced were only for those communities that had applied for participation in the program. When Tropical Storm Agnes struck the Eastern seaboard of the United States in 1972, it became clear that many communities were either unaware of the serious flood risk to which they were exposed or had been unwilling to take the necessary measures to protect residents of the floodplain. Very few of the communities affected by the storm had applied for participation in the NFIP, and even in these participating communities owners of most flood-prone property opted not to purchase flood insurance. This lack of response to the NFIP perpetuated both the reliance of flood victims on federal disaster assistance to finance their recovery and continued unsafe building in high risk flood plains.

Because of this minimal participation, the 1973 Flood Disaster Protection Act made the NFIP responsible for identifying all communities nationwide that contained areas at risk for serious flood hazard. The NFIP was also required to notify these communities of their choice of applying for participation in the NFIP or forgoing the availability of certain types of federal assistance in the community's floodplains. This NFIP effort identified over 21,000 flood-prone communities in the 50 states and the District of Columbia. It also identified one community each in American Samoa, Guam, Puerto Rico, Territory of the Pacific, and the Virgin Islands. As of May, 1999, over 19,000 communities had joined the NFIP. Most of flood-prone communities that have elected not to participate are communities whose areas of serious flood risk are either very small or have few if any structures.

The NFIP conducts two general types of flood studies, approximate and detailed. Detailed studies are conducted for communities with developing areas — that is, areas where industrial, commercial, or residential growth is beginning and/or where subdivision is under way, and where the development or subdivision is likely to continue. Communities with either minimal or no flood risk have been brought into the program without such a detailed study being conducted. Approximately 6,300 of the 19,000 participating communities are in this low risk category. Through fiscal year 1997, the cost of this massive study effort has been about $1.154 billion.

For purposes of risk identification, the NFIP uses as the standard of risk the "100-year flood," also referred to as the 1 percent flood, or the base flood. This is a degree of flooding that has a one percent chance of occurring in a given year. On most community FIRMs the 100-year flood is indicated in terms of flood elevations in feet relative to mean sea level. In coastal areas, which are exposed to the additional hazard of wave action, the base flood includes the height of the waves above the stillwater elevation. The term "100-year flood" is problematic for the NFIP. It is a term of convenience intended to convey probability but has had the adverse effect of giving floodplain residents, who tend to interpret it in chronological terms, a false sense of security.

2.2. HAZARD MITIGATION

The standards for flood plain regulation established by the NFIP are based on a nonstructural approach to floodplain regulation and are designed to supplement the federal government's program of structural flood works. While demand continues for the construction of floodworks to provide protection to floodplain residents, many are still skeptical about the efficacy of such structures. Although the NFIP has adopted a nonstructural approach to floodplain management, the NFIP rate structure gives credit to structural flood works if they are certified to protect against the base flood.

The responsibility for hazard mitigation under the NFIP is split between the federal government and the local participating community. The NFIP enters into an arrangement with the local community whereby properties built in the floodplain without full knowledge of their degree of flood risk can be insured at less than full actuarial rates. In exchange, the local community makes a commitment to regulate the location and design of future floodplain construction in a way that results in increased safety from flood hazards. The federal government has established a series of building and development standards for floodplain construction to serve as minimum requirements for participation in the program. These standards use the 100-year flood as the basis for regulation. The primary mitigation action required by the regulations is elevation of the lowest floor of a structure above the level of the base flood as determined by the Flood Insurance Rate Study and shown on the FIRM. The rates for coverage of structures built or substantially improved after the date of the FIRM are based on this elevation.

The local community is responsible for adopting and enforcing these floodplain management standards, and compliance is accomplished through the building permit process. Since local governments have jurisdiction over land use and development, it is only at the local community level that the implementation of standards will be effective. FEMA, working with the state government, conducts periodic reviews at the local level to assess the local community's enforcement of NFIP standards. A determination that a community is not adequately enforcing local ordinances can result in a period of probation, during which a surcharge is added to insurance premiums on all NFIP policies in the community. If the community fails to take corrective actions during this probation period, it can be suspended from the program, which means that NFIP coverage becomes unavailable.

While the NFIP compliance process has identified a number of violations of program standards at the local level, there has never been a comprehensive assessment of the level of compliance nationwide nor of the overall effect of program standards on local development patterns. Nevertheless, certain limited assessments have been done and a certain amount of data relevant to the issue has been collected. The Natural Hazards Research and Applications Information Center (NHRAIC) prepared a report in 1992 [1] for the Federal Interagency Floodplain Management Task Force noting a number of significant achievements in floodplain management, including more widespread public recognition of flood hazards, as well as reduced development and losses in many localities. Much of these accomplishments can be attributed to the NFIP. But the report also found that: "a considerable distance remains between the status quo and the ideal that can be envisioned". The report acknowledges the difficulty in assessing the effectiveness of floodplain management stating "that there are few clearly stated, measurable goals, and that there is not enough consistent reliable data about program activities and their impacts to tell how much progress is being made in a given direction".

In 1994, the Interagency Floodplain Management Review Committee (IPFMRC) studied the causes and consequences of the 1993 Midwest flooding. The committee's report [2], while recognizing that NFIP requirements are minimum standards that are applied to areas subject to very different flooding conditions, noted that:

> "in the Midwest, the NFIP tends to discourage floodplain development through the increased costs in meeting floodplain management requirements and the cost of an annual flood insurance premium, although this may not be the case elsewhere in the nation. Individuals and developers appear to choose locations out of the floodplain to avoid these costs. Developers have the added incentive of wanting to avoid marketing flood-prone property. Many communities visited by the Review Committee actively discourage floodplain development."

An analysis of loss experience from 1978 to 1994 also provides some indications of the effectiveness of the program's standards. The figures show that, in general, structures built before 1975, which were subject to no NFIP standards, suffered about six times more damage from flooding than those built after NFIP mitigation requirements became effective.

2.3. INSURANCE

2.3.1. Rates

The NFIP's insurance structure includes two general classes of properties: those insured at full actuarial rates and those that, because of their date of construction, are statutorily eligible to be insured at lower, "subsidized" rates — that is, rates which do not reflect the full risk to which a property is exposed. Actuarial rates for coverage are charged to property owners living outside100-year flood hazard areas and to those living within 100-year areas who built or substantially improved structures after the federal government provided complete risk information from the Flood Insurance Rate Map (FIRM).

The date of this map differentiates the two classes of properties. Properties built prior to the availability of this information (known as pre-FIRM structures) are insured at "subsidized" rates, which, while they represent some contribution on the part of the property owner to his or her financial protection, do not reflect the full risk to which the property is exposed. The 1966 HUD study considered such a "subsidy" to be important in providing an incentive to local communities to participate in the NFIP and regulate future floodplain construction.

The concept of "subsidy" as applied to the NFIP is often misunderstood. The subsidy does **not** refer to a direct infusion of taxpayer dollars to offset the shortfall in premium from properties paying this lower rate. This shortfall is currently about $500 million annually. It prevents the program from building a catastrophic reserve that can be used to pay claims in heavier loss years; instead the NFIP has been required to exercise its borrowing authority from the US Treasury to pay such claims.

While the law calls for a lower-than-actuarial rate to be charged to pre-FIRM structures in the floodplain, it does not specify the actual rate. During the 1970s, when the primary program objective was to enlist communities to participate in the NFIP, these rates were substantially lower than the actuarial rates. In 1981, the program began a process of gradually increasing the rates charged for existing construction in order to reduce the amount of subsidy and attempt to render the program self-supporting for the average historical loss year. In 1981, rates for pre-FIRM structures were increased 19 percent, and in 1983 they were increased another 45 percent. From 1983 through 1995, eight increases in rates for both structures with subsidized and actuarially-based averaged about 8 percent each. As the subsidized rates for existing structures have been raised, the number of properties requiring a subsidy has also continued to decline over the years, going from about 75 percent in 1978 to about 30 percent in 1999. The premiums paid by this group of insureds are estimated to be about 38 percent of the full risk premium needed to fund the long-term expectation for losses.

2.3.2. Private Industry Participation

Most NFIP policies are written by private insurance companies that participate in the Write-Your-Own (WYO) Program. Under the WYO arrangement, the private insurers sell and service federally underwritten flood insurance policies under their own names, retaining a percentage of premium (currently 31.9 percent) for administrative and production costs. One hundred sixty-seven companies now participate in the WYO program. At present, WYO companies write about 92 percent of NFIP policies, the remainder still being written by agents who deal directly with the federal government.

2.3.3. Community Rating System

The Community Rating System (CRS) was created in 1990 as a mechanism for recognizing and encouraging community floodplain management activities that exceed the minimum NFIP standards. The three goals of the CRS are to reduce flood losses, to facilitate accurate insurance rating, and to promote the awareness of flood insurance. When a community's activities meet these three goals, its flood insurance premiums are adjusted to reflect the resulting reduction in flood risk.

In order to recognize community floodplain management activities in this insurance rating system, they must be described, measured, and evaluated. A community receives a CRS classification based upon the scores for its activities. There are ten CRS classes: Class 1 requires the most credit points and results in the greatest premium reduction; Class 10 receives no premium deduction. A community that does not apply for the CRS or does not obtain a minimum number of credit points is a Class 10 community.

Community application for the CRS is voluntary. Any community that is in full compliance with the rules and regulations of the NFIP may apply for a CRS classification better than Class 10. A community's CRS classification is assigned on the basis of a field verification of the activities included in its application. Credit points are based on how well an activity meets the three goals of the CRS mentioned above, and are calculated for each activity using formulas and adjustment factors. The CRS recognizes 18 creditable activities organized in four categories:

- *Public information* activities include programs that advise people about flood hazard, flood insurance, and ways to reduce flood damages.
- *Mapping and regulations* activities are intended to provide increased protection to new development. These activities include mapping areas not shown on the Flood Insurance Rate Map, preserving open space, devising higher regulatory standards and storm-water management, and preserving the natural and beneficial functions of floodplains. This credit is increased for growing communities.
- *Flood damage reduction* activities involve mitigation programs where existing development is at risk. Credit is awarded for addressing repetitive loss problems, acquisition and retrofitting of flood-prone structures, and maintaining drainage systems.
- *Flood preparedness* activities include programs such as flood warning, levee safety, and dam safety.

Some of these activities may be implemented by the state or a regional agency rather than at the community level. For example, some states have disclosure laws that are creditable in the flood hazard disclosure category. Any community in those states will receive those credit points if they demonstrate that it effectively implements the law. The CRS recognizes some established methods for obtaining credits in each activity, although communities are invited to propose alternative approaches to these activities in their applications.

3. Program Funding

The NFIP is funded through the National Flood Insurance Fund, which was established in the US Treasury by the National Flood Insurance Act of 1968. Collected premiums are deposited into the fund, and losses, operating costs, and administrative expenses are paid out of the fund. The National Flood Insurance Program also has the authority to borrow up to $1 billion from the Treasury. The NFIP experienced annual negative operating results during the period 1972 – 1980 ranging from $5,400,000 to $323,228,000, resulting in the need to borrow funds regularly from the Treasury.

Congressional appropriation to repay borrowed funds was first requested in 1981 and the program received an appropriation to repay borrowing each year until 1986. The total amount borrowed from the Treasury prior to 1986 was $1.2 billion, all of which was repaid through a series of appropriations. During this same period, from 1981 through 1986, only 1983 and 1984 experienced a negative operating result. From 1987 through 1996, negative operating results were experienced in 1989, 1990,1992, 1993, 1995, and 1996.

As noted previously, before 1981, no action was taken regarding the level of subsidy being accorded existing properties in high-risk areas, the program focus being primarily on promoting community participation. Consequently, program expenses inevitably exceeded income. In 1981, the administrator of the program established a goal of making the program self-supporting for the average historical loss year by 1988. Note that the goal was that the program be self-supporting, not that it be actuarially sound.

The continuing statutory requirement to "subsidize" certain existing properties in high-hazard areas still makes actuarial soundness an unrealistic goal at this point in the program's history. "Self-supporting" means that in years where losses are less than the historical average, the program builds up a surplus to be used in years when losses exceed that average. However, even though the premiums for existing buildings have been regularly raised to the point where they are now higher, on average, than the premiums paid for new construction, they still represent only about 38 percent of the full-risk premium for these properties, which currently make up about 30 percent of the NFIP book of business. This shortfall in premiums prevents the program from accumulating the kind of catastrophic reserves that would reduce the need to borrow. The premium shortfall also makes it more difficult for the NFIP to repay funds when borrowing becomes necessary.

Nevertheless, even with the statutory subsidy, from 1986 until 1995, the program operated with a positive cash balance. However, in the four fiscal years from 1993 through 1996, the program experienced over $3.4 billion in losses, and beginning in July 1995, the program has had to borrow substantial funds from the Treasury. The level of borrowing reached $917 million in June, 1997, but by May, 1999, the level had been reduced to $730 million through repayments, including $45 million in interest. It is significant to note that from 1969 through 1997, almost $10 billion in premiums was collected from policyholders, and over $8 billion was expended by the program for loss payments and loss adjustment expense.

Until 1986, federal salaries and expenses, as well as the costs associated with flood mapping and floodplain management, were paid by an annual appropriation from Congress. From 1987 to 1990, however, Congress required the program to pay these expenses out of premium dollars without giving the program the opportunity to adjust the rate structure to reflect these costs. This resulted in a loss of about $350 million in loss reserves available to pay claims. Beginning in 1990, a federal policy fee of $25 (now $30) was levied on most policies in order to generate the funds for salaries, expenses, and mitigation costs.

4. Issues

The remainder of this paper describes certain issues confronting the National Flood Insurance Program at this writing, their background, and possible solutions.

4.1. EXISTING STRUCTURES

A major issue to be dealt with in the area of mitigation is the matter of existing structures, i.e. those buildings built in high risk flood areas before the availability of adequate risk information. The NFIP floodplain management requirements are primarily directed toward guiding future construction in the floodplain. The 1966 report projected that subsidies for existing high-risk properties would be needed for about 25 years, based on the assumption that a turnover in housing stock, particularly those structures most exposed to serious flood hazard, would, over that time, result in the large majority of structures being rebuilt in compliance with NFIP building standards and insured at full actuarial rates. However, modern construction and renovation techniques have greatly extended the useful life of buildings with the result that the replacement of this older housing stock has not occurred as quickly as contemplated by the feasibility study.

The most serious problem presented by existing flood-prone structures has to do with buildings that are damaged repeatedly but not beyond 50 percent of their value, the threshold that triggers the requirement to rebuild to specified safer elevations. These number rises about 76,000, although only about 40,000 are currently insured. These properties account for over 200,000 paid losses, almost one-third of the program's total paid losses, and the $2.8 billion paid on these losses represents about 35 percent of the $8 billion paid over the history of the program. The program is currently in the process of identifying the worst of these properties, about 9,000, and targeting them for special action such as relocation, purchase or elevation. It is estimated that a focused effort to address this universe of properties would cost about $300 million over a five-year period, but the program would recover those costs in the form of reduced losses in eight years.

4.2. EROSION

One of the major issues that faces the NFIP in coastal areas is the problem of erosion. The program provides coverage for damage caused by storm-related erosion. However, current provisions governing the NFIP do not allow the rate structure to reflect the erosion hazard, preventing the program from buttressing local coastal regulation with a financial leverage. In 1994, the Congress voted down a proposed provision that would have allowed the program to charge for the erosion hazard and to impose setback requirements on erosion-prone properties. Instead, it authorized a study to be done of the economic effects of such a provision. That study is in progress.

4.3. MAPS

One of the NFIP's important ongoing responsibilities is maintaining the accuracy of its maps and ensuring the accessibility of the risk information to its various users. The maps support a variety of activities, including local community enforcement of floodplain ordinances, the proper rating of insurance risk, and the determination by mortgage lenders of which properties require flood insurance coverage. The maintenance of accuracy means that the maps need to be regularly updated because of changes in stream channels, local development, coastal erosion, or construction of floodworks. It is estimated that the cost of modernizing the current NFIP map inventory is $800 million, an amount of money that is not available under the program's funding scheme. This modernization process would put the flood risk information into digitized format, thereby eventually making it more accessible to those users such as insurance agents and lenders who are not as sophisticated in the interpretation of risk data displayed in map form.

4.4. MARKET PENETRATION

NFIP policies in force stand at about 4.1 million as of March 1999. Single-family residences account for almost two-thirds of all policies. Non-residential (commercial) structures account for only about 4.3 percent. Definitive figures on the potential market for flood insurance are difficult to obtain. A conservative estimate is that the current 4.1 million NFIP policies represent less than half of those property owners who should carry the coverage. The reasons that have been given for this low market penetration are varied.

- Many floodplain residents underestimate the seriousness and likelihood of flooding.
- Lenders, who are required to have flood insurance placed on all loans in high risk areas, have not been especially zealous in requiring the purchase of flood insurance as the law requires.
- NFIP policies are unique, difficult for insurance agents to write, and therefore not marketed aggressively. Because they are single-peril policies, they are viewed as more costly than other lines of coverage.
- Many potential consumers expect that federal disaster assistance will adequately provide for post-flood recovery.
- Many people are simply misinformed about or unaware of the availability of national flood insurance.

4.5. SUBSIDY REDUCTION

Currently, about 30 percent of NFIP policies are subsidized to some degree, which costs the program approximately a half billion dollars in annual premiums. Subsidization was built into the program both to avoid unfairly burdening existing floodplain residents who had built without full knowledge of the risk, and as an incentive to communities to regulate the location and quality of future floodplain construction. To the extent that the mitigation components of the program are effective, the subsidy will necessarily shrink. There have been impressive examples in recent storms demonstrating that mitigation measures that meet program standards are effective in reducing losses. However, the turnover in subsidized housing stock has not occurred as anticipated. Further, a relatively small number of structures which have suffered repetitive damage over the years, but not to a level exceeding 50 percent of their value, have accounted for a disproportionate percentage of the program's total losses. These structures continue to qualify for subsidization.

Acknowledging the requirement to provide some level of subsidy to existing structures, the program has attempted, since 1981, to accelerate the reduction of the amount of subsidy on individual policies without raising premium costs past the point of reasonableness. The annual rate review continues to result in reduction of subsidies.

Any strategy to reduce the subsidy is influenced by the political climate in which the program has to operate. There are conflicting pressures to both reduce and maintain the subsidy, and the public attitude toward subsidies varies with the national mood at the moment. In the immediate aftermath of a major event like the Midwest flooding of 1993, there is often a general lack of support for allowing people to continue to occupy the floodplain with no financial consequences. However, when measures are proposed to address the problem, the public has often forgotten the storm and changed its sentiments. Further, an elimination of the subsidy that pushes premiums for many to unaffordable levels would also likely increase the demand for alternative forms of public assistance which, in the long run are more costly than the subsidy. In 1994, the Congress authorized a study to assess the economic effects of eliminating the program subsidy. This study is scheduled for completion in late spring of 1999.

4.6. RELATIONSHIP TO FEDERAL DISASTER ASSISTANCE

The availability of federal disaster assistance in the aftermath of a flood disaster is often cited as a reason why floodplain residents do not purchase flood insurance. The prevailing public impression is that federal disaster assistance is generally equivalent to the financial protection provided by hazard insurance. In reality this is not the case. In most flood disasters, the primary mode of federal assistance provided to property owners after a disaster is a low-interest loan from the Small Business Administration (SBA). Disaster victims who are deemed unable to repay an SBA loan can receive an Individual and Family Grant (IFG) from FEMA. However, the amount of these grants is limited to a maximum of $13,000, and they are intended to address reasonable needs and necessary expenses, not to make the grant recipient "whole."

A growing and more problematic form of federal disaster assistance is the buyout program authorized by Congress since the early 1990's. Buyouts are one of the uses of the Hazard Mitigation Grant Program. This program makes available an additional 15% of the total Federal disaster expenditures in a given disaster to be used for mitigation purposes. The objective of the buyout program is to use the opportunity presented by a serious flood to pay for the relocation of flood victims out of the floodplain in order to reduce the costs of future flood disasters. This program was used extensively after the 1993 Midwest floods, and it was employed in the wake of the

1997 floods in North Dakota, Minnesota, and South Dakota. The city of Grand Forks, North Dakota was especially hard hit by the 1997 floods and the buyout process is still ongoing for much of the affected area.

When buyouts are authorized, they are available to all affected residents of a targeted area, whether they are insured or not. While the NFIP can accurately claim that insurance coverage is preferable to a disaster loan or grant, a buyout results in a much larger payment to the property owner than an insurance policy. To the extent that floodplain residents begin to presume the availability of buyouts in the event of a flood, the incentive to purchase flood insurance is significantly reduced.

Another issue related to disaster assistance involves low-income individuals. The current law includes a provision which denies any future assistance to anyone who receives federal disaster assistance in the aftermath of a flood and fails to purchase and maintain flood insurance in force. While this provision is designed to promote the purchase of flood insurance, it has its greatest potential impact on grant recipients, who are generally in lower income categories than those receiving loans and less likely to be able to afford insurance. Whether the threat of denial of future federal assistance will have the intended effect of promoting insurance purchase among this segment of the population remains to be seen. Experience over the years indicates that recipients of such assistance who were required to purchase insurance seldom maintained the insurance in force. In light of this, the question must be raised as to whether insurance is the mechanism best suited to provide financial protection to a low-income population. Individuals in this income category often carry no insurance at all, let alone a single-peril policy like flood insurance.

4.7. FUTURE ROLE OF THE PRIVATE INSURANCE INDUSTRY

The original role contemplated for the insurance industry in the NFIP was to underwrite the coverage, with financial backing as needed from the government. That arrangement ended in 1977, and since that time the industry has performed in various ways in the role of fiscal agent to the program, even with the more extensive involvement of the industry under the WYO program. As we gain more and more experience with the flood program, if the subsidy continues to recede, and if the insurers remain committed to the program, it is inevitable that the feasibility of having private insurance companies again underwrite a significant portion of the coverage will surface for discussion. Lloyd's of London, for one, is now able to provide a certain level of private flood coverage because of the more favorable market environment created by the NFIP. One of the concerns expressed over the privatization of flood insurance is whether the vital connection between the availability of flood insurance and the local community enforcement of floodplain management provisions can be maintained. Unless this local enforcement continues, any significant privatization of flood insurance is not likely to be financially feasible.

5. Conclusion

An overview of a program as complex as the National Flood Insurance Program must necessarily abbreviate discussion of many aspects that warrant more in-depth explanation. One of the key elements in the program design is identifying the most effective utilization of the various governmental and private sector entities that have a role to play in flood loss reduction. Because it relies on these interrelationships to accomplish its primary objectives, the sometimes artful management of these relationships is the program's primary challenge.

6. References

1. Natural Hazards Research and Applications Information Center. (1992) University of Colorado at Boulder, CO. Floodplain Management in the United States: An Assessment Report. Prepared for the Federal Interagency Floodplain Management Task Force.
2. Interagency Floodplain Management Review Committee. (1994) Sharing the Challenge: Floodplain Management into the 21st Century. Report to the Administration Floodplain Management Task Force, Washington, DC.

FLOOD DISASTERS - INSURANCE ASPECTS INCLUDING THE 1997 ODRA FLOODS

THOMAS LOSTER

Geoscience Research Group
Munich Reinsurance Company
Munich, Germany

1. Introduction

The severe floods that affected the USA (Mississippi 1993), China (1996, 1998), India, Nepal and Bangladesh (1998) attracted great attention throughout the world. In Europe, a noticeable accumulation of, and an increase in, extreme events has also caused concerns. The floods on the Rhine and its tributaries (Germany 1993, 1995), in eastern Germany (1994), in Italy (1993) and on the Odra (Czech Republic, Poland, Germany 1997) are just a few examples of exceptional floods. The analysis of loss events worldwide reveals a dramatic increase in the loss amounts everywhere, with substantial increases in economic losses even in less developed countries. In the face of this apparent accumulation of major flood catastrophes, the call for insurance protection has become stronger. The insurance industry's role is increasingly being discussed, and in many countries of Europe this discussion focuses on the question of divesting the state of this responsibility by transferring it the private sector. It is only within the framework of a risk partnership, among the state, the affected population, and the insurance industry, that an effective attempt can be made to solve the problem of floods.

2. Losses, Trends

2.1. GENERAL ASPECTS

In economic terms, floods are the number one cause of losses from natural events. This tendency may be observed not only in terms of the major events but also in terms of the annual total of losses from many small and medium-size events. On average, floods cause more damage than any other destructive natural event. Additionally, it is important to bear in mind that the financial resources spent on flood control (sea dikes, embankments, reservoirs, etc.) by countries throughout the world is a multiple of the costs they devote to protection against other natural disasters.

As far as insured losses are concerned, the statistics are dominated by windstorms in all their various forms. Of the top nine insurance catastrophes, seven were due to windstorms. In a list of insured losses sorted by size, the costliest flood ever incurred

J. Marsalek et al. (eds.), Flood Issues in Contemporary Water Management, 197–208.

by the insurance industry, the "Great Flood of 1993" along the Mississippi, and the China Floods of 1998 rank only 19th, while the Odra flood of 1997 does not even rank among the top 25 events. The reason for the shift from "flood" to "windstorm" is the much greater insurance density involved in the case of windstorms.

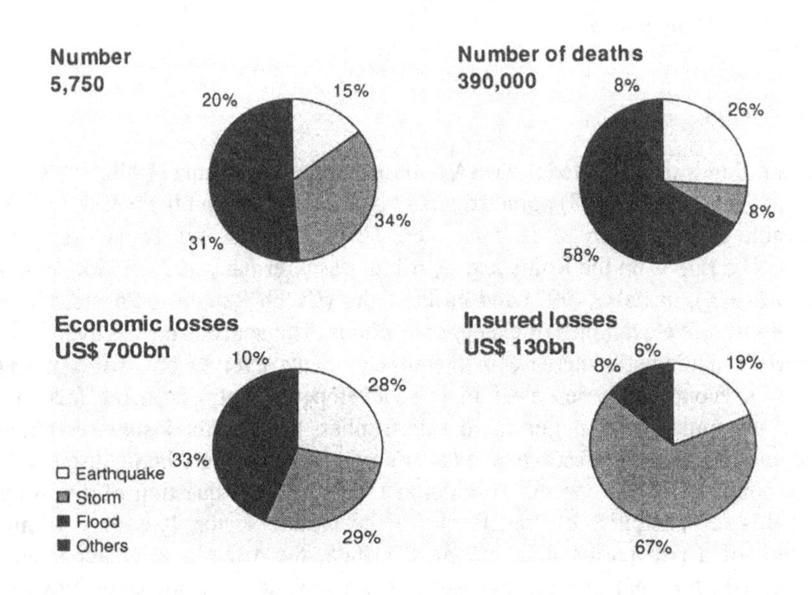

Figure 1. World-wide natural catastrophes, 1988-1997.
Percentage distribution by event type.

The main features of flood events derived from global analyses (1988-1997) are as follows:

- Floods account for about a third of all natural catastrophes.
- They cause more than half of all the fatalities due to natural catastrophes throughout the world.
- They are responsible for a third of the overall economic loss.
- Their share in insured losses is relatively small, with an average of just under 10 %, as insurance for flood risk is offered very conservatively in many markets or does not exist at all.

The following tables present the largest events of the 1990s throughout the world, broken down by the number of fatalities, economic losses and insured losses.

TABLE 1. Significant flood loss events 1990-1998

a) Deaths

Rank/Date		Country, region	Total losses*	Insured losses*	**Deaths**
1	10.6 - 30.9.1998	India, Bangl., Nepal	5,020	--/< 1[**]	**4,750**
2	May - Sep. 1998	China	30,000	1,000	**3,656**
3	21.6 - 20.9.1993	China	11,000	--/< 1	**3,300**
4	May - Sep. 1991	China	15,000	410	**3,074**
5	27.6 - 13.8.1996	China	24,000	445	**3,048**
6	Oct. - Dec. 1997	Somalia	--	--/< 1	**1,800**
7	4.9 - 2.10.1992	India	1,000	--/< 1	**1,500**

* US$ millions (original costs)
[**] -- = no data; < 1 = an estimate

b) Economic losses

Rank/Date		Country, region	**Total losses***	Insured losses*	Deaths
1	May - Sep. 1998	China	**30,000**	1,000	3,656
2	27.6 - 13.8.1996	China	**24,000**	445	3,048
3	27.6 - 15.8.1993	USA, Mississippi	**16,000**	1,000	45
4	24.7 -18.8.1995	North Korea	**15,000**	--/< 1[**]	68
5	May - Sep. 1991	China	**15,000**	410	3,074
6	21.6 - 20.9.1993	China	**11,000**	--/< 1	3,300
7	4 - 6.11.1994	Italy, N	**9,300**	65	64

* US$ millions (original costs)
[**] -- = no data; < 1 = an estimate

c) Insured losses

Rank/Date		Country, region	Total losses*	**Insured losses***	Deaths
1	27.6 - 15.8.1993	USA, Mississippi	16,0000	**1,000**	45
2	May - Sep.1998	China	30,0000	**1,000**	3,656
3	20 - 31.12.1993	Europe	2,000	**800**	14
4	5.7 - 10.8.1997	Europe	5,900	**785**	110
5	19.1 - 3.2.1995	Europe	3,500	**750**	28
6	20 - 28.9.1993	Europe	1,500	**500**	16
7	3 - 10.1.1995	USA	1,800	**470**	11

*US$ millions (original costs)

A survey of the great catastrophes shows that floods continue to cause the largest numbers of deaths in the poor and heavily populated countries of the world. In terms of economic losses, however, large loss amounts are mainly generated by the accumulation of values in the regions affected or flooded by exceptionally long-lasting events, resulting in widespread inundation. The largest insured losses are encountered, as might be expected, in the industrial countries, where the insurance density is generally at its greatest.

2.2. TRENDS

For more than 25 years, Munich Re has been gathering data on natural loss events from all over the world (property losses, bodily injury, etc.) in a special natural hazard database (NatCat*SERVICE*). The following analysis focuses on great natural catastrophes since 1950. In this context, floods are classified as great if the ability of the region to help itself is distinctly overtaxed, making inter-regional or international assistance necessary. This is usually the case when thousands of people are killed, hundreds of thousands are made homeless, or when a country suffers substantial economic losses (depending on the economic circumstances generally prevailing in that country). The advantage of this approach is obvious: an analysis of all the losses recorded might simply reveal an increase due to improved flows of information. As a result of the media revolution (global information networks, the Internet, data highways, etc.), there is a (possibly) substantial increase in the volume of information and hence in the data set itself. Great natural catastrophes can be analysed very well in retrospect, because even records that go back several decades can still be investigated today. This means that a considerably longer period of time can be covered here. Of the 550 to 700 loss events recorded each year by the Geoscience Research Group at Munich Re, an average of about three to five may be classified as great natural catastrophes (the record was in 1987 with 17 natural catastrophes, six of which were floods). In 1997 there was only one great natural catastrophe, the summer floods on the Odra.

TABLE 2. Great Flood Disasters 1950 - 1998
Trends - comparison of decades

	Decade 1950-59	Decade 1960-69	Decade 1970-79	Decade 1980-89	last 10 1989-98	Factor last 10:50	Factor last 10:60
Number	7	7	9	20	34	4.9	4.9
Economic losses	27.9	20.2	19.2	25.5	199.6	7.2	9.9
Insured losses	---	0.2	0.4	1.4	7.4	---	37

Losses in billions US$ - 1998 costs

If one compares the individual decades, it can be seen that the economic losses of the last ten years are ten times as high as in the 1950s – after adjusting for inflation. The number of loss events has also increased (by a factor of 6). As far as insured losses are concerned, flood insurance was still in the early stages of development in the 1950s, so the pertinent information can only be given for the 1960s (37 fold increase). In the future the long-term global trend towards multiple risks cover, which will often include flood losses, will push the figures up even more distinctly.

The reasons for the increase in catastrophic losses may be summarised as follows:

1. population developments globally and in exposed regions
2. the increase in values in these regions
3. the increase in the vulnerability of structures (infrastructure, building structures) and goods (stock, furnishings, etc.)
4. construction in flood-prone areas
5. often excessive trust in flood protection systems
6. changes in environmental conditions.

As far as the extent of losses caused by the Odra floods in 1997 is concerned, nearly all of the above causes occurred in the affected countries, the Czech Republic, Poland, and Germany. Construction in flood-prone areas (item 4) must be particularly mentioned. Factor No. 5 (trust in flood protection systems) must be slightly qualified. The fact that there had been no serious instances of the Odra overflowing its banks for a very long time had made the people forget the danger of flooding completely. An additional role was played by the fact that sections of river embankments in some river reaches were too old.

3. Flood Insurance, Basic Issues

Subject to the absolutely indispensable preconditions of 1) no anti-selection, 2) insurance of sudden and unforeseeable events only, and 3) substantial deductibles, it would be reasonable to provide flood coverage at a flat premium rate for a large segment of the mass business portfolio, i.e. for those objects that are not regularly exposed to river flooding. That sounds simple at first, but a closer look will reveal that the difficulties are in details.

1) **Anti-selection**

Flood insurance is usually sought only by the owners of property that is frequently flooded or obviously exposed to flooding. This means that flood, unlike many other natural hazards, presents insurers with the particular problem of what is called "adverse selection" or "anti-selection":

- There is no spatial spread of the insured risks; the community of insurance holders is a relatively small group of people who are all exposed to a high flood risk and the losses incurred are normally high.
- A balance of losses over time is not possible because flooding occurs too often, so that loss events cannot be regarded as unexpected events. The element of the unexpected, however, is prerequisite to any insurance coverage.

Premiums based on past loss experience and future loss expectations would have to be so high that policyholders would normally find them prohibitive. All attempts to combine flood with other natural hazards (windstorm, earthquake, etc.) in the coverage have failed to date. Clients that pay large premiums for life, health, car, and liability insurance see no reason to pay premiums for risks that can hardly affect them.

2) **Unforeseen and sudden events**
It is one of the principles of insurance that coverage can only be granted for unforeseen, sudden events. In the case of many river floods, however, this simply does not apply. For many rivers, the only question is when flooding will occur. And then there are sections of the river that are hit at almost regular intervals (e.g. the Middle Rhine in Germany). Flash floods present a different picture: in this connection, flood damage is insurable because flash floods are caused by heavy rain - often during local storms - and can occur almost anywhere. The necessary geographical spread of risks is given and adequate premiums can be calculated with a relatively high degree of reliability. The community of insurance holders is large, and the frequency of people being hit by extreme events is low. Therefore, the premiums can be also kept low.

3) **Deductibles**
The incorporation of substantial deductibles in the insurance terms and conditions is one of the most effective precautionary measures. It has a number of advantages not only for the insurer but also and in particular for the insured. The insurance company has to settle a lot fewer claims and the total reimbursement for an event makes up only a fraction of the total loss. At the same time the insured will make a great effort to improve his preparedness measures, because he will always have to bear any loss that may occur up to the limit of the deductible. Consequently, the risk situation improves and the premiums decrease quite substantially, but the protection against major losses is maintained in full. In spite of these obvious advantages for all concerned, the opportunities available to insurers are often heavily restricted by the competitive situation. It is only when the loss burden explodes that the companies in the affected market react with the introduction of deductibles.

4. The Floods on the Odra in the Summer of 1997

4.1. GENERAL ASPECTS: LOSSES AND LESSONS LEARNED

A strong low-pressure system which had developed over the Mediterranean at the beginning of July, moved over to the Czech Republic and Poland. The extensive rain fields generated precipitation amounts of up to 570 L/m^2, resulting in flash floods of devastating force in many places. In the Czech Republic, mainly the rivers in the east of the country were affected, while the worst hit area in Poland was in the south. The torrential rain in the mountains ran off into the Odra. There was extensive flooding throughout the upper reaches. Numerous other rivers in Poland (the Nysa, Vistula, etc.) burst their banks. A total area of about 6,000 km^2 in 14 Polish provinces was flooded, with more than 100,000 houses in 2,500 towns and villages under water. Large areas in cities like Opole and Wroclaw were submerged in flood waters. Infrastructure was severely impaired, with 2,000 km of railway lines, 3,000 km of roads, and 900 bridges impassable, and gas, electricity, and water supplies interrupted for weeks. Business and commerce were hit hard. Numerous industrial losses resulted in lengthy business interruption and problems with deliveries. The agricultural sector was hit particularly hard: 40,000 farms and 1,800 km^2 of arable land were inundated, 8,000 head of livestock were drowned.

The flood was of historical dimensions. Never before had the Odra experienced such a large discharge of water – exceeding 4.2×10^9 m^3. An essential role was played not only by the immense amount of rain that fell but also by the natural sealing of the ground surface that resulted from this copious precipitation. Classifying this flood in statistical terms (return period) is very difficult as the majority of observation instruments were torn away by the deluge, and consequently it is possible to make a rough estimate of the discharge only. The return periods for some discharge peaks in the upper Odra and the tributaries were said to be as high as 10,000 years. All bodies of water in this region, both large and small, burst their banks and inundated large areas of land. This had a positive effect, however, in the lower reaches, where the river forms the border between Poland and Germany. Had it not been for the floods in the Czech Republic and Poland, the water levels there would have been about 150 cm higher and would have given those defending the embankments no chance whatsoever. These embankments were put to a severe test; they had to withstand enormous pressure from the water for weeks on end and were near to bursting just about everywhere. Gigantic efforts on both sides of the Odra averted a catastrophe at least in this region.

TABLE 3. Summer floods 1997*
Overall balance

Country	Date Houses damaged/destroyed	Deaths	Economic losses (US$m)	Insured losses (US$m)
Czech Rep.	5-30.7. 2,150/26,000	60	1,850	310
Poland	5.7-10.8. 68,200	55	5,500	450
Germany	17.7-10.8. hundreds damaged	---	360	35

* The Odra river and its tributaries; the Czech Republic data includes those for the Odra and the Morava River and its tributaries

This brief survey of losses shows that the extent of loss varied widely from country to country. Among other things, this was due to the duration of the event. The analysis of these floods allowed important conclusions to be drawn with regard to the extent of loss and flood defence. These conclusions may be summarised as follows.

- Flood defence functioned relatively well in many of the places affected – considering the extreme magnitude of the floods. The work performed by the fire brigades and the voluntary rescue services was particularly good. But there were shortcomings too.
- In the Czech Republic there was a lack of expertise in many places, particularly at the municipal level. Consequently, there was no co-ordination of skilled workers and sometimes there were no task forces on hand. There were also deficiencies in the flow of information, and the people were not given proper instructions.
- In Poland, suitable equipment like boats, pumps, power generators, etc., was not available in some places. These shortcomings were partly offset by the enormous efforts of volunteer helpers on the spot. They proved that, given sufficient involvement and purposeful efforts, almost inconceivable protection measures are possible and they prevented an even worse catastrophe. For example, large parts of Wroclaw's old city were successfully defended.
- Germany was fortunate enough to be forewarned about the approaching waves of water. This made it possible to take quite effective measures at an early stage (evacuations, dike stabilisation, etc.). Nevertheless, serious problems surfaced here too. Improvements are needed in terms of the compatibility of telecommunication systems and in terms of the co-operation between the authorities and volunteers, and other helpers in disaster management.

Reporting in the media, particularly in Germany and the Czech Republic, was sometimes sensationalistic and misleading. Lucid information was often scant.

- Rivers need water meadows and flood plains. On the Odra, flood plains measuring a total of 3,000 km^2 have been settled in recent decades.
- The settlement of known flood plains is critical and in the long term this is a source of damage, even if the protective measures (dikes, flood retention basins, etc.) are good and nothing happens for years. This was manifested in many parts of the countries affected (the Czech Republic: Mohelnice, Olomouc; Poland: Wroclaw, Opole; Germany: Ziltendorfer Niederung).
- Systematic investigations are needed on the rates of discharge, exposures, loss potentials, and flood management if these countries are to be in a better position to cope with similar events in the future. It must be borne in mind, however, that there is always a residual risk in the case of floods.
- The use of early flood warning systems must be optimised. This applies both to meteorological equipment like data collection and radar networks, and to hydrological reporting networks and discharge models. Some of these problems have already been tackled effectively by the responsible authorities.
- The events on the Odra triggered an international dialogue between politicians, scientists, and authorities in the neighbouring countries. The inclusion of the flood risk in the range of subjects dealt with by the international commission for the protection of the Odra (IKSO) is an important initial step towards getting the problem of floods and flooding on this river under control.

One positive aspect to be mentioned in passing is that there was a great demonstration of solidarity with those affected, with private donations coming from people both at home and abroad, and this contributed to providing speedy help in coping with the loss.

4.2. INSURANCE ASPECTS

For the insurance industry there are also conclusions to be drawn from the floods on the Odra. The situation in each of the countries affected was different.

- **Germany**
  In the area of the former GDR (i.e. until 1990), buildings and contents were insured against floods. After reunification, the insured portfolios were assumed or purchased by west German insurance companies. On account of the problems mentioned above, however, flood cover was excluded in western Germany at that time and was granted only in isolated cases. Unknown to many people in the new federal states in the eastern part of Germany, the insurance contracts were therefore changed and the flood risk was excluded in most cases. In this particular case, however, the insurance industry made a goodwill gesture by settling many of the losses, involving a total payment of around DM 35 million.

- **Czech Republic and Poland**
  The political situations in the Czech Republic and Poland posed some unusual circumstances. It was only in the beginning of the 1990s that the state monopolies had been converted into joint stock companies, although in some cases the state still maintained a major share in them (e.g. PZU in Poland). In view of the fact that both countries were badly hit by the catastrophe and these young markets had little experience in this field, it is remarkable how well the catastrophe was dealt with in the final analysis.
  - In the Czech Republic more than 100,000 policies were affected. After only a few weeks more than half of the claims had been settled and almost all of them had been settled within the period of a year.
  - In Poland, where about twice as many insurance policies had to be processed, claims settlement was fast and effective. The regional offices set up claims settlement offices, manned by experts (engineers, etc.). It is conspicuous just how many cases were subjected to close scrutiny and inspection, surpassing even the norm in other countries of central Europe.

All in all, the insurance industry demonstrated that it is able to act quickly and effectively, and that it has an important role to play in the process of coping with losses.

Action to be taken immediately by the insurance industry

Besides providing purely financial protection for policyholders, insurance companies perform two further tasks that are vital for their own existence:

- **Loss analysis**

  It goes almost without saying that insurers have to analyse their claims data meticulously and draw conclusions from these analyses for the purpose of risk-commensurate rating. The loss patterns of the Odra floods present an excellent opportunity to deduce loss parameters that are specific to particular areas and lines of business, and which in turn can be incorporated in analyses of the loss potential and eventually in the rating process itself.

- **Accumulation control**

  Accumulation control involves a detailed analysis of concentrations of liabilities. Large accumulations can lead to major loss burdens in the case of catastrophes. Accumulation control is an important tool that enables an insurance company to define its business policy objectives – in terms of production targets, underwriting guidelines and the establishment of reserves – and to determine its reinsurance requirements. It is one of the most important tools for underwriting natural hazards successfully. Uniform standards have already been defined for earthquake and windstorm accumulation control in many countries, but there is an immense backlog of work to be cleared as far as the flood risk is concerned.

5. Risk Partnership for Flood Damage Prevention

Disaster preparedness and coping with disaster once it has struck only works, however, if the three main groups affected – state/public agencies, the affected population, and the insurance industry – work together. They are called on to join forces and enter into a risk partnership, in order to make the most of the opportunities that present themselves. Only effective co-operation will lead to developments from which all those concerned, and in particular the flood-prone population, can benefit. Insurers and reinsurers can make a valuable contribution beyond performing their core function of indemnifying losses by making their knowledge available. They have extensive experience in the field of loss analysis and reinsurers in particular also have scientific and engineering expertise and can thus offer information and advice.

Some of the areas in which close and methodical co-operation between insurers and scientists and authorities may be expected to produce fruitful results are given in the following examples.

- **Generation of exposure analyses and exposure maps**

  The goal is not only to provide politicians or municipal authorities with planning aids. Insurers are an important link in the chain of loss minimisation. They should be involved in this work at an early stage so that their requirements can be properly considered. After all, they should be in a position to use exposure zoning for assessing and rating the risks they insure.

- **Loss potential analyses**

  Loss potential analyses, such as water level/loss functions, must be carried out in the light of the insurers' requirements. These analyses should be conceived in such a way that they consider features specific to particular lines of business and thus provide the insurance industry with a useful aid.

- **IT systems, Geographical Information Systems (GIS)**

  The data kept by scientists and users are to be as uniform as possible. IT systems should be geared to each other. In the case of the flood risk, for instance, GIS should be employed. Gearing the data and their platforms to each other will make it possible to work quickly and effectively. This means that the insurers store their liability information (portfolio data) in a GIS and can at the press of a button make use of risk-relevant exposure information provided by scientists.

The public, the scientific world, and the insurance industry must join together in formulating their requirements, preparing them in such a way that politicians can derive clearly recognisable options for action (e.g. land use restrictions) from them.

6. Future challenge - climate change

The increase in the frequency of major flood catastrophes observed in recent decades is a matter of increasing concern to the public, politicians, and the economy, which means first and foremost the insurance industry since it covers a substantial proportion of the losses. The influence of climate change on the frequency and intensity of flood catastrophes has yet to be proven, but there are numerous indications of such an influence. In addition to the rise in sea levels forecasted for the coming century by the Intergovernmental Panel on Climate Change (IPCC), the rise in humidity levels as a result of increasing condensation will have a particularly grave effect as it will exert a decisive influence on all precipitation and convection processes in the troposphere. Generally speaking, an increase is to be expected in torrential rainfalls at regional level. That has already been confirmed by an initial analysis of measurements in Europe and America. There is also firm evidence of shifts in the seasons in some countries, e.g. in Germany (drier summers, wetter winters). In Europe, milder winters, which are likely to occur in a warmer climate, can also have a grave effect on the flood risk. They lead to an increase in the natural ground sealing during the winter months as a result of precipitation in the predominantly liquid form. Winter storms and low-pressure systems combined with copious precipitation produce surface runoff, which quickly results in floods. It is also conceivable that storms with torrential rain will penetrate further into the continent of Europe. The lack of snow results in a loss of the blocking effect of the cold high-pressure systems over eastern Europe, so that winter storms are no longer diverted and can penetrate far into the continent, as was the case in the unprecedented series of storms in central Europe from January 25 to March 1, 1990 (eight destructive storms in close succession). If the IPCC's predictions are confirmed, the storm surge risk could also increase on many coasts as rising sea levels in concert with an increase in the circulation of the atmosphere might boost the number of storm surges.

7. Conclusions

In the future we must reckon with more – and larger – flood events and catastrophes. For this reason it is important to take appropriate steps without delay. On one hand, the efforts that are being taken to combat climate change must be stepped up, while on the other hand we must pay more attention to flood protection and loss minimisation or prevention in the ways described above. There are many well-known technical, organisational, and financial measures designed to reduce the flood hazard and many of them are available for use. Long-term and short-term precautions play just as large a part in this as catastrophe aid and measures taken after loss events. One thing is certain, however: flood is a subject that involves everyone, which means that everyone has to work together in combating it. The authorities, the field of science, and the insurance industry have clearly defined roles. The insurers must join together in formulating their requirements and designing them in such a way that politicians can use them to derive clearly recognisable policy options (e.g. land use restrictions). The Odra floods in the summer of 1997 have set important developments in motion. In spite of the many promising ideas being formulated, we must always remember that in the end we have to live with floods and that there is no possibility of one hundred percent safety. We must never lose sight of the flood risk even if we are tempted to do so by a series of years without any losses. In principle, the insurability of floods is already a given fact. This applies primarily in the case of flash floods, but it can also apply to river floods if the principle of solidarity and the risk partnership among the state, the victims, and the insurance industry function properly.

RECENT EXPERIENCE WITH THE USE OF MATHEMATICAL MODELS IN FLOOD ACTION PLANNING AND FLOOD FORECASTING & WARNING

KARSTEN HAVNØ and JACOB HØST-MADSEN

Danish Hydraulic Institute (DHI)
Hørsholm, Denmark

1. Introduction

In the past, catchment and river engineering has often been based on procedures and design methods, which were too narrow in focus. In many places river training and urbanisation have caused unforeseen negative impacts on groundwater, water quality and flooding conditions. It is increasingly acknowledged that catchment and river engineering needs to be based on a more holistic approach and technology than those traditionally used. Many rivers in Europe flow across international boundaries, and harmonisation of methodologies between the countries is required.

In recent years, severe floods have occurred in several parts of Europe, both as localised flash floods and as basin-wide floods on major river systems. By their nature floods are generated by the coincidence of several meteorological factors, but man's use of the river catchment also has an important contribution to the severity and consequences of the events. Even though the investment in flood control structures has steadily increased during the second half of this century, the damage caused by floods has continued to increase dramatically.

After the severe flooding on the Rhine, the Rhone, the Po and other rivers in the early 1990s, the European Commission organised an expert workshop to discuss the current state-of-the-art of river flood management and the needs for improvement in this area. It was decided to launch a Concerted Action on River Basin Modelling, Management and Flood Mitigation, RIBAMOD, with the goals to:

- identify difficulties arising from past management practices,
- identify the state-of-the-art in this area,
- identify best practice,
- prepare an overview of the current EU research projects in this area,
- identify research needs, and
- provide a scientific basis for decision makers.

The Concerted Action has identified a clear need for international development and collaboration within:

- risk assessment procedures and terminology,

J. Marsalek et al. (eds.), Flood Issues in Contemporary Water Management, 209–218.

- integrated modelling and decision support,
- long term assessments of the environmental impact of structural and non-structural measures, and
- needs for common databases and standards.

These issues are now being addressed through the EUROTAS project (European River Flood Occurrence & Total Risk Assessment System) funded also by the European Commission and being implemented from 1998 to 2000.

2. Flood Management Model (FMM)

The traditional approach to flood control has been reshaped in recent years to a flood management oriented concept. Such a concept operates with different levels of flood risk for different areas - the lowest level for the urban areas and the highest for agricultural areas. The implementation includes flood forecasting and emergency planning tools, which facilitate effective management during the flood event.

It is regarded essential, that new flood control and management concepts are properly evaluated on a regional basis by using integrated models. These models should be used to evaluate both short-term and long-term the impacts.

2.1. EXAMPLE: A FLOOD MANAGEMENT MODEL FOR BANGLADESH

Bangladesh is a flat delta at the confluence of three of the world's major river systems: the Ganges, the Brahmaputra and the Meghna. With more than half of the country under the 12.5 m contour, about 30% of the Bangladesh area suitable for cultivation is flooded in a normal year. An estimated 50% is vulnerable to either monsoon or tidal floods. Only 20% of the vulnerable area is protected. Flooding in Bangladesh is commonly caused by a combination of several factors such as the over-bank spilling of the main rivers, runoff generated by heavy local rainfall, and cyclone tidal bores or storm surges. The 1987 and 1988 floods were two of the most severe on record, caused widespread damages to crops, urban areas and infrastructure, and resulted in more than 3,000 lost lives.

2.2. THE CONCEPTS BEHIND THE FLOOD MANAGEMENT MODEL

Flood management is concerned with decision making based on policies reflecting the needs of communities and the environment. It is a complex task and often without solutions which fully satisfy all concerned parties. The many important components which need to be considered, such as land use, environment, infrastructure, flood control structures, irrigation needs, agriculture, economics, society, fisheries, flood preparedness and flood forecasting, render decision making and policy formulation extremely difficult.

The modelling of floods in Bangladesh has been conducted using DHI's MIKE 11 computer based flood models. The models provide outputs such as flood levels along the rivers and on flood plains and, more importantly, simulate the impacts of interventions on the plains. Flood models do not, however, produce the flood maps needed for identifying and prioritising flood management zones, nor do they produce maps of impacts on flood levels which greatly assist in assessing alternative solutions and carrying out multi-sectoral flood impact analyses.

To produce these maps and to perform multi-sectoral impact analyses, GIS technology is required. Flood depths and levels are represented as layers of data in the GIS, which can be geographically related and analysed with data from other flood management components. The maps and results of multi-sectoral analyses are easily assimilated using a combination of graphical and statistical formats.

The Flood Management Model (FMM) is an integrated MIKE 11 GIS modelling system. FMM has the potential to assist in clarifying and disseminating information through enhanced mapping of impacts on flood levels in relation to communities, agriculture, fisheries and the environment. The maps can also help provide project design specifications, monitor and assess the performance of flood control and drainage structures, and help distribute flood forecasts in a readily acceptable form to the general public.

2.3. THE FLOOD MANAGEMENT CYCLE

Flood management follows a cyclic process linking ideas, proposals, consultations, adopting proposals, preparing guidelines, design, construction/implementation, and operating and maintaining finished schemes. It is this process of flood management planning, design, implementation and operation cycle, in which FMM plays a useful role. Fig. 1 shows the structure of a Flood Management Model.

At the planning level, FMM helps assess proposed flood mitigation options and prepare environmental impact assessments. At a design level, FMM functions as a tool for determining civil works design criteria, designing structure operation rules, and providing inputs to flood preparedness programmes. At the implementation stage, FMM may be useful for a range of needs from scheduling flood prone construction works to a flood preparedness training aid. Real-time FMM operation linked with flood forecasting can help guide structure operators and assist emergency relief operations. FMM can also help present the consequences on flooding due to repair and maintenance of structures.

2.4. EXAMPLES OF FMM OUTPUTS

2.4.1 *Flood Depth Maps*

Flood depth maps show in graphical detail the variation in inundation depth over the floodplain, along with the flood-free areas. They give a clear picture of the depth and extent of flood inundation. The maps are produced using the results from a flood model simulation and can be produced for any instant in time, or be based on the maximum flood levels over the entire simulation. Fig. 2 illustrates a flood depth map.

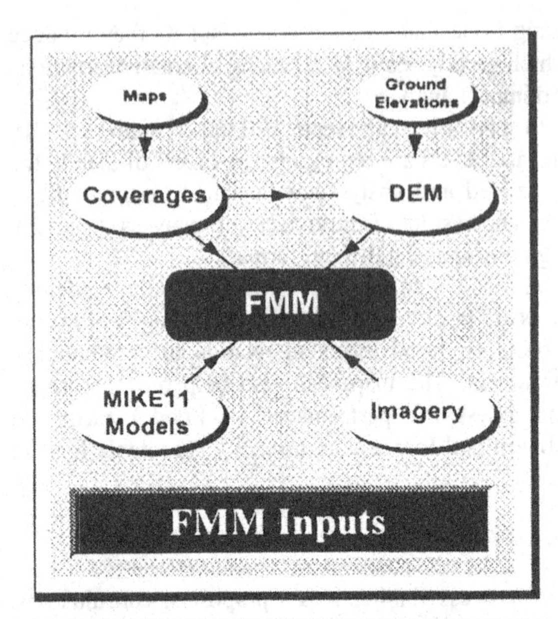

FMM requires inputs of coverages (roads, rivers, settlements), a DEM, flood models and, if available, satellite and photo imagery.

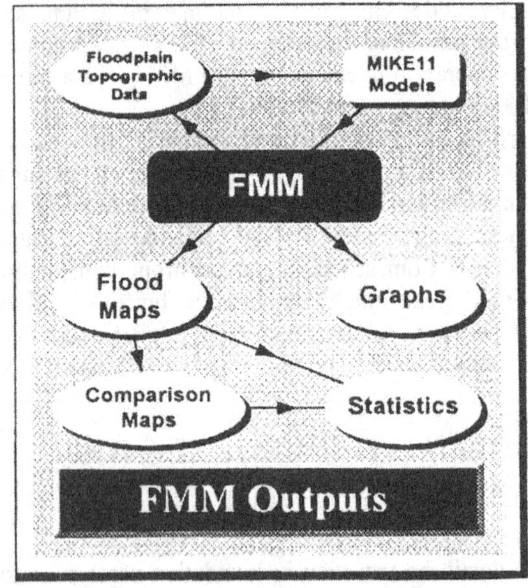

FMM outputs topographic data for use in flood models and post-processes flood model simulation results into flood maps, comparison maps and graphs. Statistics are produced from the flood and comparison maps.

Figure 1. Inputs and outputs of the Flood Management Model.

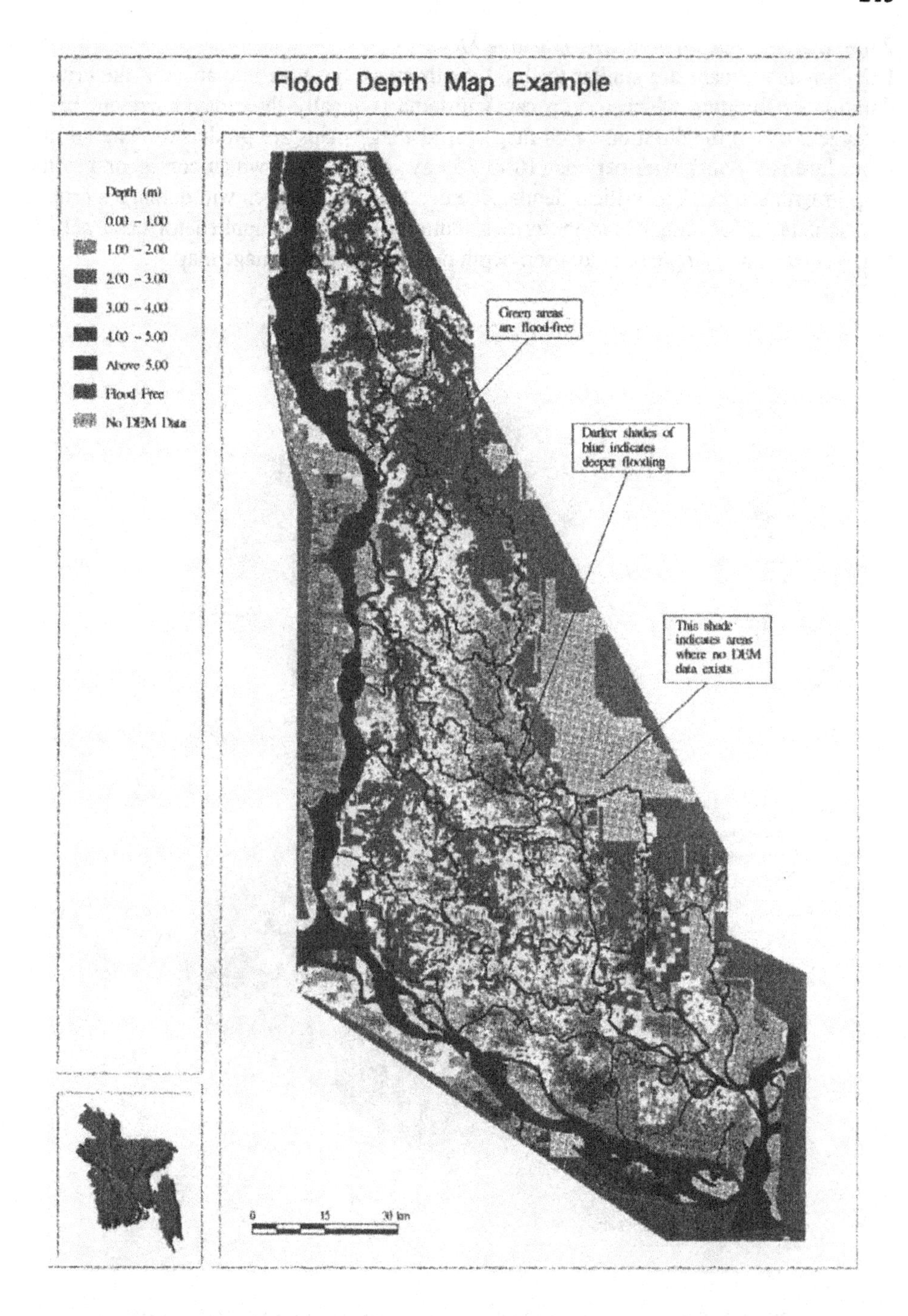

Figure 2. Example of a flood depth map.

2.4.2 *Duration/Depth and Crop Damage Maps*

Duration/depth maps are similar to flood depth maps, but take into account the critical duration of flooding which a crop can withstand (typically three days) without being damaged. From the duration depth map crop damage maps are produced. The normal procedure is to work with periods (10 to 15 days or monthly) which correspond to the crop growth stages. The critical depths (the depth of water which will damage a crop if it is inundated for longer than the critical duration) must be supplied for each period. Fig. 3 shows an example of a duration depth map and a crop damage map.

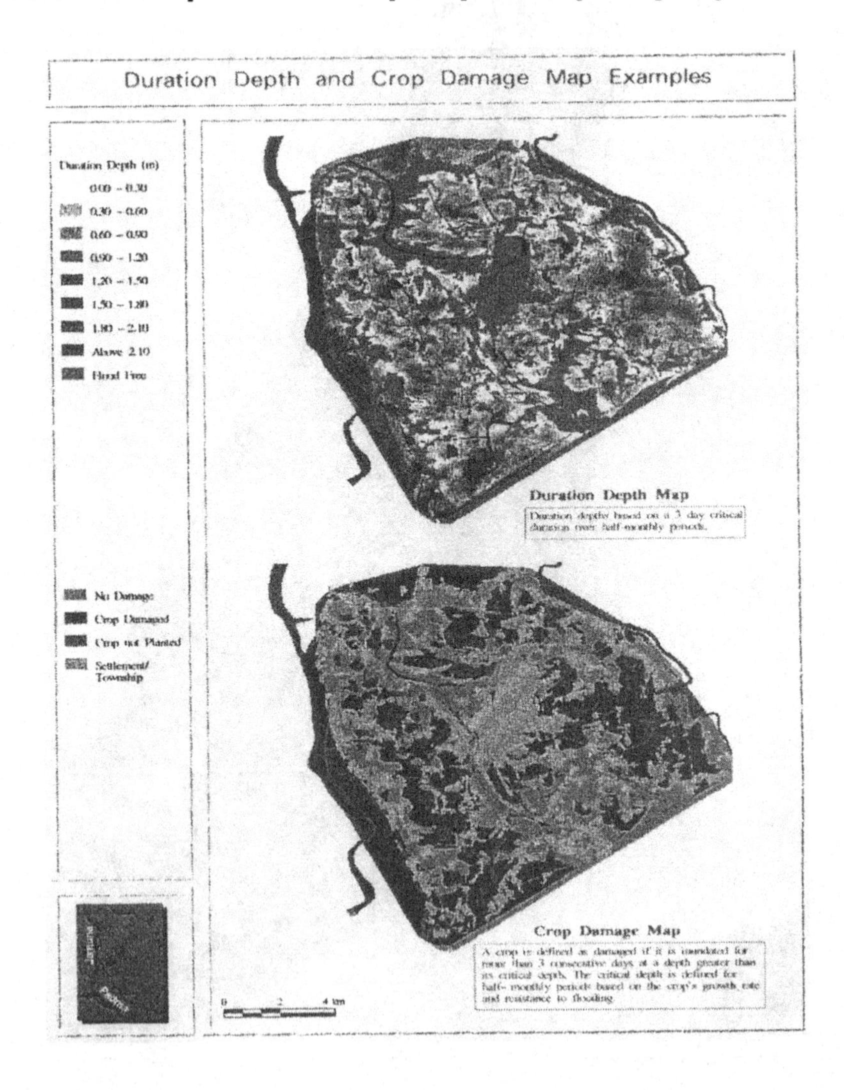

Figure 3. Examples of duration depth and crop damage maps produced by the FMM.

3. Combining Meteorological and Hydrological Models

As most floods are caused by meteorological events, a potential for improved flood forecasting accuracy is to combine meteorological and hydrological forecasting models for real-time forecasting. Although operating on different scales in time and space, recent experience shows that an operational combination of the two types of models can improve the forecast accuracy significantly. It requires a high resolution meteorological model which can incorporate a satisfactory resolution and representation of the individual hydrological catchments. Operating meteorological models at a local scale for quantitative precipitation forecasts requires super-computer computational facilities and access to boundary information for large-scale weather system models.

3.1. EXAMPLE: REAL-TIME FLOOD FORECASTING WITH METEOROLOGICAL AND HYDROLOGICAL MODELS

The Yangtze River in China is the longest river in Asia and one of the largest in the world. The middle part of the river is very flat and in the riparian valley areas flood disasters are frequent and severe. The flood-threatened areas along the mid-Yangtze river are economically important as they include some of the most developed industrial areas in China and are heavily populated. During the devastating 1998 flooding the river waters reached unprecedented levels in several parts of the river network, endangering the lives and property of millions.

The Changjiang Water Resources Commission (CWRC) is responsible for flood forecasting and flood control along the Middle Yangtze River Valley. The Danish Hydraulic Institute (DHI) and the Danish Meteorological Institute (DMI) have since 1996 been co-operating with CWRC to extend the capabilities of CWRC's proven flood forecasting procedures. A flood forecasting system has been established for the Middle Yangtze River combining both river modelling and meteorological modelling to provide water level forecasts for lead-times up to 5-days. The rainfall-runoff and river model now in operation at CWRC is based on DHI's MIKE 11 hydrological and hydrodynamic modelling system (Fig. 4). Quantitative precipitation forecasts are produced by the High Resolution Limited Area Meteorological Model (HIRLAM) jointly developed by a number of European meteorological institutes. The flood forecasting model covers an area of around 200,000 km^2, of which large parts are flood prone, and includes over 1500 km of the Yangtze River as well as the major tributaries.

During its first operational year in 1998, the MIKE 11 flood forecasting model produced daily flood forecasts for a number of locations in the river system. Every night during the flood season detailed weather and precipitation forecasts were produced by the HIRLAM model on DMI's super-computer in Copenhagen. The meteorological model results have then been downloaded by the forecasters at CWRC's headquarters in Wuhan and used for evaluation of the developments in the severe weather situations in the area and as direct input to the MIKE 11 Flood Forecasting Model.

The results from the 1998 flood season have demonstrated that the HIRLAM model can predict the development of important rain bearing weather systems in the Yangtze River Basin 48 hours ahead. The generated quantitative rainfall forecasts, when introduced into the river model, significantly improve the water level predictions particularly for lead-times larger than 24-hours. The combination of hydrological and meteorological modelling has brought about a significant improvement in the calculated river level forecasts particularly during the unusually rapid rise in the Yangtze water levels in 1998. Another advantage of physically based river models is the integrated representation of the whole river system allowing for a wide range of presentations in the form of maps, graphs and water level profiles giving overviews of complex situations.

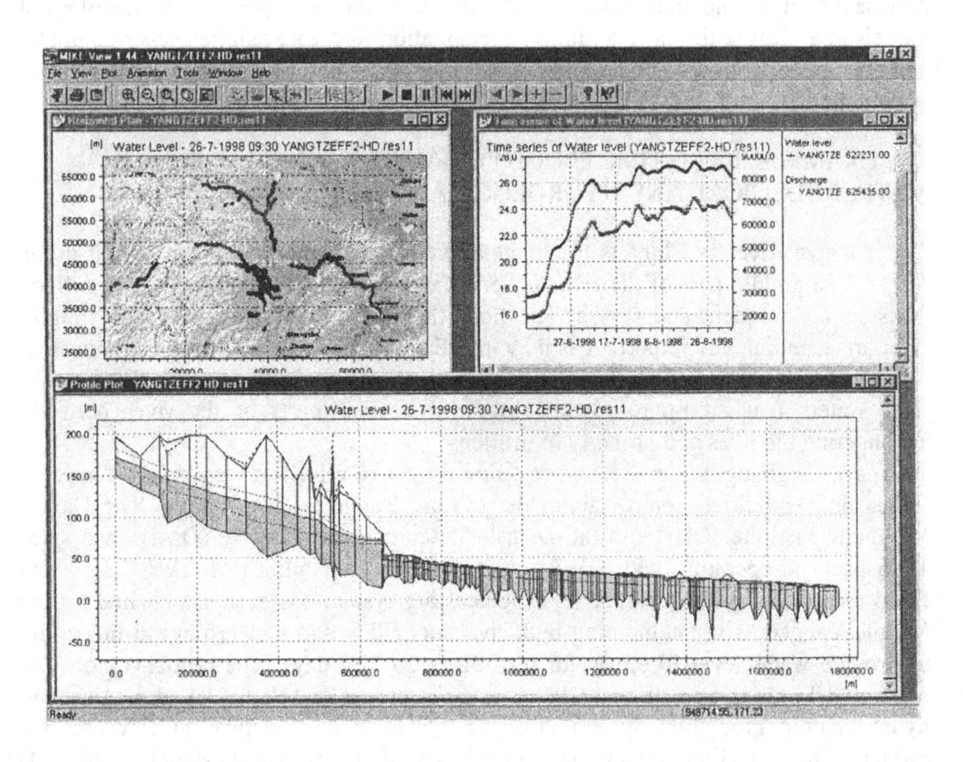

Figure 4. Overview of the MIKE 11 Yangtze forecast model set-up.

The 1998 flood in the Yangtze Valley was exceptionally severe. Millions of people were involved in safeguarding dykes and the flood protection capacity of existing reservoirs was utilised to the maximum. However, with the river rising to levels where the stability of the embankments safeguarding large cities was threatened, managers were forced to consider opening the embankments around predefined flood protection polders and flood these less sensitive areas in order to protect vulnerable cities. Evacuations were carried out moving several hundred thousand people in order to prepare for the inundation. The flood control authorities finally decided against a major forced emergency flooding and prevented the inundation of the largest of the flood protection areas. The analyses and forecasts by CWRC contributed positively to this decision.

Presently, the forecasting model includes only the river channels and catchment areas. However, the occurrence of dike collapses and flood control inundation in critical situations call for the establishment of a Flood Management Model. The extension of the model to flood plains will further improve the forecasting under extreme conditions, and furthermore provide a powerful flood management instrument for quick and reliable analyses of the downstream effects of various flood control scenarios.

With financial support from The World Bank, DHI has initiated the modelling of the entire Yangtze River with the purpose of making overall quantitative and qualitative water resources assessments. This will be accomplished by joining the forecasting model with existing MIKE 11 models of the upstream and downstream river reaches and by supplementing these with a DHI tool for basin-wide water use and resource allocation, MIKE BASIN.

The extended MIKE11 model will further benefit the 3-5 day forecast on the flood prone middle part of the river, since it will eliminate the present dependency on uncertain inflow forecasts along the upstream model boundary.

4. System Requirements and the use of Internet Technology for real-time flood forecasting

Most flood forecasting systems implemented in the past have been tailored to specific hardware and software platforms and very few generic system elements have been developed and implemented. Rapid technological development has consequently outdated many systems with significant consequences for flexibility and necessary resources for upgrade of systems.

Standalone hardware solutions have in the past been replaced by Local Area Network configurations but considering requirements for dissemination and warning there has always been a clear need to share and distribute flood information to a much larger number of recipients.

DHI is presently implementing a new flood forecasting system in East Anglia in the UK with the use of the industry standard hardware and software, Internet technology and generic software elements, in order to reduce system limitations, improve flexibility and provide powerful dissemination facilities.

4.1. EXAMPLE: FLOOD FORECASTING SYSTEM IN THE UK BASED ON INTERNET TECHNOLOGY, INDUSTRY STANDARD HARDWARE AND GENERIC SOFTWARE.

The Anglian Region of the Environment Agency (UK) covering 27,200 km^2 or 20% of England & Wales has purchased a state-of-the-art flood forecasting system designed to improve the timeliness, accuracy and reliability of flood forecasts. A consortium including the Danish Hydraulic Institute (DHI), W.S. Atkins Consultants Ltd and Litton Weather Services International is implementing the system with PB Kennedy & Donkin acting as Engineer to the contract.

The system will be proved on two test catchments – the Welland-Glen and the Witham – before being extended to other catchments across the region. Measured rainfall, river flows and river water level data from 650 points across the Anglian region will be supplied by the already established Anglian Region Telemetry System. Radar-based rainfall and rainfall-runoff forecasts will also be used. The system will be 'open' to a range of model types from different suppliers including existing hydrological and hydraulic models and DHI's MIKE 11 river modelling system, which is widely used in the UK. The forecast system is designed to deliver timely estimates of flows and levels for up to 900 forecast points on rivers throughout the Anglian Region.

Forecasts will be made available to Agency staff across the Region using a web browser interface (Fig. 5) and a set of purpose-designed data visualisation tools. Industry standard hardware has been selected to ensure independence.

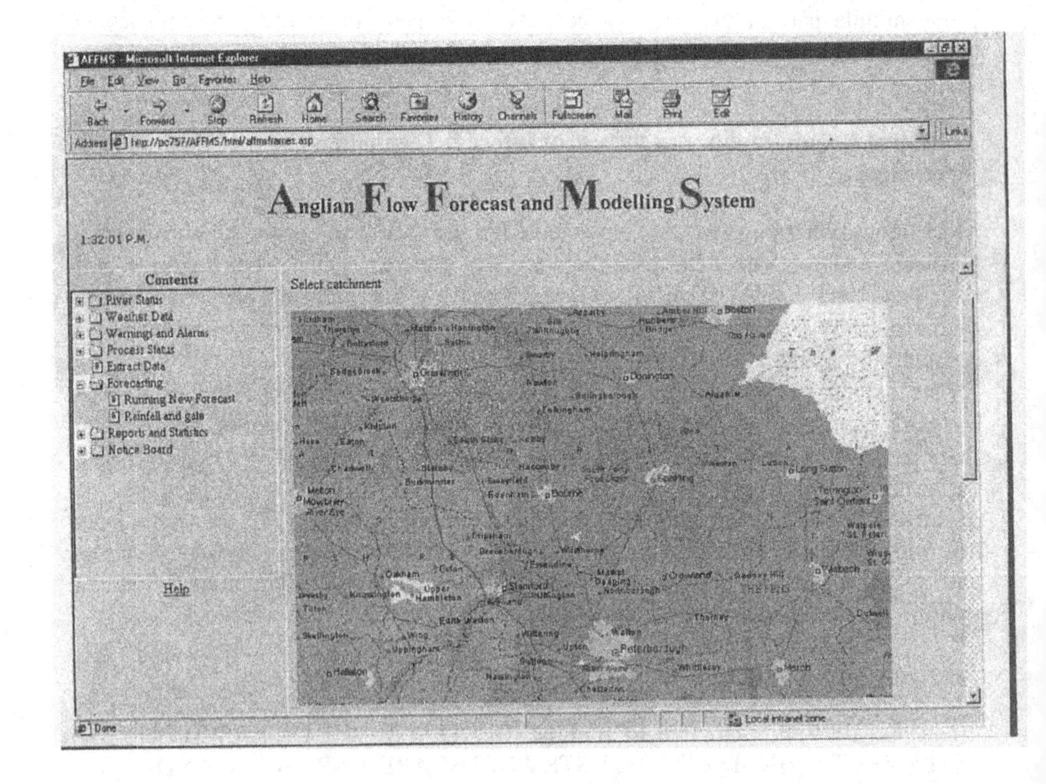

Figure 5. Web browser interface from forecast system in East Anglia (UK).

RECENT TRENDS IN HYDROLOGICAL FLOOD WARNING SYSTEMS DEVELOPMENT: THE CZECH EXPERIENCE

JIRI ZEZULAK

Czech Agricultural University, Dept. of Construction
165 21 Praha 6, Czech Republic

1. Introduction

Hydrological models based on the distributed or semi-distributed approach work equally well as tools of operational hydrology in real-time mode or provide for decision support in simulation analysis based on historical data. It might appear that both sorts of applications are identical. There is a remarkable difference in their codes, however, particularly in the way they communicate with the man-in-charge. Nemec in his pioneering work [1] identified hydrological forecasting as "...the prior estimate of future states of hydrological phenomena in real time, not to be confused with hydrological design-data computation (prediction) ...". The substance of the technical activities connected with each is basically different. Both these activities require historical data (time series) and may use the same or similar methods of hydrological analysis and modelling. However, hydrological forecasting comprises additional technical activities connected with other hydrological and non-hydrological subjects, such as network design, data processing, remote-sensing techniques, tele-communications, operational use of computers, etc. In view of this, the subject of hydrological forecasting should not be viewed as one particular hydrological technique, but as an economic activity using many technological developments, both hydrological and non-hydrological. The objective of the paper is to indicate developmental problems.

2. Real-time Aspects of Hydrological Modelling

It is inevitable that such a system be designed in a modular form, open for further development, and that it provides a wide selection of modelling techniques, and of principle hydrological, hydraulic and environmental processes such as

- snow melt and ablation
- rainfall-runoff relation
- flood routing in a river system
- reservoir operation and control
- biochemical and other processes related to the above.

From hydrological aspects and taking into account the water cycle genesis, these processes require decomposition of the investigated hydrological system (a *case*

J. Marsalek et al. (eds.), Flood Issues in Contemporary Water Management, 219–227.

system) into a number of smaller elements, (the sub-systems), much easier handled by the tools of mathematical simulation. In this section, we discuss topological aspects of hydrological system decomposition and algorithmic structures related to *'event oriented'* and *'case oriented'* mathematical formulations.

An establishment of the analogy between data structure and program architecture of the code is one of the prerequisites of the properly structured program system. It is generally required that the program follows the case system environment via its data structures. We can go even further on in hydrological modelling and design such that the form of the data structures conform to the structure of investigated *case system*. Thus, the existing analogy in structures

HYDROLOGIC SYSTEM ⇒DATA ⇒COMPUTATIONAL SYSTEM

certifies a well designed algorithm, conforming with the norms of recent programming technologies. Such analogy usually appears in two aspects:

- in spatial (topographical) decomposition of the *case system* based on principles of topology structures (e.g. graph theory), and
- in the context of time-dependent computation-control statements, computerised image of the genesis of runoff, the *'event oriented'* aspect.

2.1. BASIN DECOMPOSITION AND RELATED ALGORITHMS

The need for decomposition of a complex hydrological system follows e.g. from the scheme of the reservoir as a part of the water resources system of three co-operating, power producing reservoirs on the river Drin in Albania (Figure 1). Several categories of hydrological processes are included in the model. The computation-control code uses the strategy of an *event-driven* algorithm based on Petri-net computation, [2]. Several hydrological processes are mathematically formulated, as shown in the figure.

While leaving out (not necessarily) the domain of hydrological processes, we limit our concern to the topology of the reservoir. As indicated in the upper part of Figure 2, the reservoir model consists of four basic subsystems, of the water balance equation AA-RES, associated with two flow-control units (CC-MB3 and DD-MB3, the spillway and water release controls) completed by the power house unit, EE-ELN. One will obviously recognize complexity of the topological structure of such a system, far more accentuated when real-time data inputs and thus changes in the algorithm of the data model come into consideration. This is also why interactive data processing procedures are inevitable in such systems.

2.2. TIME-RELATED COMPUTATION CONTROL, SIMULATION VERSUS FORECAST

While the mathematical *simulation* usually aims at some sort of 'reconstruction' of past events or at event-design scenario, in the real-time *forecasting* mode the model is used to cast the future event which will be tested by reality in a few hours or days, possibly and thus a matter of possible criticism.

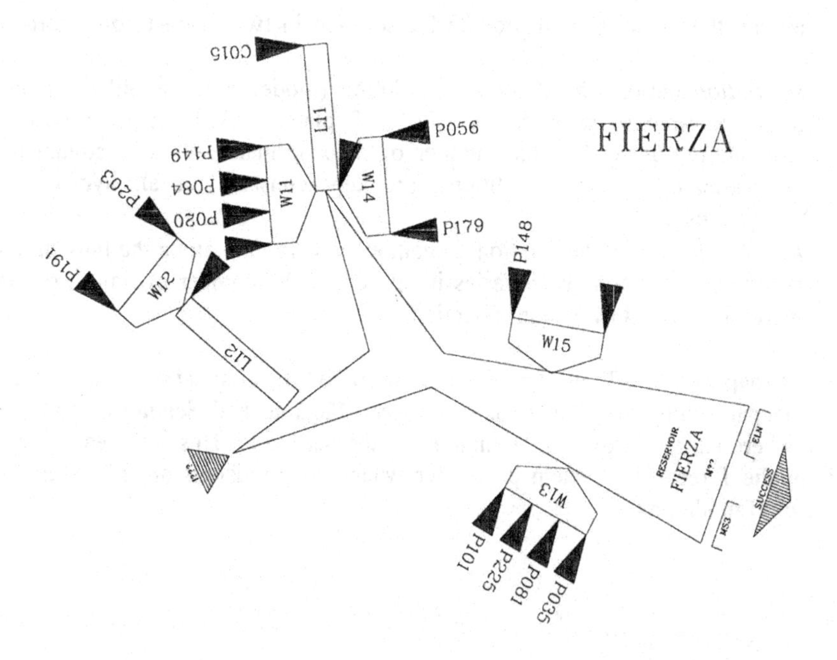

Figure 1. Definition sketch of the reservoir and adjacent hydrological sub systems.

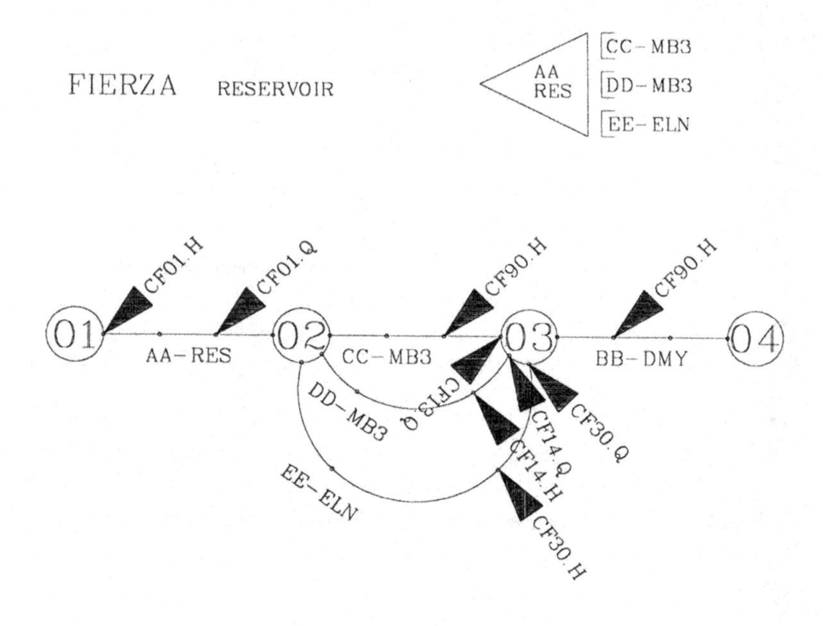

Figure 2. Three control units reservoir: definition sketch.

This is why the model should provide for solution in two distinct computational options:

- *simulation* option: the domain of solution (model state variables) is built above a *rectangular* region of the X-T plane, where the time base of computation as well as the number of time ordinates at the upstream and downstream boundary conditions, are kept constant for all hydrological subsystems; and
- *forecasting* option: the solution domain extends in time along the longitudinal co-ordinate, the time base varies in space and the number of time ordinates increases in the downstream direction.

The 2D snap of the X-T plane in Figure 3 shows a typical situation of flood wave propagation path along six river reaches between Sagaing and Seikhta of Irrawady (Burma), where lateral inflows to the main river are negligible. This is the easiest case indeed for the forecasting system developer where no provisions are necessary for surface-runoff modelling.

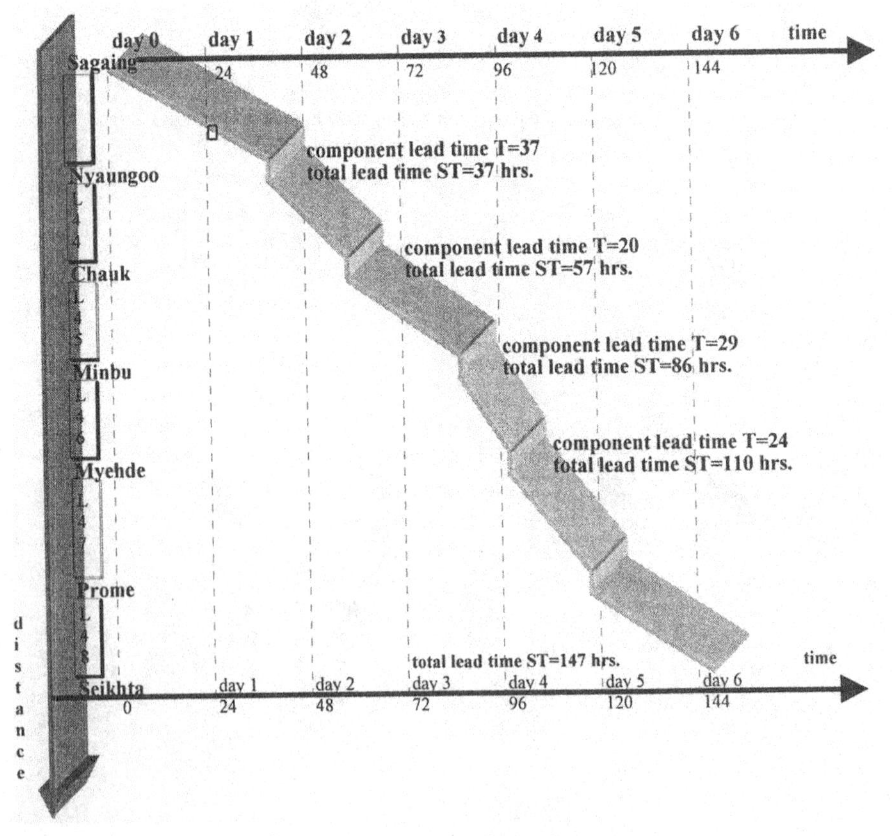

Figure 3. X-T plane in the forecasting mode.

The 3D solution surfaces over X - T plane along three model components are shown in figures 4 and 5. Figure 4 shows the data transfer (for *hot* start) between two subsequent model operations in *simulation* mode. This is a typical situation in simulation analysis where all boundary conditions are known to the full extent of the time-base.

In *forecasting* mode, Figure 5, concern is given to computation control mainly with respect to the following aspects:

- extent of lateral inflows, outflows, real-time state variables observations and flow control,
- allocation of the key-time (actual-time) on the time axis, thus subdividing the X-T plane into sectors of *past* and *future*,
- possibility of extrapolation of the upper inflows and lateral inflows, and
- determination of next model *hot* start call and storage of relevant initial conditions for the next run.

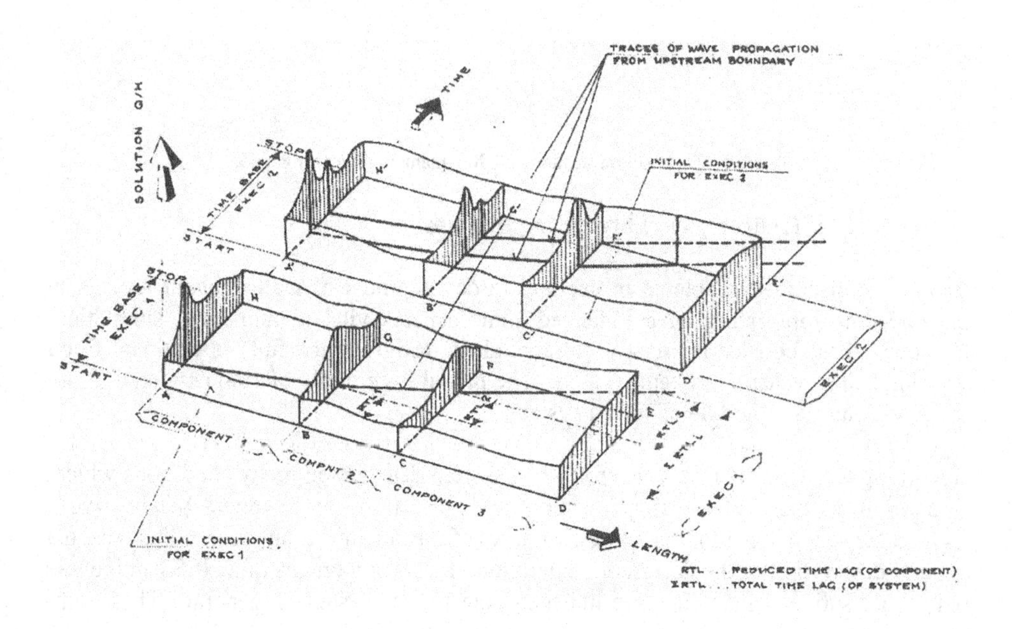

Figure 4. The solution surface over X-T plane in simulation mode.

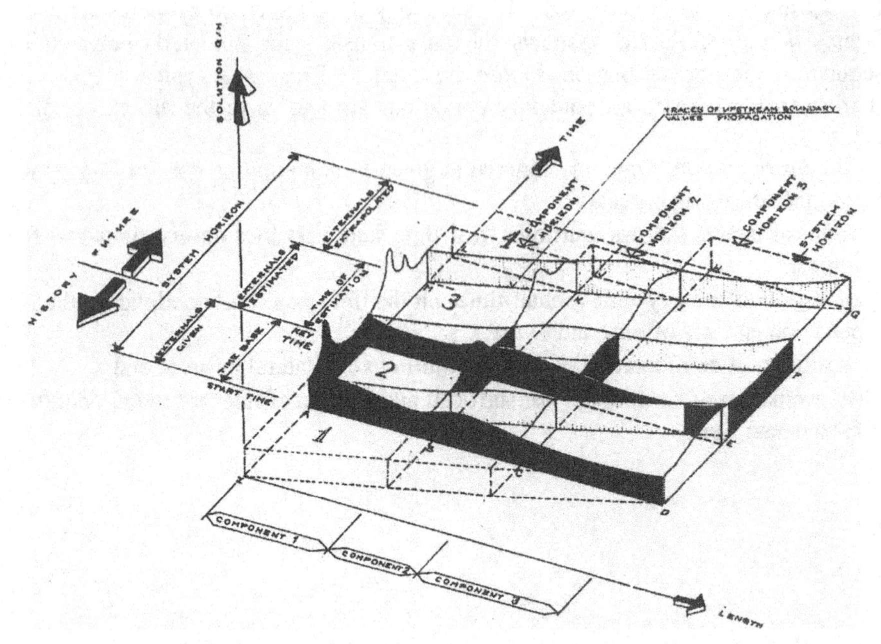

Figure 5. The solution surface over X-T plane in forecasting mode.

3. Modules of Hydrological Forecasting Systems

The above-discussed problems in use and in development of mathematical models in the real-time forecasting have indicated some aspects vital to numerical simulation. However, these considerations are not enough in designing the fully operational flood warning and forecasting systems (HFS). The model is certainly of vital concern but not the most important module of such a system.

According to Nemec [1], any concrete HFS system depends on many conditions which, in addition to those of a technical nature and the natural environment (basin, river), include the social and administrative structures of the specific country. For this reason a general description of the design and operation of the components of a HFS can include only those sub-systems which are indispensable for the system in general. These sub-systems are:

- historical and real time data collection,
- data transmission,
- data base management,
- forecasting procedure (modelling) in development and in operational mode,
- forecast dissemination services, and
- forecast evaluation and updating.

The different sub-systems and their interdependence specific for organisation of Czech Hydrometeorological Institute (CHMI) are illustrated in Figure 6. The selection of any particular sub-system is dependent not only on the conditions mentioned above but also on the other sub-systems, which in many cases may already exist.

The software components of HFS cover an even wider scope of data processing, as pointed out by Fread et al. [3]:

- *calibration* system, working on historical data to estimate basin parameters. This system does not essentially coincide with real-time models
- operational *forecast* system of three major components: data entry, pre-processor and forecast
- *extended stream-flow prediction* ESP, enabling to generate probabilistic forecast of longer lead time upon the deterministic
- *interactive* forecast program, enabling user intervention to all phases of data processing. Today, the graphical user interface in MS Windows or Unix environment is a must for all sophisticated HFS.

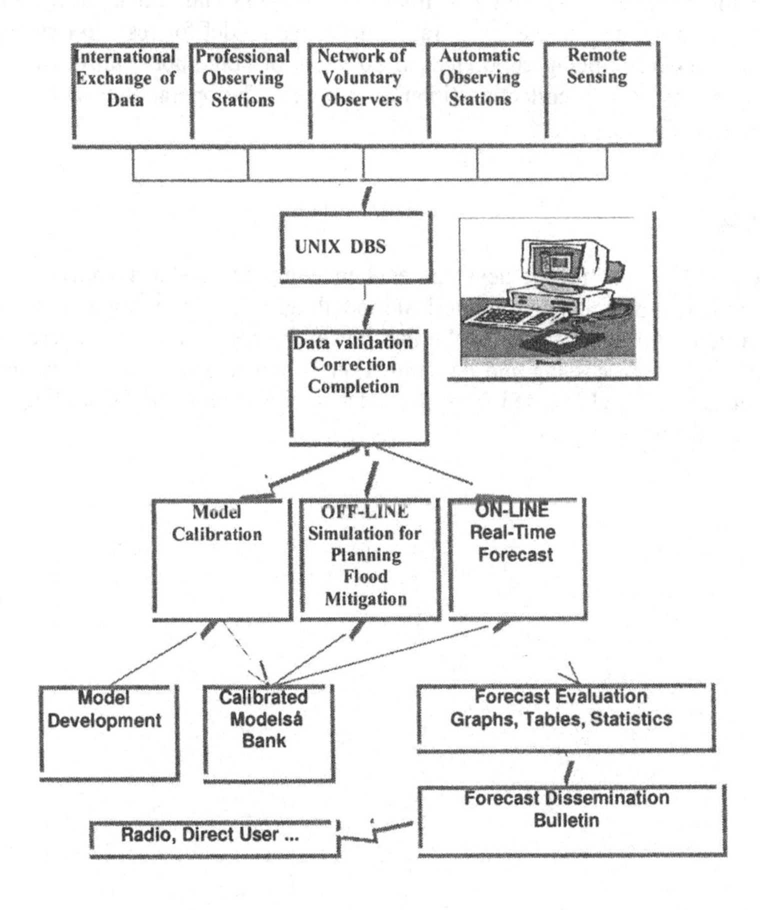

Figure 6. Modules of hydrological flood warning and forecasting system of CHMI.

4. Selected applications

At present, the CHMI operates the HFS for the lower Elbe basin, which consists of 20 river reaches and 13 headwater- and intermediate catchments, including snow melt and ablation models. The HFS originated from the modelling system, AquaLog [4], and the operational database, AquaBase. In view of the 1997 and 1998 floods, it has been decided to enhance the system particularly in flood prone regions of upper parts of the Elbe basin. It is planned to conform to the original system but, in the same time, to replace the original APIc modelling technique by the Sacramento model and to install a hydrodynamic model for selected river reaches of lowland rivers. The system is under development and some preliminary results of model calibration are already available.

Figure 7 shows an example of the forecasting model components in Orlice region. This version consists of 7 catchment components linked by 4 river reaches and one multipurpose reservoir. The Sacramento model and the snow melt and ablation models will be used for catchment modelling, a transport diffusion and full hydrodynamic model for river routing, and no-wedge storage interactive model for reservoir routing. The Sacramento model is being calibrated on 30 years of continuous daily data and selected flood events of data collection (interval 1 hour). The operational model will use an automated updating procedure.

5. Conclusions

Advanced technology and recent developments in computer systems enhanced the climate for automated data analysis, numerical modelling, post-processing and forecast dissemination. One should be aware of the fact however, that even under severe time limitations in forecast processing and dissemination, a human factor should be never underestimated and fully eliminated from procedures of hydrological forecasting and from subsequent decision making.

Figure 7. Topography of Orlice model.

6. References

1. Nemec, J. 1986. Hydrological Forecasting, Wat. Sci. and Technology, Dordrecht.
2. Zezulak, J., Krejci, J., 1998. The Real-time Modeling of Water Resources Systems using Petri-nets: Proceedings of Hydroinfomatics '98, Copenhagen.
3. Fread, D. L., Smith, G.F., Day, G.N. 1991. The Role of Real-time Interactive Processing in Support of Water Resources Forecasting in the Modernized Weather Service, 7th Intl. Conference on Interactive Information and Processing Systems, New Orleans, La.
4. Zezulak, J., Krejci, J., Smolik, M. 1991. AquaLog, the Decision Support System in Field of Water Management, Water Pollution and Hydrology. Como, Italy: Proceedings of 3rd Intl. Software Exhibition for Environmental Sciences and Engineering.
5. Zezulak, J. 1992. Information System for Automated Data Collection and Processing for Energy Power Development on River Drin. Geneva: WMO/UNDP Project ALB/86/003.

OPERATIVE CONTROL AND PREDICTION OF FLOODS IN THE RIVER ODRA BASIN

M. STARÝ[1] and B. TUREČEK[2]

[1]*Hysoft Brno*
Rolnická 9,Brno,CR
[2]*Povodí Odry*
Varenská 49,Ostrava,CR

1. Introduction

The company Povodí Odry (The River Odra Basin) is in charge of the river network system and the control of water runoff from the basin of the area of Northern Moravia, the total area of which is approximately 6252 km^2. This part of Moravia is situated in the upper part of the Odra River basin. This is the area where decisive flood passages originate. The management of the Povodí Odry company is aware of potential risks of the occurrence of catastrophic floods and their consequences. In its effort to forecast the runoff from the basin from rainfalls, and in the effort to utilise the transformation reserves contained in the system of reservoirs, the management began as early as 1991 to extend the water-management dispatching by the possibility of the operative control.

2. Basic Approaches

Hereinafter are the basic approaches which represent the foundation of the operative control system.

A. The control system is located within the framework of a certain basin. Together with this basin forms a macro-system of the area of hundreds to thousands of square kilometres. It is necessary to find an acceptable extent of simplifying the solution of a given problem.

B. The system under consideration cannot be controlled continuously for many reasons (to adjust control valves continuously). It is necessary to apply the "on-line" method for the control, and to utilise the computer facilities in a form of the operator's adviser, and to leave final decisions to the human factor.

C. The basin has its own dynamic (transfer) properties. These may be changed only by changing the positions of control valves (spillways, outlets). The problem of the optimum control may be formulated at every decision point as a repeated finding of such positions of control valves (finding such dynamic or transfer properties of a

J. Marsalek et al. (eds.), Flood Issues in Contemporary Water Management, 229–236.

system) at which the subsequent flood flow through the system will be transformed in a required way.

D. It is necessary to select such methods that make it possible to simulate the behaviour of the system continuously in time steps, and at the same time, to offer the snap of the actual state of the system (discrete point simulation of the continuous process).

E. Considering the large area of the basin, the spatial and temporal distribution of precipitation cannot be neglected. A sufficient number of precipitation gauging stations with remotely-transmitted data (i.e. on-line transmission of the data from gauging station to the centre) should be available in the basin that would be capable of securing the distribution of the precipitation on the surface.

F. Automated information system of measuring and also on-line transmission of data should be available in the watershed which are destined for measuring the intensity of precipitation in precipitation gauging stations, the positions of water levels in reservoirs and for measuring water stages in all selected profiles of streams in the river system.

G. The temporal variation of the historical precipitation data and the prediction of future precipitation must be available for the operative control in all precipitation gauging stations. The prediction must not be shorter than the lag time of water flow through the system, the longer, the better.

H. The subsurface runoff from the basin cannot be excluded from the simulation. It has been proved that this runoff is a constituent of the precipitation - runoff process that must not be neglected, and which is markedly manifested especially in cases where several floods immediately follow one another. This runoff considerably affects the rate of the decrease of receding limbs of hydrographs, and in the floods to follow, also the intensity of the maximum peak discharges. It is necessary to maintain the balance between the volume of rainfall on the basin, the groundwater storage, the volume of water stored in reservoirs and river beds and the volume of water run-off through the closing profile.

3. Control Algorithm

Equipping a basin with measuring components with a remote-data transmission is the first step of the solution. For the needs of the operative control, the measured data must be verified. This reduces the risk of errors caused by failures of measuring apparatuses or by data transmission.

3.1. SIMULATION OF THE SYSTEM BEHAVIOUR

The schematization of the location considered must be carried out in order to satisfy the needs of the simulation of the system behaviour and the calculations referring to the operative control. In this case, the schematization is based upon the division of the basin into river elements, surface elements suspended on these river elements, and reservoirs. The flow of water through the system is a combination of the hydraulic and hydrological methods. The surface runoff and the stream runoff are solved with respect to the necessary fast simulation by the kinematic approximation; the subsurface flow from the suspended areas by a conceptual regression model, the basic equation of reservoirs by the Runge-Kutta method. The relation of the precipitation totals on the basin in the preceding week to the initial infiltration rate is solved by a regression model.

3.2. PRINCIPLE OF CONTROL

The process of the operative control is a sequence of decision - time points mutually shifted by a time step, usually of a constant duration (see Figure 1). It is necessary to know the state of the system at each time point. A part of the information is acquired from measurements (the course of the past and present precipitation and the current filling of reservoirs). The remaining values (the surface runoff on the surface of the location under consideration, the stream runoff in the river network and the groundwater level in the underground reservoir) must be estimated by calculation. For this purpose, the calculation in time comes back to the moment when a steady state of the system is assumed (in the immediate foregoing period with no precipitation, and no manipulation of the outflow from the reservoirs).

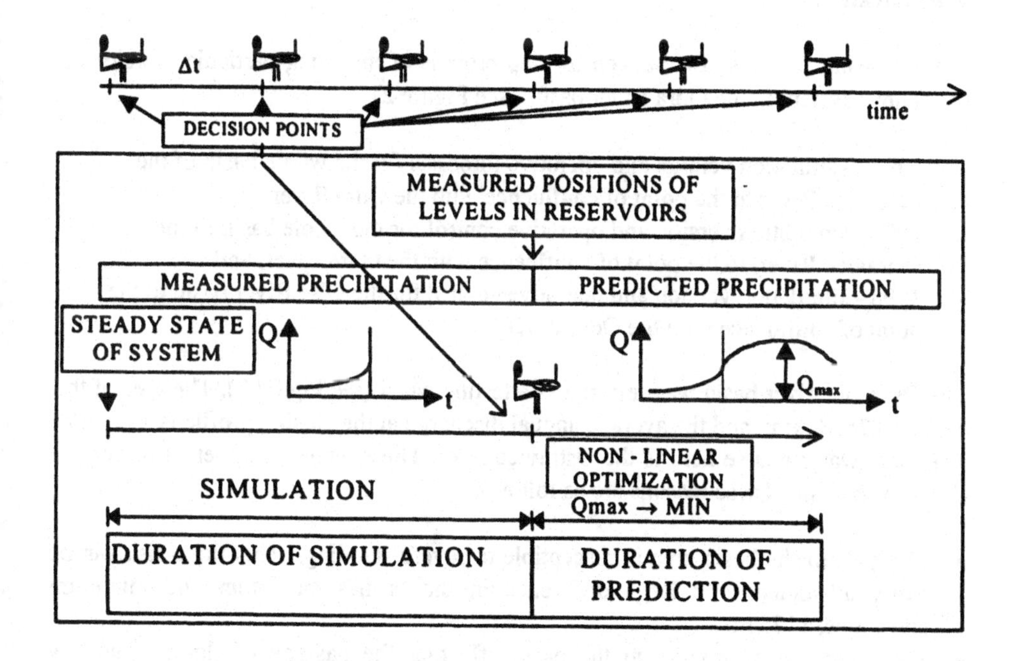

Figure 1. The control algorithm.

The calculation of the steady state flow in the river network has been carried out here. The corresponding position of water level in the underground reservoir has also been found for this state. Subsequently, a simulation of the behaviour of the system up to the assumed decision time point is started. At this point the correction of the system state is carried out. The water level in the reservoirs and the stages in the selected profiles are corrected.

At this moment, taking into account the precipitation forecast, the non-linear optimization starts. This is solved in a form of a simulation model with the optimised selection of parameters. The parameters are represented by unknown positions of control valves. The sum of squares of deviations between the simulated maximum peak discharges and the required values (these values may be zero) in the control profiles in the river network is minimized. This sum is the criterion of optimization. The computed positions of control valves are adjusted on the real system, and their positions are not changed till the next calculation. The process mentioned above is based on the principle step-by-step adaptation of the position of valves depending on the changing state of the system.

The mentioned procedures were programmed in a HYDROG program which enable either to simulate the behaviour of the system, or to predict discharges in an arbitrary profile, or to optimise the control of the flood passage as cited in [1].

4. Application

The following points show the step-by-step process of installing particular versions of the control system in the Odra River basin (see Figure 2):

- 1994 - simulation version and operative forecasts for the whole basin of the Ostravice River to the point of confluence with the Odra River,
- 1996 - simulation version and operative control for the whole basin of the Ostravice River to the point of confluence with the Odra River, and
- 1998 - simulation version and operative control for the Olše River basin to the point of confluence with the Odra River.

The Ostravice river basin was selected as a testing basin (see Figure 3). The area of this basin is 826.78 km^2, and the average annual discharge at the closing profile is 15.5 m^3/s. The main water course here is the Ostravice river. There were a number of reasons to choose this basin. These reasons are as follows.

- This part of the basin area is susceptible to severe flooding. This flood risk can be partly attributed to a large difference in the basin's maximum and minimum elevations, 1125 m.
- There are five reservoirs in the basin affecting the passage of floods. The key hydraulic structures are Šance, Morávka and Žermanice.
- The basin was already partly equipped with precipitation gauges with remote - data transmission.

The only thing to do was to add some new gauging stations. At present, 17 gauging stations are available in this watershed. Each of them has the possibility of remote - data transmission to the control centre by means of radio signals. The operative control utilises the hourly precipitation totals. The measured water stages in reservoirs and in selected profiles are transmitted in a similar way. The prediction of future precipitation is still a problem. For this purpose, under conditions of the Ostravice River basin, a numerical model ALADIN from METEO FRANCE is being used since October 1997. This model gives the precipitation prediction for 24 to 48 hours ahead, with the time step of changes from 3 to 6 hours.

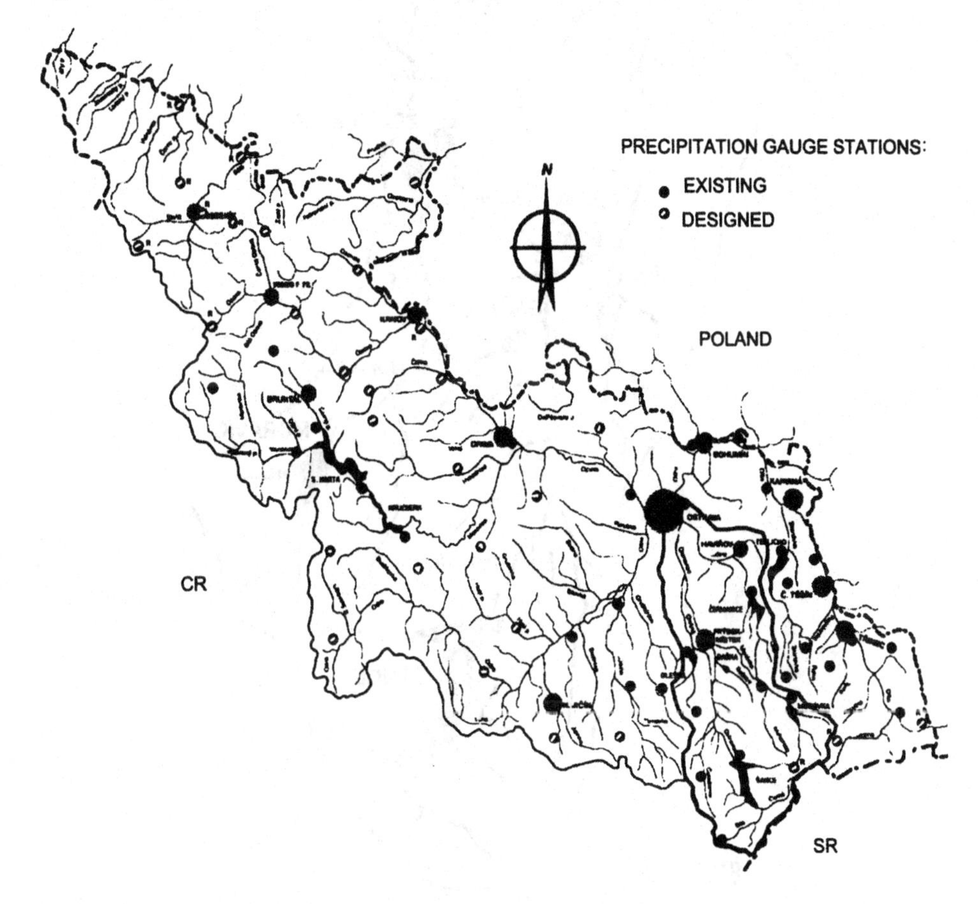

Figure 2. The Odra River basin and the Ostravice River subbasin.

The control system was used for the first time in July 1997 when the extreme regional flood took place as cited in [2]. Analyses of the situation were made when the floods were over. The attained effects of the transformation of floods by the system of reservoirs simulating the traditional control rules were compared with the results obtained by the operative control. The analyses resulted in a number of facts. Below are some of these facts which reflect the results obtained in two important profiles: Frýdek-Místek town and Ostrava town.

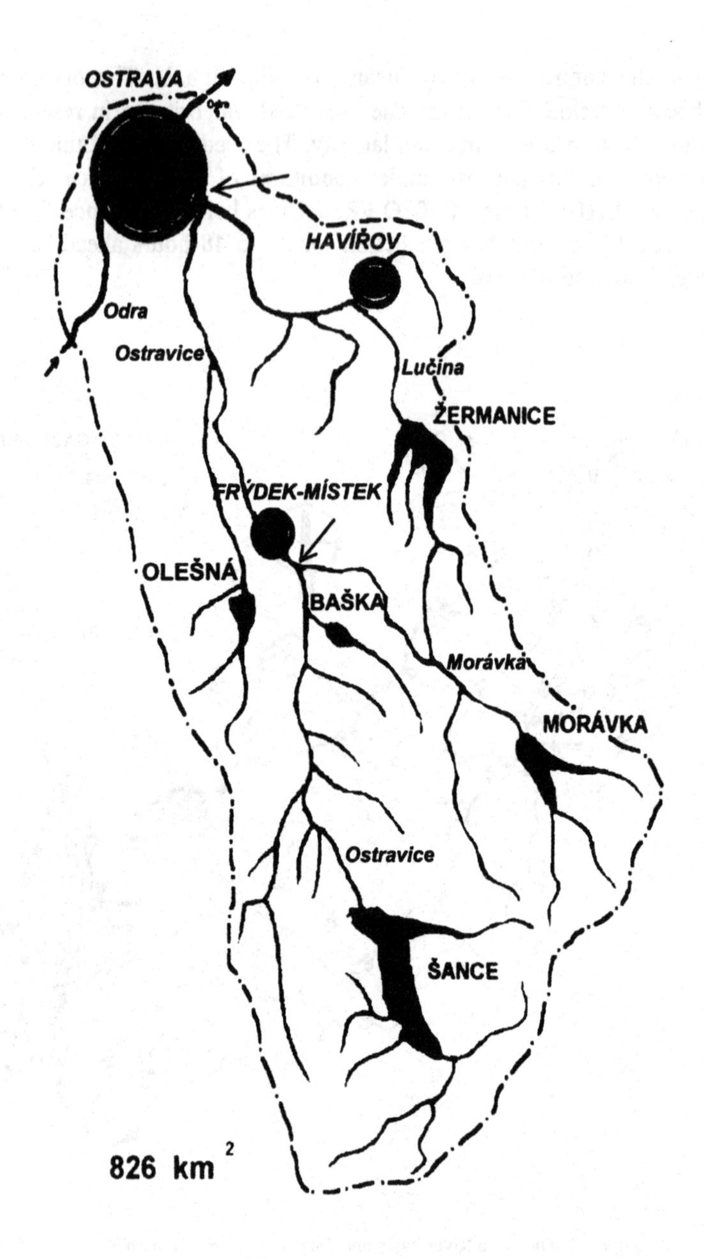

Figure 3. The Ostravice River Basin.

TABLE 1. A comparison of maximal discharges using the traditional control rules with the results obtained by the operative control

Type of discharge	Discharge (m^3)	
	Frýdek-Místek	Ostrava
Q_{max} (with operative control)	735	880
Q_{max} (without operative control)	800 to 850	950 to 980
Theoretical capacity of the river channel	750	950
Q_{max} (basin without reservoirs)	1000	1080
Q_{100} value (basin with reservoirs)	550	842

Potential consequences in Frýdek-Místek: continuous overflowing of embankments by jets on both sides at height of 30 to 50 cm, destroying of embankments, flooding Frýdek-Místek, and the advance of water in backwater areas towards Paskov and Vratimov could be expected.

Potential consequences in Ostrava: local overflowing of retaining walls of the systematic river regulation of the Odra River to city quarters of Moravská Ostrava and Slezská Ostrava (historical centre of Ostrava).

5. Discussion and Conclusions

From both above-mentioned examples, it is evident that the effect of the operative control was somewhere around 10 per cent from the point of view of the evaluated maxima. However, it is necessary to realise that both evaluated profiles are located at a relatively long distance from the reservoirs (23 and 43 km). The interbasin from which the runoff is not controlled is large, and highly affects the course of discharges.

A number of other analyses and tests revealed the following. Let us define the effect of the operative control in an arbitrary profile of the river system in the following way. Let this be the difference between the maximum peak discharge without the operative control of the WS and with the operative control. Then, based upon the resulting analysis of the problem, it is possible to state that this effect ranged between 10 to 30 per cent depending on the position of the profile in the river system. This effect was higher close to the reservoirs and it gradually decreased downstream together with the increasing area of the interbasin.

It has been confirmed that in the basins where reservoirs with the protective function were constructed, a considerable transformation effect may be attained. The operative control is one of other potential ways of increasing this effect. This way of control enables the use of whole transformation potential from the system. Some preconditions exist for this statement, whether it refers to a system of reservoirs or a single reservoir. Neither the present project documentation nor the present and the older standards and regulations took this solution into account. This is a reserve.

From the beginning, program HYDROG has been drafted to be applicable for an arbitrary basin. This is a consequence of using the non-linear programming methods for operative control. So, no use of rules or heuristic methods are applied for the control the use of which always link the program with a particular basin. The good results have stimulated us and we continue making improvements. This year we are engaged in supplementing the HYDROG program with the modulus of melting of the snow cover and by the modulus of simulation of the controlled inundations and simulation of the overflowing the river embankments.

The operative control is gradually applied in other subbasins of the Odra River. In 1999, it will be installed for Opava River basin. In 2000, we are planning to do the same for all subbasins of Odra River. The connection of the above-described control system in real time to the ALADIN model was the first one of this type in the Czech Republic, that is, the first connection of the numerical prediction meteorological model with a hydrological model focused on operative control.

6. References

1. Starý, M. (1996) HYDROG version 8.3. Software for the simulation and operative control of water-management systems during the passage of floods, Hysoft Brno.
2. Starý, M., Šeblová, H. and Tureček, B. (1998) *The operative control of the passage of floods*, Proceedings of the third international conference on hydroinformatics, Copenhagen-Denmark, 831-836.

HYDROINFORMATICS FOR RIVER FLOOD MANAGEMENT

ROLAND K. PRICE

Professor of Hydroinformatics
INE Delft, Westvest 7, PO Box 3015, 2601 DA Delft, The Netherlands

1. Introduction

Hydroinformatics is about making use of advanced information and communication technologies to handle information concerning the aquatic environment so that we can improve our understanding and management of that environment. Information here should be viewed as having the sense of the capacity to impart knowledge rather than the more narrow sense of classical information theory. The management of flooding in rivers, that is, what happens when the river flow exceeds the capacity of the designated channels, whether they have evolved naturally or been artificially enhanced, is one of the more important areas that can benefit from hydroinformatics. Indeed, because flooding can affect so many communities, sometimes with devastating, life-threatening consequences, it forms an important focus for hydroinformatics generally.

2. Information Management

Consider the information needed for river flood management. We can identify a number of different aspects. Firstly we have to *acquire* information about past and present flooding in the river. This involves not only data concerning the physical domain past and present, but also the human dimension: the historical, sociological, legal, economic and even political aspects. Much of this information is geographical and time dependent. Some of it is *archived* in databases and documents, while much resides in the heuristics and experience of human beings. One of the tasks of the organisation is to *analyse* the information it has available and to determine what new information needs to be collected, archived and made *accessible.* Invariably some form of *modelling* is required to make sense of the acquired data and to generate new data representative of the river system, particularly for the future. On the basis of an analysis of acquired and generated information the organisation generally provides its employees with systems for *decision support*, particularly to assist in times of emergency when human beings are under stress or pressure to get things done. Finally, there is the need to *communicate* decisions at all levels and to provide *feedback* from the information system to the real world, the river, its environs and the people associated with it (Fig. 1).

J. Marsalek et al. (eds.), Flood Issues in Contemporary Water Management, 237–250.

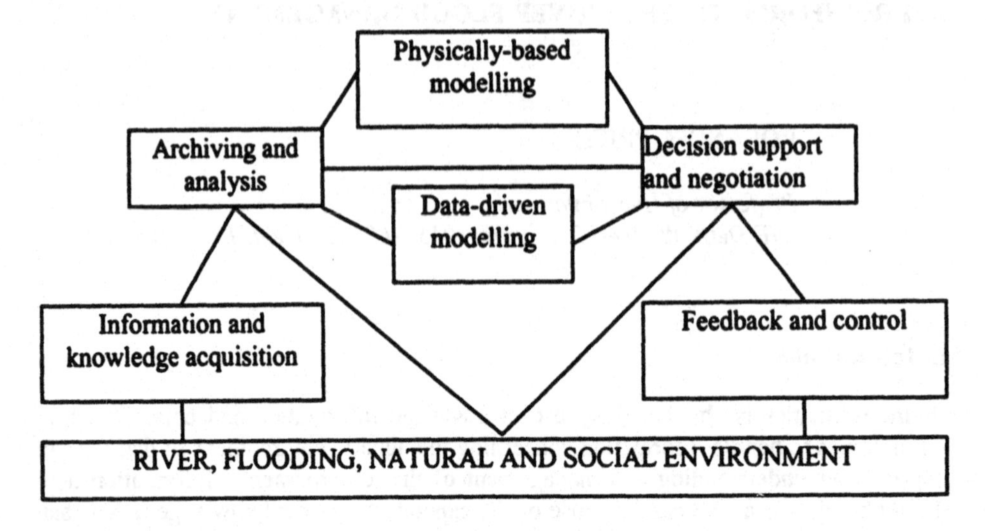

Figure 1. Information management

3. Information for the River

Consider now the river as a water-based asset. The flows in a typical river are largely stochastic in nature. We interfere with the river to get it to behave in ways that we perceive are more convenient for us and to satisfy our sense of responsibility for the environment. We protect agricultural land and urban areas from extreme events. We seek to ensure adequate depths for navigation. We try to minimise pollution and its effects. The river is viewed both as a generic object with properties that pertain to all or some groups of rivers, and as a particular object with its own unique characteristics. We bring to bear on a particular river our generic knowledge and understanding gained from elsewhere. So, for example, we have generic knowledge of the behaviour of flow in natural and artificial channels and over associated flood plains. Much of this knowledge is stored in databases and documents. Increasingly we are introducing different technologies such as weather radar, remote sensing, and other monitoring systems to gather data we think are important for management. But there will always be a limit to the data we can collect. For this reason we seek to complement the acquired data with modelling, such as digital elevation and land use models. In fact, we encapsulate much of our generic knowledge of the hydraulic performance of rivers within our simulation modelling software. The software is instantiated for a specific river through the inclusion of surveyed and monitored domain data for the river.

4. Modelling Software

The process of creating reliable and robust modelling software is a skilled task that few engineers have the knowledge and experience to accomplish in a safe and reliable way. Producing the software itself is one task; applying modelling software for a particular river is quite another. It is important to recognise therefore that other kinds of experience and skill are needed when developing a model. In the 1960s and early 1970s it was largely the software developers who also built their own models. Now, however, the software is widely available, giving other engineers the opportunity to apply modelling software for a wide variety of rivers, sometimes in circumstances not envisaged by the original software developers. This opportunity is being enhanced through the development of the sophisticated and better, user-friendly interfaces that are now available for modelling systems. For example, in the last few years there has been considerable progress made in linking modelling interfaces with databases, GIS and data visualisation tools. In addition, decision support systems are being introduced with encapsulated experience of modelling experts.

5. Problems with Modelling

If the 'traditional' technology is progressing quickly there is also growing awareness that there are nevertheless outstanding problems with modelling that have not been properly addressed. During a formal review of the Easter 1998 floods in the UK [1] it was found that the computer models of some of rivers grossly under-predicted the extreme water levels that occurred. The predictions of a computational model used to estimate flood levels through critical reach at Leamington Spa did not correspond to what actually occurred. This is partly due to bridge and weir structures of unusual form, some in close proximity, as well as complicated variations in channel and flood plain geometries and surfaces. Errors the order of 0.5 to 0.8 m were recorded. These were significant. The review called for national guidelines on computational hydraulic modelling to be prepared for use by the Environment Agency and its consultants. In turn, this raises the issue of validation of software. Further, the review concluded that computational modelling should be limited to situations where the theoretical analysis is valid and sufficient data are available for proper calibration and verification. In particular, it suggests that contract terms for the engagement of consultants on specialist computational modelling should be changed to bring them into line with those for consultants' services on other aspects of flood defence.

- Responsibility for adopting a scientifically sound approach should be placed on the consultant.
- Professional indemnity insurance should be required in an amount appropriate to the possible consequences resulting from the use of unsound science or incorrect analysis (£5-10M cover is suggested).

Certainly, such requirements will place a yet greater load on the modelling consultant to ensure that the right experience is available and in use.

Besides these criticisms there is an underlying awareness that uncertainty associated with model predictions is a growing issue. For example, modelling studies of

Northampton done in 1980s and early 1990s were used to design defences up to 100-year return period. But the hydraulic model was established using data from a flood with an estimated return period of 18 years. The conclusion is that the reliability of the model when predicting the 100-year flood is uncertain. Interestingly, the reviewers were drawing attention to issues that were simultaneously being raised elsewhere.

6. Validation of Software

The issue of validating modelling software is one that has vexed engineers for a long time, particularly in computational hydraulics. It is, in fact, impossible to validate every path through modern sophisticated software products such as MIKE11, TELEMAC, SOBEK and ISIS, to mention but a few. What was suggested by Dee [2] is that one of the best ways forward is to produce a paper-based validation in which the developers of the software attempt to substantiate the claims made for their model, possibly enhanced by case-studies of other independent experts using the software.

7. Dangers of Improper Calibration

In 1998, Cunge [3] issued a call to reconsider the problems and even dangers of improper calibration, that is, the evaluation of model parameters such that the model predictions closely match observations. His warning was that sophisticated modelling software in the hands of inexperienced or naïve engineers can be the cause of dangerously wrong conclusions on the basis of generated results. This warning has been issued by others before him, but only recently has there been a growing awareness that something now needs to be done about it.

The problem is that calibration parameters for a model can be forced beyond reliable limits or, put in another way, the model can assimilate the monitored data incorrectly. The model therefore cannot then be assumed reliable for events other than the calibration events, particularly for those events that are extreme for the river. This analysis has lead Cunge and others to address the problem of uncertainty in modelling. The argument continues that it is insufficient for physically-based (deterministic) models to provide results that do not carry with them a 'health warning' coupled with some measure of the damage to 'health'. Such misinterpretation seems to be a peculiarity largely of the models that claim to have some physical basis in their formulation rather than the statistically-based models for catchment runoff, say. The temptation is to regard the physically-based models as being capable of being 'forced' in terms of their calibration parameters, or being applied to events outside the calibration events. There is now a growing amount of research being done on the whole issue of modelling uncertainty, whether using standard sensitivity or Monte Carlo analyses or working directly with uncertainty in the original algorithmic equations.

8. Deducing and Explaining Short-comings in Models

Another, complementary line of research is to explore the deficiencies of physically-based models. Consider, for example, the Saint-Venant equations. The applicability of these equations to river modelling has been analysed in detail by many researchers. The general conclusion is that they can embody the principle physical processes involved. Even then however, it is questionable whether the application of the equations to irregular natural catchments and rivers has been properly explored. Their acceptance is largely based on a cumulative experience of engineers in different circumstances, sometimes with questionable results. Invariably however, researchers are aware that some processes are imperfectly, or even incompletely described. So, for example, the complex nature of the boundary texture in the channel or flood plain and the nature of the various shear flows generated at say, the channel flood plain boundary, make it difficult to accommodate the various length scales involved. Some researchers go to extreme lengths to try and incorporate the energy losses that are indicated, using formulae developed on the basis of particular experimental data for each category of energy loss. Usually these formulae are brought together in a form that is questionable in terms of their integrated effect. For these and other reasons attention is being given to see if it is possible to find other, less objective ways of understanding the deficiencies in an instantiated model.

The idea is that the difference between the predictions of a model calibrated using parameter values estimated by engineering experience and monitored data is not an error to be lamented but to be welcomed as the container of valuable information. The problem is to explore the structure of the residual difference using other tools. Such tools that seem appropriate for the task are the new technologies emerging from other disciples. These include artificial neural networks, fuzzy logic, and genetic programming. If, say, an artificial neural network (ANN) can be trained on the residual in terms of the original input to the model (rainfall or flows) then it may be possible to examine the weights of the ANN. Such an examination would explore the effects of inadequate representation of the input or domain data, and possible missing processes from the governing equations. This can be done using Monte Carlo or other similar techniques.

9. Modelling Application

One of the conclusions of the UK Easter Floods report [1] was that modelling should be rationalised and founded on a small number of state-of-the-art techniques, relevant to the range of basic catchment characteristics found across the regions of the UK. This step has the benefits of restricting the number of modelling software products to those approved in some sense by the responsible authority and ensuring that proper training programmes are in place. It also means that a body of experience can be accumulated to the benefit of new or inexperienced modellers. However, this strategy also runs the risk that it may exclude better products in the future, unless the authority keeps under regular review developments in the field.

10. Knowledge Management

We can now go further and explore the implications of knowledge rather than simply information management. Indeed much of what we have already been talking about concerns knowledge as well as information. Knowledge management is concerned with managing the life-cycle of knowledge relevant to areas that are mission-critical for an organisation. In our case we are concerned with a river basin management organisation or a flood warning enterprise. The life-cycle includes the process of acquiring, generating, archiving, analysing, applying, disseminating, assimilating and discarding knowledge. Through such a view we are now regarding knowledge as a kind of product. This is associated with a transition from a modern to a post-modern culture within society. There is a corresponding change from *knowing* to *consuming knowledge;* for more information see Abbott [4]. The following arguments are derived for engineering researchers for whom knowledge management is a specific issue. However, their concerns are reflected in the needs of groups brought together to manage flooding in specific rivers.

It is increasingly being realised that the key to the future of developed societies is the way in which they handle knowledge. This is changing rapidly. Gibbons [5] has presented arguments in support of this observation. He suggests that in the past, academic institutes were key to the generation of new knowledge of a generic nature. They defined and conducted their own fundamental research with funding provided by central government and with minimal accountability. Now however, with the expansion of higher education since the Second World War there are large numbers of people trained in universities who are doing research in a variety of different non-academic organisations. Some of these are government laboratories; others are organised commercially to compete in the international market place. These organisations are concerned primarily with knowledge production directly in the context of applications rather than with knowledge for its own sake. For them, knowledge is a commodity that can be bought and sold; that is, the market forces of supply and demand have an important role in the knowledge production process.

In the past, engineering researchers could also choose the topics of their research without outside reference. Now however, in line with the needs of the market, research, and especially engineering research, is required to be socially accountable. Whether funded directly by government or by industry, results are communicated increasingly to stakeholders and participants during the course of the knowledge production rather than at the conclusion of the project such that they can effect some interaction with the knowledge producers. In addition the nature of the problems addressed is changing with the move away from traditional disciplines, such as physics and civil engineering, to more hybrid topics that reflect the inter-dependence of one aspect of a problem on another. For example, consideration of the social and environmental impact of most civil engineering construction projects, including river flood management, is now mandatory. The integrated nature of these problems means that multi-disciplinary teams of people are being brought together from different locations and organisations to find integrated solutions. Flexibility and response time are crucial factors. Quality control on research and development is imposed by competition in the market, cost effectiveness and social acceptance. These are all important pre-requisites for effective river flood management.

Although there continues to be a significant need for original basic research, that is, knowledge generation in basic topics such as flood generation and hydraulics in river basins, it is known that over 90% of the knowledge produced globally is not produced where its use is required. Consequently, besides the continuing need for the generation of new knowledge on, say, how to handle flood events, the primary task is to acquire and *reconfigure* existing generic and case history knowledge that is produced locally and elsewhere. In addition, the demand will be more for problem identification and solution in terms of specific applications rather than the development of a general underlying theory of some fundamental aspect. All of this requires an efficient means of gathering intelligence.

Because future knowledge production, whether off-line for flood defence or on-line in a flood emergency, will depend more on multi-disciplinary teams rather than specialised small groups with a common expertise, the efficiency of these teams will depend increasingly on the relationships between the members. In turn, such relationships will require excellent communication facilities and, in particular, communication networks. In effect the teams will work as virtual organisations. They will have access to comprehensive documentary information acquired globally and built up from local experience to minimise their replication of knowledge produced elsewhere and to take advantage of the research results and knowledge production of others, and, in the case of emergency management, of previous situations.

In view of the extreme complexity of flood management more emphasis will be put on acquiring good quality and reliable data both off-line and in real time to effectively monitor the integrated systems. Data mining techniques will be available to extract new knowledge on these systems from the resulting large databases with better opportunities for real-time forecasting and control.

Computational modelling will continue to be necessary to comprehend the complex interaction of various processes within the flooding situation. The modelling will be applied not only to the physical, chemical and biological processes affecting or affected by the environment of the river basin but also to those management and business processes that reflect human society's interaction with that environment. It is anticipated that access over a communication network to generally, even publicly, accepted *instantiated* models will be vital for effective research and application by knowledge workers in different centres responsible for flooding in a given river.

Knowledge generation is known to occur better within the context of a social environment that is conducive of *innovation.* For example, most engineers find informal gatherings in workshops and conferences an important source of new ideas and initiatives. Consequently the ability to create a similar environment through a communication network that provides opportunities to share and exchange ideas (chat lines, computer and video conferencing, white boarding, etc.) will greatly enhance the innovation of the engineering teams. This can be especially important in an emergency event when the team is under pressure.

Knowledge produced by these teams will need to be retained and archived in an accessible form for a wide range of people, particularly as the engineers move and as they change their skills to meet the differing demands of their successive jobs.

These forms of knowledge management will enhance the knowledge production process of the individual engineer and the associated teams. However, because the work done will be socially accountable it is important that the knowledge is disseminated effectively. All of the facilities described above will become significant in the successful dissemination of the acquired knowledge to those responsible for implementing that knowledge at government, municipality, consultant, contractor and operator levels, as well as by pressure groups and the general public, to improve the management of the river basin. The responsibility to provide advice will involve access to

- acquired and maintained document archive as a navigable repository of the river basin organisation's knowledge and experience [6]
- relevant databases and on-line monitoring systems
- instantiated computational models of the physical, management and business processes and associated modelling tools
- problem identification and solution service through expert advice
- net-based learning as a form of continuous learning and as a means of assimilating knowledge within the work culture or the individual's world view (cf the Anglian Water University). The Easter 1998 Floods review [1] reported that there is a scarcity of senior staff (in the UK Environment Agency) with advanced academic training and qualifications in hydrology, open channel hydraulics and computational hydrologic and hydraulic modelling

For further elaboration on knowledge management in civil engineering see Delft Cluster [7].

The elements of this form of knowledge management system are dependent on effective communications that are increasingly Internet-based and make extensive use of the recent technologies such as, Java, Active X, Flash, etc. There are a number of projects that are looking at extensive use of on-line access to modelling systems, databases and monitoring/sensing systems. One such project is ELTRAMOS. This is focussed on giving remote users access to sophisticated modelling systems such as MIKE11 and TELEMAC for rivers. The idea is that access to such software is supported by decision support systems that identify the problem, select the software and support instantiation.

These ideas are being explored also in another telematics project, TELEFLEUR, that is developing management of flood emergencies in urban areas where there is very short lead-time between rainfall event and the flooding. This is the case for cities on the north Mediterranean coast, such as Genova and Athens. The solution to the problem is perceived as forecasting rainfall over these areas using meteorological models supported by human interpretation of the model results and available monitored data (with radar in the future). Hydrological models then use the predicted rainfall to estimate the timing, severity and duration of flooding. The issuance of warnings is supported by a decision support system, as is the management of the emergency and the handling of restoration operations.

The communication of information from models and databases is increasingly done using images rather than text. GIS-type graphics displays are therefore seen as an important means of communicating information on such varied aspects of river flood management as topography, land use, rainfall intensity distribution, stream flow, flood zone mapping, flood extent mapping, crisis point identification, damage assessment, emergency relief routes, and so on; see Gourbesville [8]. Indeed, Parker in the Easter

1998 Floods review [1] stressed the need for high quality, high resolution flood plain mapping.

11. Risk

Another important area that is receiving considerable attention is assessing the risk of flooding. Such assessments take into account the whole range of contributions to the risk of failure of the flood defence system to provide its service of protecting a region against flooding. This situation arises in different situations around the world, ranging from low lying land such as in the Netherlands and Bangladesh, to many large cities, particularly in developing countries, such as Bangkok, Shanghai, Manila, and Dhaka. At the heart of the systems developed in the Netherlands are sophisticated simulation models; for example, see Stelling et al. [9]. For a review of research into hydrological risks in Europe, see European Commission [10].

12. Modelling using Data-driven Techniques

It has already been indicated above that physically-based models form the best means of trying to estimate what happens when a flood event greater than previously experienced occurs. This is because such a model has built within it some knowledge of the physics, such as the dimensions of the flood plains, even if values for the parameters: runoff coefficients, roughness values and discrete headlosses, under such extreme conditions have not been 'calibrated'.

When there is a need for very rapid predictions of water levels, say, in a forecasting situation, or when long time series are being run, or a Monte Carlo analysis is required, then the physically-based model is cumbersome, even with powerful computers. For these, and other reasons, researchers are turning to other technologies for solutions. There are now a number of situations where artificial neural networks (ANNs) are being used to predict and forecast extreme events in rivers. To be successful the ANNs have to be trained on reliable data. Often this is not available directly from monitoring. In this case the physically-based model becomes the generator of a database of events for the ANN; see Gautam [11], Stary [12] and Khindker [13]. Where some good monitored data for rainfall and runoff is available, then ANNs can be used to fill in missing rainfall data as well as to extend catchment runoff predictions.

Besides ANNs there is also the possibility of using a fuzzy logic generator to do the same thing as an ANN. Because a fuzzy rule-based model uses distributed representation of knowledge in the form of membership functions it has good fault tolerance. This has benefits when using field data with low precision and accuracy. Imprecise linguistic information can be accommodated. Like neural networks fuzzy rule-based systems should not be used to extrapolate beyond the training data. There is also the 'curse of dimensionality' for fuzzy logic generators in that the relationship between the number of inputs and rules is exponential. For examples of this connectionist technology see See and Openshaw [14], Abebe [15] and Bazartseren [16].

More recently, others have been experimenting with chaos theory to understand the underlying structure of (chaotic) time series, such as a water level series, in order to make forecasts based only on the time series. In some cases, where there is only an upstream level gauge for reference the use of chaos theory to improve the forecasting can be important. For example, see Liu and Lin [17], Rahman [18] and Huang [19].

Genetic algorithms are increasingly used with these technologies to optimise the selection of parameters, while genetic programming has its value in helping to identify the underlying algorithmic structure.

13. Flood Forecasting

The UK Easter Floods Review Report [1] recommended that flood forecasting models should be used more widely. To facilitate the implementation of flood forecasting there should be adequate rain gauge coverage for key catchments. Simple things should be observed such as siting flow measuring and telemetry equipment above likely extreme flood levels. A more demanding recommendation in the Review is that 'at least one station in each sub-catchment associated with a significant risk area should be capable of measuring extreme flood discharges'. Again, the Review recommends that 'a standardised recording of key information, facilitating quick appraisal, should be introduced nationally'. Finally, the reviewers suggest that 'flood forecasting operating procedures should ensure that, in addition to using model predictions, close attention is paid to monitored flood levels'. This caution exposes a concern to limit the trust placed in models while stressing the need for better appreciation of the value of monitored data. This places a greater demand on improved telematics and archiving and analysis of the acquired data.

14. Warning Systems

The operational management of floods has different requirements depending on such factors as the response time, opportunities for control, existence of forecasting tools, monitoring systems, warning systems, and so on. The generation of a flood in the lower parts of the Rhine for example, can be tracked over a period of days. There are some options for control, sometimes with difficult choices to be made over whether to allow land to be flooded in order to reduce peak discharges downstream. Also, in the Rhine forecasting can be done with reasonable accuracy due to the large amount of historical data and the availability of reliable modelling systems. The situation in a city such as Genova, Italy, however is at the other end of the spectrum. Here it is virtually impossible to have an accurate forecast of rainfall and the response time of catchments flowing through the urban areas is the order of half to two hours. This means that the opportunities for providing reliable forecasts giving the population ample time to react are limited. The present system is very dependent on meteorological modelling based on the global atmospheric model run daily at Bracknell, UK, a European model run at Rome and a corresponding model for Northern Italy run at Genova. The smaller scale models depend on the larger models to provide boundary conditions. Predictions of rainfall over the catchments running through the city have to be amended by an expert rainfall forecaster. The predictions can then be run through a hydrological model to estimate location and severity of flooding. The warning system builds on these data to generate warnings both for the initiation of flooding and any restoration activities. Such a system also uses monitored data. In the near future it is hoped that weather radar will also be available to provide some input to now casting procedures to complement the existing forecasts. Through the TELEFLEUR project the system is being developed further and a decision support system installed in Athens and Genova to provide better management of flood emergencies in the cities.

Another similar but different system is available in Taiwan. There the problems are again different. The island is frequently in the path of typhoons that track across the area from east to west. There is a mountain range that runs north-south on the east side of the island. Intense rainfall in these mountains generates runoff that flows west through densely populated areas. The streams are extensively embanked some distance from the main channels down to the coast. The embankments can be very high in order to provide a high degree of protection to the population behind them. The storms are tracked by satellite and radar. The information is sent to a central control where emergency flood managers monitor the situation and issue warnings as appropriate. As in the Netherlands and other regions where the rivers are extensively embanked the risk for those living in the flood plains behind the embankments is generally greater than before the construction of the embankments. A dilemma for flood warning systems is that they have the tendency to even enhance this confidence further such that the overall risk may increase rather than decrease when such a scheme is implemented. Consequently, the Taiwan Provincial Government has set up, not only a flood warning service, but a complete hydrologic information centre. The aim is to provide information on

- Flood stages of major streams
- Major precipitation stations
- Important reservoirs
- Groundwater and land subsidence
- Flood prevention and emergency repairs
- Flood control command for the Tan-shui river
- Typhoons from the Weather Bureau

Careful attention has been given to providing clear, concise and unambiguous messages supported by recognisable graphics. Satellite images of cloud cover and storm movement are supported by maps showing rainfall and vivid graphics displaying flood levels behind levees. See the web site *http://www.tpgwrd.gov.tw*

15. Using Internet, and Fact and Judgement Engines for Decision Making within the Community

Flooding is a human community problem because it endangers life and property. It is not clear whether it is a problem to the environment in that floods are a natural phenomena anyway, and the environment left to its own devices, without interference from man, would find its own way of coping with or accommodating flood events.

What could be at issue here is whether environmental health in the state that it is managed by society is prejudiced by flooding. This raises other issues such as decision making and education.

If the society has a definite interest in flooding it is questionable how far down the chain the decision making process may go. Certainly, there is a more open attitude to flood zoning for warning and insurance purposes in countries like the USA than in many other countries. In some cities in Europe, for example, there would be severe political consequences if flood zoning were introduced. This means that there are limitations in most countries to the number of people involved in decision making. This however is changing through the introduction of Internet. More people can have ready access to information, and therefore to decision making.

Abbott and Jonoski [20] have researched the implementation of network distributed decision support tools as Multi Agent Systems (MASs). The idea is that fact engines (the simulation models for flooding in a river, say) are accessible remotely over the Internet in a standardised form for a range of people to use as the basis for their decision making. The decision making process, however, may involve a number of stakeholders. Therefore the authors suggest that another, 'judgement' engine should be supplied to enable those involved make their decision iteratively, taking into account the decisions of others and the emergence of new 'facts'. Such a system could be used, for example, to trade-off the benefits of reducing the frequency of flooding in an urban area, say, for the costs incurred through flooding farming land upstream. Such a system would include tools for negotiation. This could be even more important for problems involving flooding in trans-national rivers. For example, Switzerland, Germany, Belgium are all upstream of the Netherlands. There is a need to settle not only issues of river management at the planning level, but also at the operational level. It is claimed that this Internet-based type of decision making system could help in areas such as this.

One of the responses of the Environment Agency to the Easter Floods in 1998 has been to take more seriously the communication with the public about the possible risk of flooding. With more than 1.3 million homes in recognised flood plains the Agency has a large task in educating the inhabitants. Whereas a variety of means can be used for education purposes, a growing use will be made of Internet and WWW pages in order to put flood warning onto the national agenda. This highlights the fact, as is evident from other aspects, that flooding is very much a sociological as well as technical problem. It is people who live in the flood plains, who are affected by flooding, who demand development in the flood plain and who exploit it to their benefit. Consequently, a proper treatment of any river in terms of its flooding potential has to include an understanding of the people directly affected in terms of their economics and politics as well as their culture and use of the river.

16. Decision Support Systems

There are a number of decision support systems that are being developed recently, especially in EC projects. Such projects include AFORISM (SUPREME) focussed on the Reno river in Italy, and TELEFLEUR (see above) for flood warning in Athens and Genoa. The aim of these DSSs is to provide structured advice to assist decision makers, particularly when acting under pressure. Reasoning based on factual knowledge is used in these systems, depending on input from different centres and facilities, and augmented by a general knowledge base. Dependencies between tasks determine how the reasoning progresses. This provides a means for navigation through the knowledge and causal relationships. What such a DSS provides, given adequate knowledge on which to draw, is a system that reinforces best practice, and minimises the risk of unsafe decision making. For an extensive treatment of safe hydroinformatics systems see Ahmad and Price [21]. Ammentorp et al. [22] describe a GIS-based DDSS for flood warning. See also Catelli et al. [23].

17. Conclusions

Hydroinformatics, with its focus on handling the information associated with the aquatic environment, is well placed to make significant contributions to the management of flooding in rivers. The key areas include

- Acquisition of domain and performance data for a river liable to flooding
- Archiving and analysis of such data to yield additional knowledge and information on flooding, its causes and consequences
- Physically-based and data-driven modelling of rivers in flood for planning, design, prediction, forecasting and control
- Implementation of decision support systems for reliable management of flood defence schemes and flood emergencies
- Telematics for warning and communication during flood events
- Internet-based tools for negotiation and education on flood protection schemes and awareness
- Recognition that flooding in a river is as much a sociological as well as a technical issue (flooding is a socio-technical problem)

Research and development in all of these and other areas is continuing, and promise to open up new, more cross-technological ways of managing rivers in flood.

18. References

1. Bye, P. and Horner, M. (1998) Easter 1998 Floods. Environment Agency, UK.
2. Dee, D.P. (1993) A framework for the validation of generic computational models. Tech Rep X109, Delft Hydraulics, PO Box 177, 2600 MH Delft, The Netherlands.
3. Cunge, J.A. (1998) From hydraulics to hydroinformatics. Proc 3rd Int. Conf. On Hydroscience and Engineering, Cottbus, 1998.
4. Abbott, M.B. (1999) Editorial. In the first edition of the (new) Journal of Hydroinformatics, (to be published), IAWQ, London.
5. Gibbons, M. (1998) Higher Education Relevance in the 21st Century. Association of Commonwealth Universities.
6. Price, R.K., Ahmad, K. and Holz, P. (1998) Hydroinformatics concepts. Proc. NATO ASI on Hydroinformatics Tools for Planning, Design, Operation and Rehabilitation of Sewer Systems, Harrachov, 1996, Kluwer Academic Publishers.
7. Delft Cluster (1998) Investeringsvoorstel Delfts Cluster. Available from Grondmechanica Delft, PO Box 69, 2600 AB Delft, The Netherlands.
8. Gourbesville, P. (1998) MIKE11 GIS: Interest of GIS technology for conception of flood protection. Hydroinformatics '98, Copenhagen, Babovic & Larsen (eds), Balkema, Rotterdam.

9. Stelling, G.S., Kernkamp, H.W. J., Laguzzi, M.M. (1998) Delft Flooding System: a powerful tool for inundation assessment based upon a positive flow simulation. Hydroinformatics '98, Copenhagen, Babovic & Larsen (eds), Balkema, Rotterdam.
10. European Commission. (1998) Hydrological risks: Analysis of recent results from EC research and technological development actions, DG12, Environment and Climate programme.
11. Gautam, D.K., (1998) Regressive neural networks for the modelling of time series MSc thesis, IHE Delft.
12. Stary, M., Selova, H. and Turecek, B. (1998) The operative control of the passage of floods (on the Ostravice river). Hydroinformatics '98, Copenhagen, Babovic & Larsen (eds), Balkema, Rotterdam.
13. Khindker, M.UL-H, Wilson, G. and Klinting, A. (1998) Application of neural networks in real time flash flood forecasting. Hydroinformatics '98, Copenhagen, Babovic & Larsen (eds), Balkema, Rotterdam.
14. See, L. and Openshaw, S. (1998) Using soft computing techniques to enhance flood forecasting on the river Ouse. Hydroinformatics '98, Copenhagen, Babovic & Larsen (eds), Balkema, Rotterdam.
15. Abebe, A.J. (1998) Application of fuzzy logic for knowledge representation in hydroinformatics MSc thesis, IHE Delft.
16. Bazartseren, B. (1999) Use of artificial neural networks and fuzzy adaptive systems in controlling polder water levels in the Netherlands. MSc thesis, IHE Delft.
17. Liu, C.L. and Lin, S.C. (1998) Artificial neural network forecasting for chaotic dynamical process. Hydroinformatics '98, Copenhagen, Babovic & Larsen (eds), Balkema, Rotterdam.
18. Rahman, S.M.M. (1999) Analysis and prediction of chaotic time series. MSc thesis, IHE Delft, The Netherlands.
19. Huang Yan (1999) Application of data driven methods for flood forecasting. MSc thesis, IHE Delft, The Netherlands.
20. Abbott, M.B. and Jonoski, A. (1998) Promoting collaborative decision -making through electronic networking. Hydroinformatics '98, Copenhagen, Babovic & Larsen (eds), Balkema, Rotterdam.
21. Ahmad, K and Price, R.K.(1998) Safe hydroinformatics. Proc. NATO ASI on Hydroinformatics Tools for Planning, Design, Operation and Rehabilitation of Sewer Systems, Harrachov, 1996, Kluwer Academic Publishers.
22. Ammentorp, H.C., Jorgensen, G.H. and van Kalken, T. (1998) FLOOD WATCH - A GIS based decision support system. Hydroinformatics '98, Copenhagen, Babovic & Larsen (eds), Balkema, Rotterdam.
23. Catelli, C., Pani, G. and Todini, E. (1998) FLOODS: FLood Operational Decision Support System. Hydroinformatics '98, Copenhagen, Babovic & Larsen (eds), Balkema, Rotterdam.

FLOODSS
A FLOOD OPERATIONAL DECISION SUPPORT SYSTEM

EZIO TODINI

University of Bologna
Department of Earth and Geo-Environmental Sciences
Via Zamboni, 67
40127 Bologna, Italy

1. The Need for a Flood Planning and Management DSS

The EU is extremely sensitive to the flood problem: several research projects have already been funded on the subject and more specifically, after the report of an Expert Meeting, held in Padua in September 1997, clearly emerged the definition and the need for a holistic flood management.

"The mitigation of flood damage and loss does not only depend upon the actions during floods but is a combination of flood preparedness, operational flood management and post-flood reconstruction and review. Pre-flood activities include:

- flood risk management for all cause of flooding and disaster contingency planning,
- construction of physical flood defence infrastructure and implementation of forecasting and warning systems,
- land-use planning and management within the whole catchment,
- discouragement of inappropriate development within the flood plains, and
- public communication and education of flood risk and actions to take in a flood emergency.

Operational flood management can be considered as a sequence of four activities:

- detection of the likelihood of a flood forming (hydro-meteorology),
- forecasting of future river flow conditions for the hydro-meteorological observations,
- warning issued to the appropriate authorities and the public on the extent, severity and timing of the flood, and
- response by the public and the authorities.

Depending upon the severity of the event, the post-flood activities may include:

- relief for the immediate needs of those affected by the disaster,
- reconstruction of damaged buildings, infrastructure and flood defences,
- recovery and regeneration of the environment and the economic activities in the flooded area, and
- review of the flood management activities to improve the process and planning for future events in the area affected and more generally, elsewhere."

Given the large number of high risk situations as well as the extremely high costs in terms of casualties and damages involved in flooding [1,2], the implementation of all

J. Marsalek et al. (eds.), Flood Issues in Contemporary Water Management, 251–260.

the structural and non-structural measures aimed at its reduction, becomes essential in this holistic view of the problem.

While structural measures can be implemented with a sufficient degree of reliability in areas where the anthropic pressure is low, when dealing with heavily anthropised areas, and in particular in urban areas, the need for non-structural measures becomes extremely high. There are essentially four different non-structural measures to be realised for flood control [3]. The first one is the identification of the areas of high flood hazard and the development of flood risk maps. The second one is the setting up of reliable real-time flood forecasting and warning systems that allow for an operational forecasting lead time (not less than 6 hours possibly 12). The third one is the definition of flood risk emergency plans, with the preparation and the distribution to the Authorities in charge of a sort of unified Manual of Procedures describing, for each area at risk, the delineation of the flood threaten area, the risk thresholds, the Authorities in charge, the action to be taken, who to alert, etc. etc.. The last non-structural measure is the development of an integrated Decision Support System aimed on the one at centralising the information in an Operational Emergency Unit, and on the other hand at monitoring and co-ordinating the activity of all the groups involved on the basis of the Manual of Procedures, given that the main problem during flood events has been identified as the lack of co-ordination among the plethora of Authorities and organisations operating for flood damage alleviation.

2. Aims and Structure of FLOODSS

Following the non-structural measures requirements, the FLOod Operational Decision Support System (FLOODSS) system under development is aimed at analysing and anticipating catastrophic flood events and at preventing and mitigating their effects on the economical, social, environmental and cultural heritage. FLOODSS responds to the need of an integrated tool that, taking advantage of available High Performance Computer platforms allows, (1) for planning purposes, to locate areas at risk and to estimate expected damages; (2) to forecast floods and inundation phenomena on the basis of real-time analysis of the present meteorological situation and of forecasts available at different time and space scales; (3) to evaluate the effects of decisions aimed at reducing social, economical and environmental damages on the basis of planned or real-time forecasted scenarios; (4) to allow continuous training of personnel.

FLOODSS is based upon the experience gained in national and international R&D projects such as EFFORTS (European Flood Forecasting Operational Real Time System; 1988-1991) ([4,5]), which produced a real time flood forecasting package already operationally installed on several rivers in Italy , in Germany and China; AFORISM (A comprehensive FOrecasting system for flood RISk Mitigation and control; 1990-1993) [6] which produced a feasibility study for what is being proposed under the FLOODSS project; ODESSEI - Open architecture DEcision Support System for Environmental Impact assessment, planning and management (EUREKA-EU487; 1994-1997) [7], which allowed for the development of an open architecture framework under which FLOODSS will be structured; PRIMAVERA (Previsione Regionale Idro-Meteorologica Accoppiata per la Valutazione delle Emergenze e del Rischio di Alluvione; 1997-1998) [8] and TELFLOOD (Forecasting Floods in Urban Areas Downstream of Steep Catchments; 1996-1998) [9] that are studying the link between Limited Area Meteorological Models quantitative rainfall forecasts and flood forecasting models.

Following the findings of AFORISM, FLOODSS will allow quantitative impact analyses on the basis of environmental, social, economical criteria by combining flood maps (generated by simulation or real-time forecasting models) with geo-referenced data (land use maps, cadaster maps, road maps, traffic information, etc.) using several tools developed in the ODESSEI project FLOODSS will also allow for the link to real-time hydro-meteorological data acquisition systems and to real-time flood forecasting systems.

Figure 1 shows the overall schematic diagram of FLOODSS, that will include at full development:

- A set of relational and mathematical tools;
- A statistical model of rainfall extremes (GEV – Generalised Extreme Values distribution) [10], combined with a Probability-Weighted Moments estimator [11];
- A hydrologic model known as ARNO which has already been used for flood forecasting in several basins in different countries [12];
- A hydraulic one-dimensional river routing model, the PAB [13];
- A combined one/two dimensional flood plain inundation model based on the control volume finite element approach (CVFE) [14];
- A real-time flood forecasting and simulation system based upon EFFORTS [4,5];
- A GIS which allows for automatic catchment identification, river identification, setting up and calibration of rainfall-runoff models as well as of one and two-dimensional hydraulic models;
- A KBS which helps the user at three different entry level to navigate into the system and to find the answers to the queries;
- An advanced and interactive GUI which enables the user to obtain an immediate insight into the problem to be solved by setting up in an interactive mode all the necessary data and conditions and by looking at the results in a one, two, or three dimensional high definition graphics;

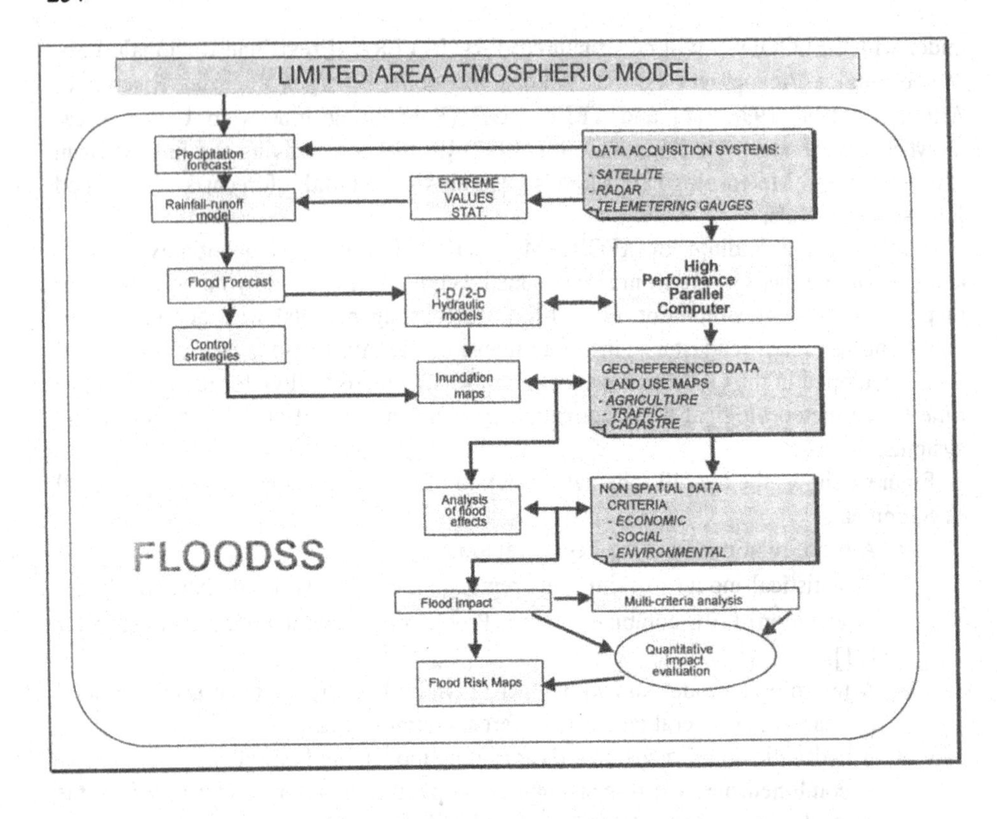

Figure 1. Schematic representation of FLOODSS.

The DSS allows high level graphic manipulation of data thanks to its 2D and 3D rendering modules based upon AVS (Advanced Visual System) - EXPRESS. It is possible to visualise these data geo-referencing them on the territory thanks to the GIS modules of the system (a modified version of GRASS where several vector routines have been corrected and extensively re-written at ET&P). The software supports typical GIS operations. These operations may help the user in using the data in order to:

- visualise maps
- retrieve alphanumeric information regarding a specific graphic element (i.e. road, river, railway etc...)
- generate and print thematic maps
- modify existing maps
- graphically visualise the result of a query.

In addition, the GIS capabilities are fully integrated with the models and can be used for instance to identify from the Digital Elevation Map a specific catchment boundary, its geo-morphological characteristics, the drainage network, and to set-up a rainfall-runoff model) by linking the precipitation data at geo-referenced stations to the selected catchment area (Figure 2). All objects are managed and integrated so as to create a logical sequence of operations to answer the user requirements and the user is guided in his choices by a set of expert system rules and controls which help him in solving the proposed problem.

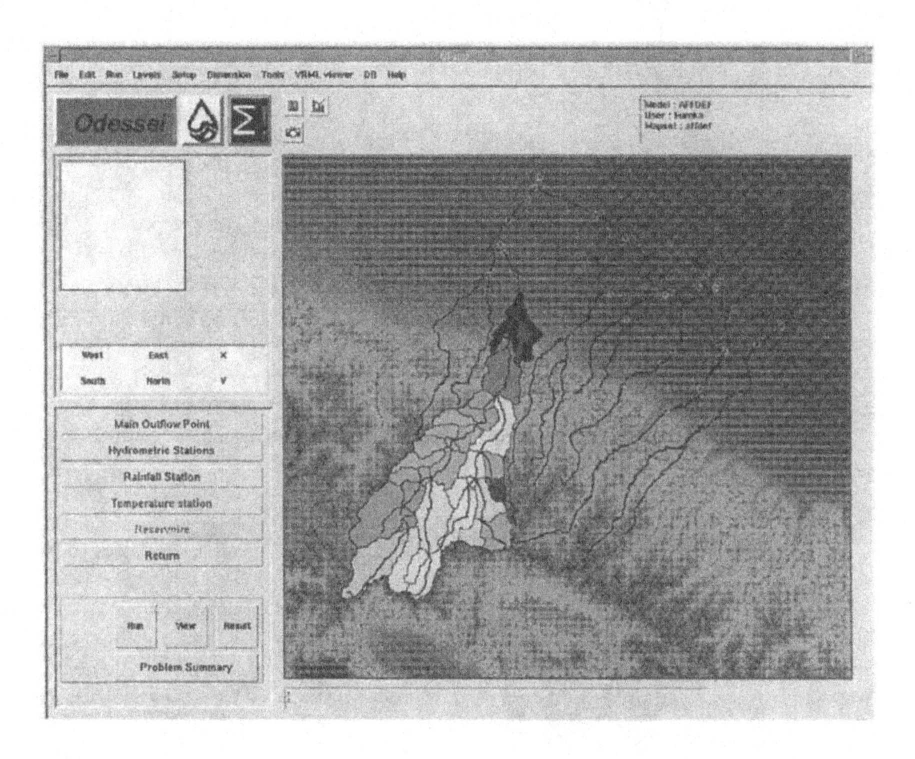

Figure 2. Automatic identification of catchments and drainage network in the Reno river.

The available tools and the structure of ODESSEI is kept in FLOODSS, where a flooding problem may be viewed as a cascade of the following operations:

- Estimation of the precipitation intensity for given duration and return period;
- Estimation of peak flow and hydrograph shape using a rainfall-runoff model (Fig. 2);
- 1-D routing and identification of over-banking areas (Figures 3, 4 and 5);
- 1-D/2-D simulation and evaluation of the inundated areas, together with flow velocity and residence time, and production of a flood hazard map (Figures 6, 7, 8 and 9);
- Assessment of vulnerability of the area and production of flood risk maps.

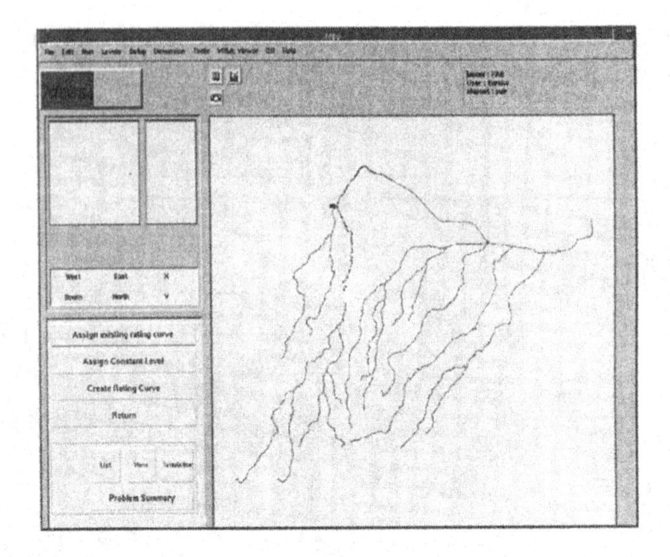

Figure 3. Selection of a specific river reach for 1-D flood routing.

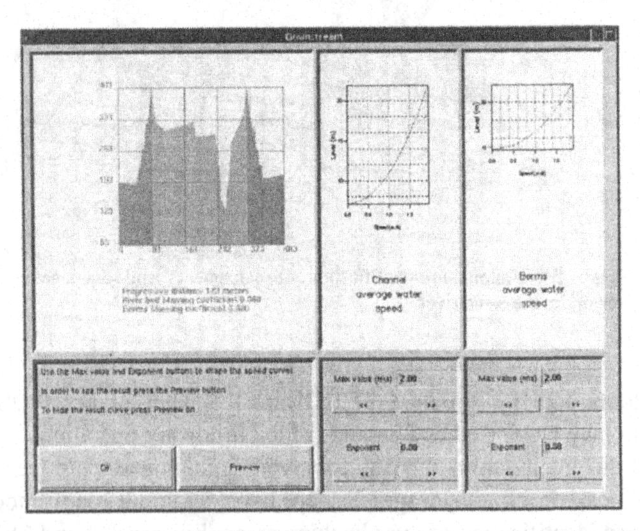

Figure 4. A simplified procedure for setting up of downstream boundary when a rating curve is not available.

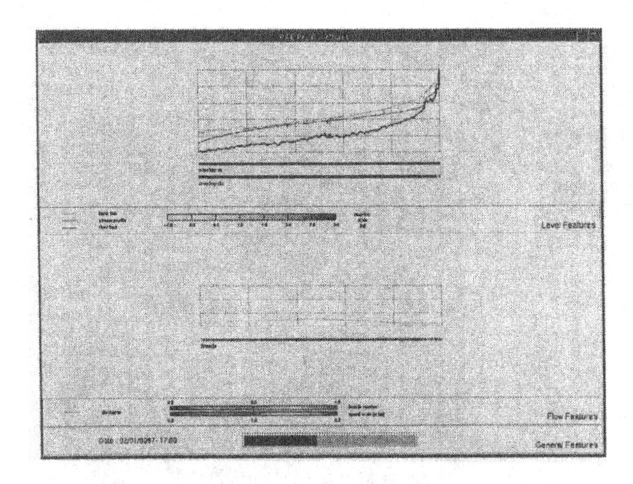

Figure 5. Analysis of 1-D flood routing results.

FLOODSS fully integrates the CVFE [14] model with the Expert System and the GIS so that all the operations to set up the model of a particular area are simplified and can make full use of the available geo-referenced data such as river cross sections; digital terrain model; land use maps; dykes, roads, railways position and elevation, etc.

The GIS automatically searches for the closed cells in the selected area (Figure 6) and the generation of the FE mesh, conditional upon the break lines, is also automatic while land use maps are used to determine a first guess of the friction coefficients to be used in the flood plain model (Figure 7). Special tools have been developed to allow for the automatic identification of buildings from the cadaster maps in order to appropriately generate the FE meshes (Figure 8).

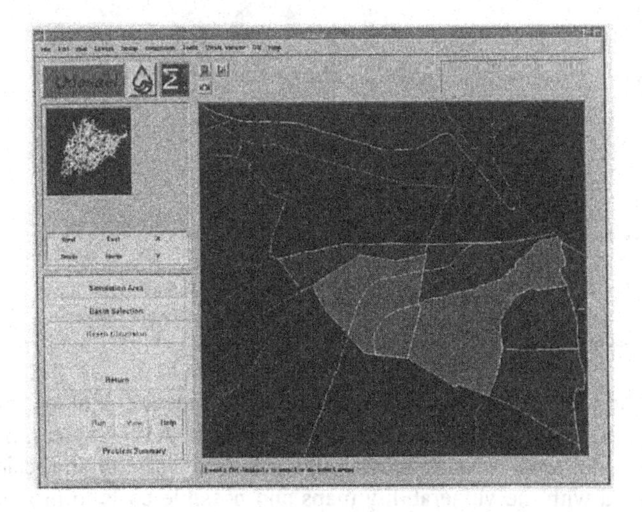

Figure 6. Selection of an area in the flood plain: automatic identification of the hydraulic cells divided by breaklines (rivers, canals, dykes, roads, railways, etc.).

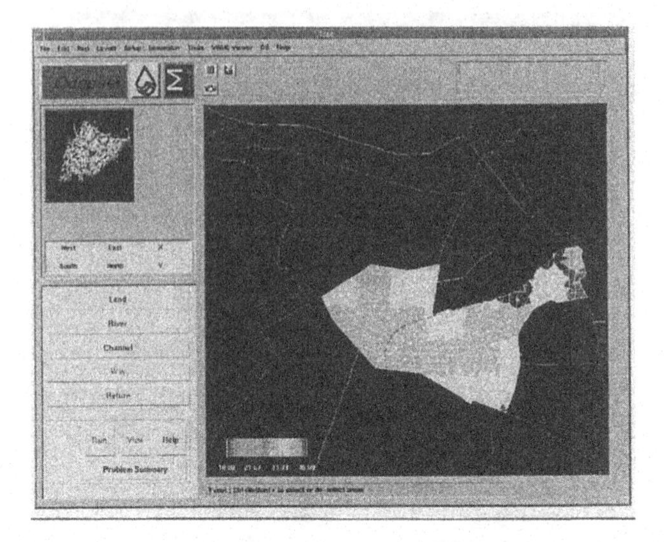

Figure 7. Generation of the triangular irregular mesh for the CVFE and friction coefficient estimation from land-use maps.

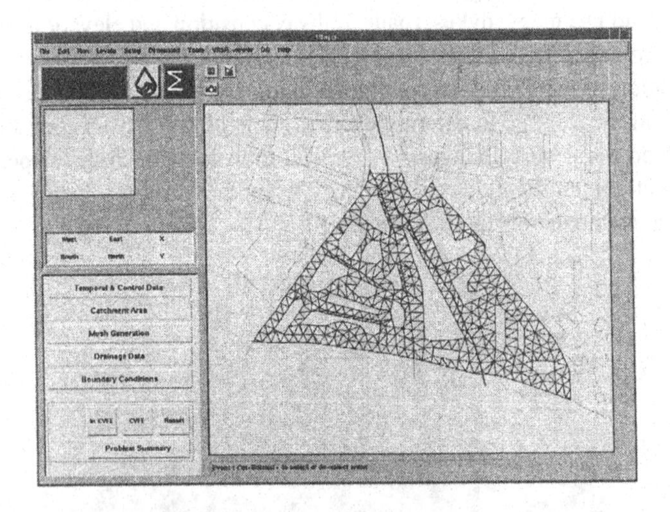

Figure 8. The identification of the flooding domain and the automatic generation of the triangular irregular network from a cadaser map.

After simulation, the results can be displayed in a technical 2-D mode (Figure 9) or in a more realistic 3-D Virtual Reality mode through AVS-Express. Finally the simulation results, in terms of water elevation, water flow velocity and residence time may be intersected with the vulnerability maps and possible costs to produce the risk maps as described in [15].

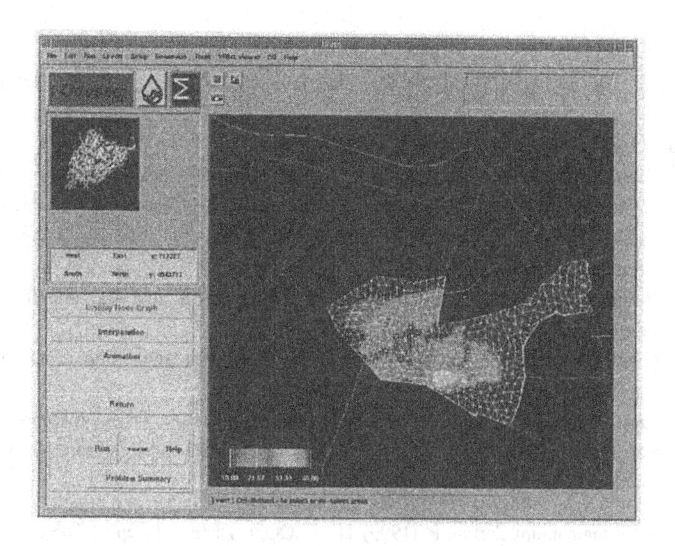

Figure 9. 2-D analysis of simulation results in terms of water depth.

3. Conclusions

Flood forecasting today can be performed with a substantial margin of reliability, for time horizons of twelve hours, in both large and intermediate-sized basins, while experimentation is well under way aimed at using the precipitation forecasts generated by LAM's directly as input in flood forecasting models.

Downstream of the meteorological and flood forecasting, however, the question of the planning and coordination of emergency situations is still unresolved: in fact it is essential to have not only the flood forecast, but also a detailed emergency plan and the necessary co-ordination among the different authorities involved in the management of risk. The latter is probably the most critical and frequently overlooked aspect of real-time flood forecasting and control, given the multiplicity of different Institutions and Authorities involved in the decision making processes during flood emergencies.

This problem can probably be solved within the frame of a holistic approach to flood management, which requires the study and the development of decision supporting tools such as FLOODSS, where the integration of the GIS capabilities in the system together with the parallelisation of the flood plain inundation model will allow for extensive risk analyses and the preparation of flood risk maps and their use in real time in association with its flood forecasting and flood simulation capabilities. For all these reasons, it is the belief of the authors that the availability of a system allowing for the hydrological and hydraulic analyses; the evaluation and the preparation of the flood risk maps; the planning of structural and non structural measures; the real time flood forecasting with the possibility of analysing the advantages of the different intervention scenarios, will constitute a major breakthrough in the four areas related to flooding: planning, real time management, dissemination and training of personnel.

4. References

1. National Land Agency Japan, (1994) *Natural Disasters in the World* - Statistical Trend on Natural Disasters.
2. Munich Re (1998) *World Map of Natural Hazards*, Munich.
3. Todini, E. (1996) Interventi non strutturali per la difesa delle alluvioni, in U. Maione and A. Brath (eds.), *La sistemazione dei corsi d'acqua naturali*, Editoriale BIOS, Cosenza, pp.417-441 (in Italian).
4. ET&P (1992) The Fuchun River project - A computer based real-time system. EC-China Cooperation. Final Report Research Contract CI13-0004-I
5. Todini, E., Marsigli, M., Pani, G. and Vignoli, R. (1997) Operational real-time flood forecasting systems based on EFFORTS. Proc. RIBAMOD Workshop.
6. Todini, E. (1995) AFORISM - A comprehensive forecasting system for flood Risk mitigation and control - Final Report of Contract EPOC-CT90-0023.
7. Todini, E. and Bottarelli, M. (1997) ODESSEI: Open architecture Decision Support System for Environmental Impact: Assessment, planning and management, in Refsgaard & Karalis (eds.), *Operational Water Management*, Balkema, Rotterdam, 229-235.
8. Todini, E. (1998) Un progetto che fa PRIMAVERA. ARPA, n. 0, 23-25 (in Italian)
9. Todini, E. and Bongioannini Cerlini, P. (1999) TELFLOOD: Technical Report. DISTGA Università di Bologna.
10. Jenkinson, A.F. (1955) The frequency distribution of the annual maximum (or minimum) of meteorological elements, *Quarterly Journal of the Royal Meteorological Society*, **81**, 158-171.
11. Hosking, J.R.M., Wallis, J.R. and Wood, E.F. (1985) Estimation of the generalized extreme-value distribution by the method of probability-weighted moments, *Technometrics*, **27**, 251-261.
12. Todini, E. (1996) The ARNO rainfall-runoff model, *Journal of Hydrology*, **175**, 339-382.
13. Todini, E. and Bossi, A. (1986) PAB (Parabolic and Backwater) An unconditionally stable flood routing scheme particularly suited for real time forecasting and control, *Journal of Hydraulic Research*, **24**, 405-424.
14. Di Giammarco, P., Todini, E. and Lamberti, P. (1996) A conservative finite elements approach to overland flow: the control volume finite element formulation, *Journal of Hydrology*, **175**, 267-291.
15. Di Giammarco, P., Todini, E., Consuegra, D., Joerin, F. and Vitalini, F. (1994) Combining a 2-D flood plain model with GIS for flood delineation and damage assessment. in: P. Molinaro and L. Natale (Eds.), *Modelling of flood propagation over initially dry areas*, Proc. of the Specialty Conf. ENEL-DSR-CRIS, Milan, pp. 171-185.

REAL-TIME FLOOD FORECASTING: OPERATIONAL EXPERIENCE AND RECENT ADVANCES

EZIO TODINI

University of Bologna
Department of Earth and Geo-Environmental Sciences
Via Zamboni, 67
40127 Bologna, Italy

1. The Predictability of Flood Events

As far as the predictability of flood events is concerned, it is necessary first of all to debunk a common misconception. Whilst it is true that there are hundreds of hydrological rainfall-runoff models, it has however been abundantly demonstrated that, for the purposes of flood forecasting, the most important effect is the dynamic of the overall soil filling and depletion mechanisms and the resulting variation in the size of the saturated basin area, as a result of which the rainfall that falls in the basin has a direct effect on the flood discharges which is not attenuated by the soil's absorption capacity. These concepts have spawned models which are very widely used today and have been extensively described in the literature, such as the model developed by Zhao [1], by Moore and Clarke [2], the TOPMODEL [3] and the ARNO [4].

Maybe the most widely applied real time flood forecasting system (apart from the old ALERT system based upon the Sacramento model widely used in the USA) is EFFORTS (European Flood Forecasting Operational Real Time System) ([5], [6]) based on the ARNO model. Table 1 gives a list of the real-time systems based upon EFFORTS that have been installed and have become operational in the last ten years together with the calibration results, expressed in terms of explained variance, the size of the catchment for which they were developed and the installation date. All the systems, except for the Po which is only based upon a hydraulic flood routing model, combine rainfall-runoff hydrological forecasting to hydraulic flood routing. Due to the presence of an additional real-time error correction term, actual EFFORTS are generally better than those expressed by the calibration explained variances given in Table 1.

J. Marsalek et al. (eds.), Flood Issues in Contemporary Water Management, 261–270.

TABLE 1. Installation date, size of basins and EFFORTS performances expressed in terms of explained variance

Basin	Inst. Date	Area (km^2)	Expl. Var.
Po at Pontelagoscuro	1992	70,091	n.a.
Fuchun at Lan Xi	1991	18,236	0.96
Tiber at Rome	1996	16,545	0.96
Tiber at Corbara	1993	6,100	0.98
Adda at Lago di Como	1993	4,572	n.a.
Arno at Nave di Rosano	1994	4,179	0.98
Danube at Berg	1994	4,037	0.94

An explained variance of the order of 94-98% shows that the hydrological forecasting problem has in itself actually been resolved: the existing problems are frequently upstream and downstream of the hydrological model. As a matter of fact, there is a lack of reliable and up-to-date rating curves and the geometrical descriptions are frequently approximate and/or rarely controlled; at the same time, very few authorities responsible for flood management have thus far seemed to show any real interest, during flood events, for "uncertain" discharge forecast up to 8-12 hour in advance.

2. Operational Requirements and Predictability Horizon

It has been pointed out in the American literature that it is necessary to have real-time flood forecasts with an advance warning time of six-eight hours in order for these forecasts to be of any use and to allow measures to be taken to reduce the danger level or the damage connected with the flooding. In general, given the stress of emergency conditions it is reasonable to set the forecasting lead time at twelve hours so as to enable an organisational early warning phase to be mounted before the actual operational emergency is triggered. Only for very small catchments for which this is not possible at the present time one should at least aim at three to four hours of lead time.

In order allow a quantitative forecast up to 12 hours in advance, the basins can be subdivided into four broad classes according to their size and concentration times:

- Large basins (>10,000 km^2), for which the flood forecast with up to 12 hours advance warning can basically be made on the basis of the water levels (or discharge quantities) recorded in one or more upstream sections.
- Intermediate basins (1,000-10,000 km^2), for which the flood forecast up to 12 hours advance warning can be made on the basis of the precipitation measurements alone; in other words, the future rainfall has a negligible effect on the discharges which will occur in the next 12 hours as a result of the interception, infiltration, soil dynamics and transfer of runoff on the slopes and through the drainage network.
- Small basins (100 - 1,000 km^2), for which the flood forecast up to 12 hours advance warning can be made only when a precipitation forecast is available for a number of hours.

- Very small and urban basins (< 100 km^2) for which 12 hours in advance warnings are not possible, but for which generally 3-4 hours are sufficient for issuing an effective alert.

As can be seen from these observations, the problem of the short-term predictability of a flood event (up to 12 hours) can already be resolved for medium-large basins provided one has a real-time tele-metering network in place. However, the forecasting problem remains unresolved for a large variety of minor basins which would require quantitative future rainfall forecasts out of the range of what is known as "nowcasting" with a quality far superior to that provided by the present models.

As an example of systems implemented for large basins, various water level real-time monitoring stations have for some years now been installed on the main stream of the Po from Becca to Pontelagoscuro. For the latter section an automatic flood early warning system has been designed and in operation for over five years. The system, installed with the Po Water Authority, is based on the EFFORTS package using the PAB program [7] for hydraulic flood routing. Downstream of the hydraulic model is a Kalman Filter based upon the MISP algorithm [8] which allows the model to be continuously adjusted to the recorded measurements. The analysis of the last 5 years of operations are extremely satisfactory as can be seen from Figure 1, which compares the 24 and 36 hours in advance forecasts with what has actually occurred.

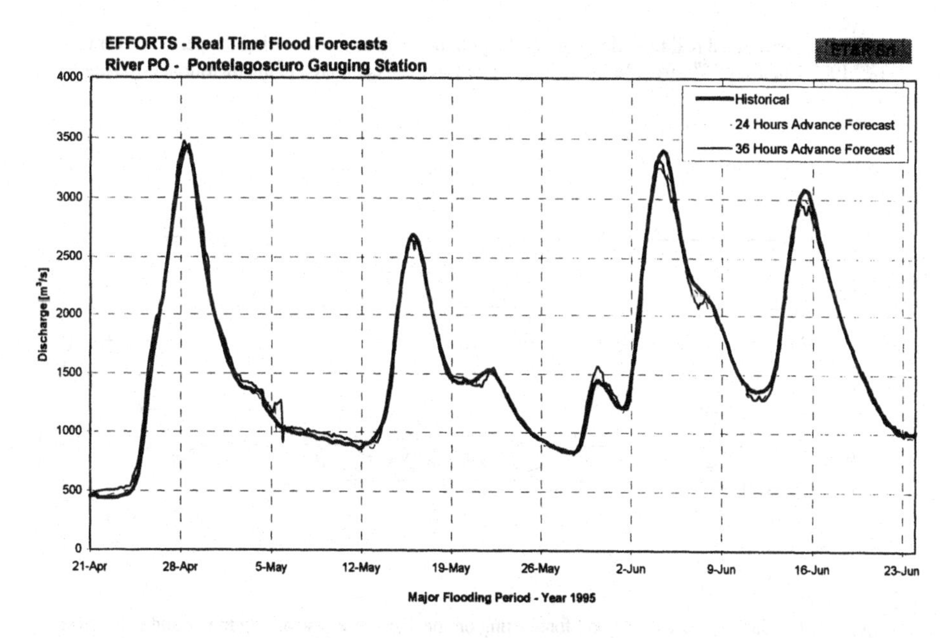

Figure 1. Comparison of real time flood forecasts in the Po river and observed flows.

Good quality results have also been reported by several users systems which include both the rainfall-runoff models and the flood routing models: in 1993 on the Fuchun a peak flow forecast of 11,100 m^3/s was issued 36 hours in advance which compared very well with the actually recorded 11,200 m^3/s; on Lake Como, 12 hours in advance lake level forecasts have been reported to be exceptionally accurate, with errors of the order of 1 cm; Figures 2 and 3 show the succession of 12 hours in advance forecasts compared to the actual recorded flows on the Tiber for several flood periods and one can observe that the system operational forecasting reliability and accuracy are in line with the needs for implementing river control and flood remediation strategies.

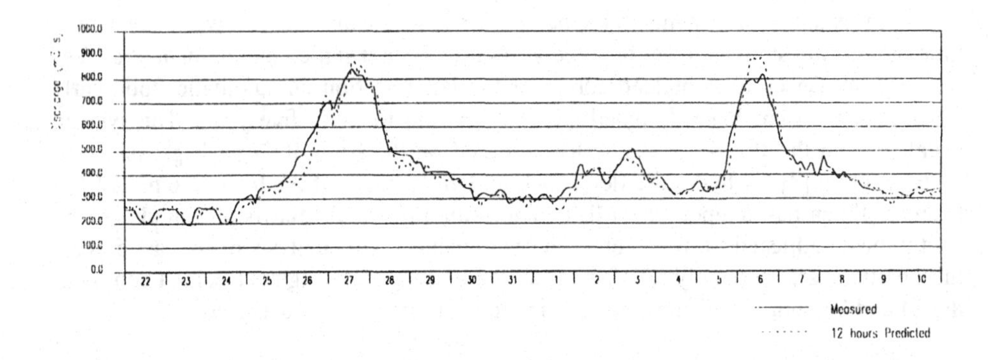

Figure 2. Operational real time flood forecasting on the Tiber at Rome during the period December 22nd 1993 - January 10th 1994. 12 hours in advance forecasts compared to the measured discharges.

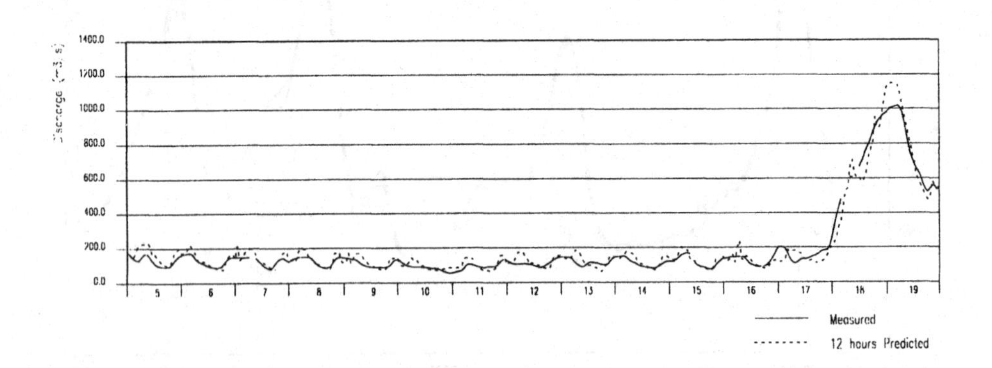

Figure 3. Operational real time flood forecasting on the Tiber at Rome during the period November 5th 1996 - November 19th, 1996. 12 hours in advance forecasts compared to the measured discharges.

3. Extending the Forecasting Lead Time

With regard to the problem of sufficiently long forecasting lead times in smaller basins, research is still under development: two projects PRIMAVERA [9] and TELFLOOD [10] recently dealt with the operational use of the rainfall forecasts, provided by limited area meteorological models, better known as LAMs, as inputs to rainfall-runoff models.

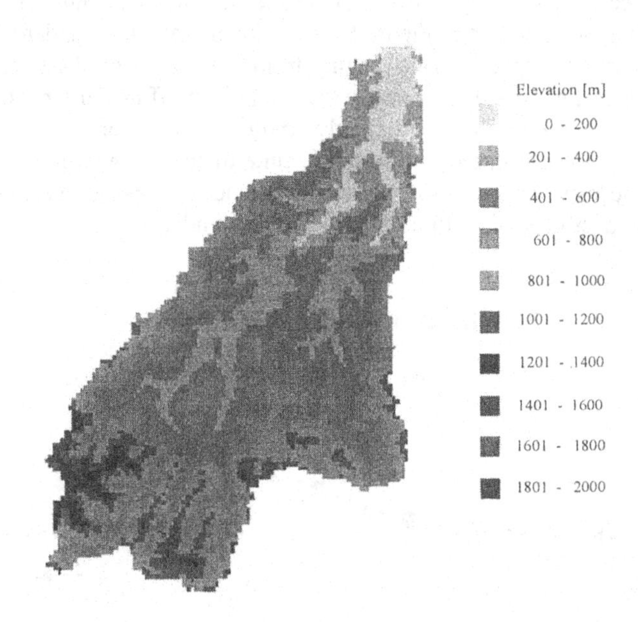

Figure 4. Digital elevation map of the Reno river at Casalecchio.

TABLE 2. Flood periods used as case studies in the PRIMAVERA project

Event	Date	Year
1	18-25 September	1994
2	10-17 June	1994
3	2-11 December	1992
4	27-2 October/November	1992
5	29-9 September/October	1993
6	4-11 November	1994
7	30-3 December/January	1993-94
8	13-18 April	1994
9	18-22 October	1993

Studies were carried out on a sample catchment: the upper Reno closed at Casalecchio (Italy), which catchment is of approximately 1,000 km^2, shown in Figure 4. Six years of hourly data were collected for 26 rain-gauges, 10 temperature gauges and 2 level gauges. Nine flood events, listed in table 2 in descending order of importance, were identified and used as case studies.

A first experiment was performed without adding a "noise model" to the output discharge forecasts. The noise model is normally added in real-time flood forecasting schemes because it can improve the forecasts at very short lead times. In order to assess the potential benefit of the LAM forecasts, at each step in time, flood forecasts were simulated introducing as input to the ARNO rainfall-runoff model the LAM operational forecasts available in Bologna on meshes of 20x20 km^2. The statistical analysis of the results is shown in Figure 5 where the bias and the standard deviation of the forecasting errors is given at lags ranging from 1 to 12 hours. One can appreciate that 50 m^3/s is a sort of base error, namely the rainfall runoff model error that does not vanish even if the future rain is perfectly known. This error remains practically unchanged for four hours due to the response time of the basin, while from the fifth hour on both the bias and the standard deviation tend to increase, reaching –23 m^3/s and 100 m^3/s respectively when forecasting 12 hours in advance.

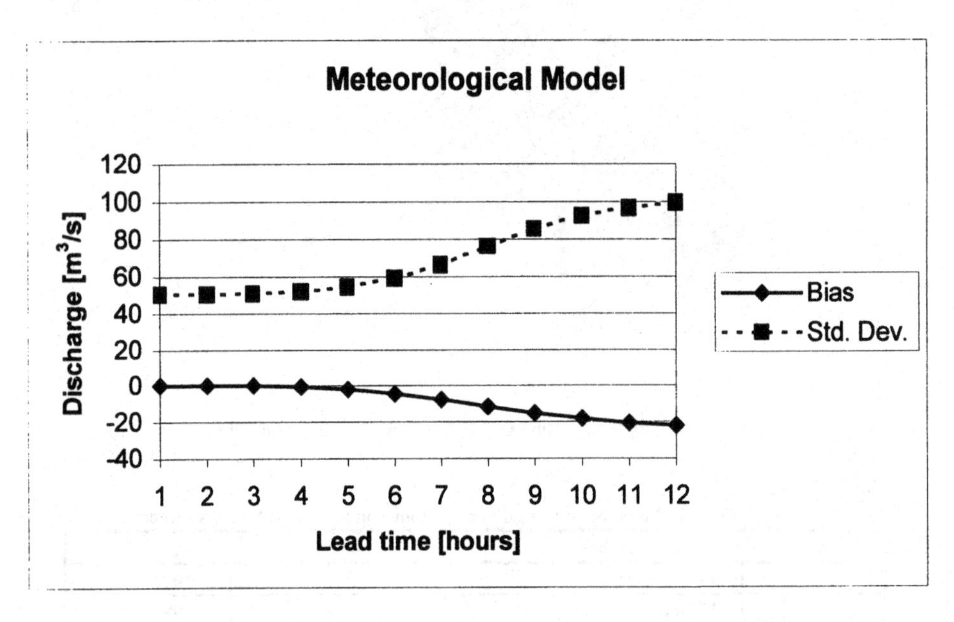

Figure 5. Bias and standard deviation of forecasting errors for 1 to 12 hours in advance flood forecasts using the LAM forecasted rainfall.

A second approach was to generate an ensemble of 20 future rainfall traces conditional upon the latest measures using a phase space model ([11], [12]) with a formulation similar to the one proposed by Yakowitz [13] and Yakowitz and Karlsson [14] better known as Nearest Neighbour. The model used, which may be interpreted as a locally stationary model, is described in [10]. By using at each time step all the generated "scenarios" as inputs to the rainfall-runoff model, 20 possible flood forecasts were issued and averaged. This required more than 32,000 simulations with the ARNO model. The results are given in Figure 6 in terms of bias and standard deviation: using this approach, the deviation from the base error of 50 m^3/s starts at least three hours later and the bias is now reduced to –5 m^3/s at a lead time of 12 hours. This means that the rainfall forecast provided by the Nearest Neighbour approach is not only more realistic than the one produced by the LAM, but on a short horizon (3 hours) the forecasted rainfall is of the same quality of the observed one.

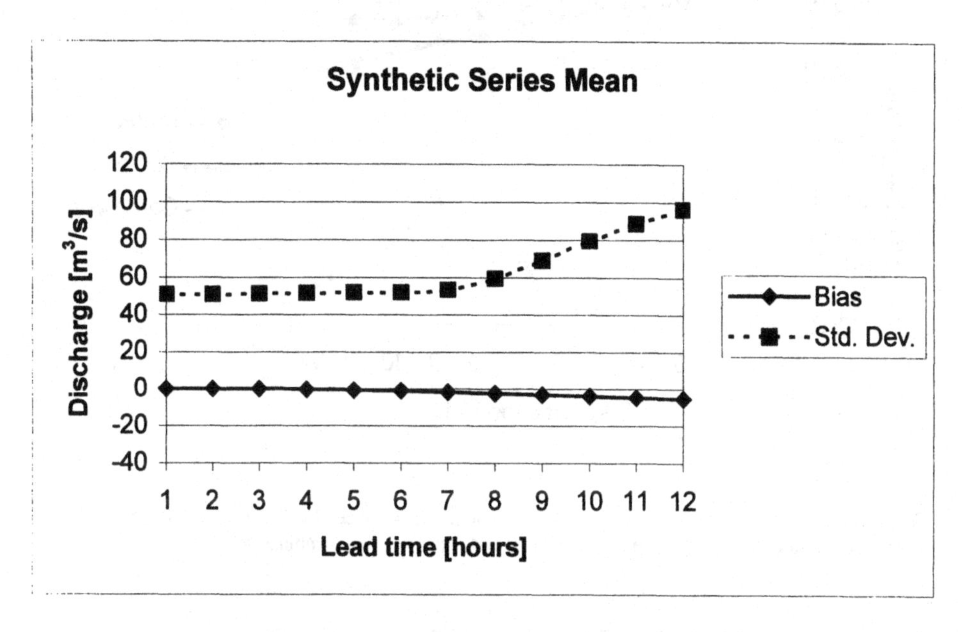

Figure 6. Bias and standard deviation of forecasting errors for 1 to 12 hours in advance flood forecasts using the nearest neighbour scenarios.

The following two figures show a comparison in terms of bias (Figure 7) and standard deviation (Figure 8) of the flood forecasting errors when using the LAM data, the conditional Nearest Neighbour ensembles and their Bayesian combination. It is possible to notice that the Bayesian combination is better in terms bias but a little poorer in terms of standard deviation, in a range between 6 and 10 hours, while improvements can be expected for longer lead times.

Successively, a second experiment was performed by adding to the previous forecasts a noise model based upon an auto-regressive representation of errors and using the MISP algorithm. No substantial changes can be seen in terms of bias (Figure 9), while a noticeable improvement can be observed for the first 4-5 hours for the three types of forecasts: this result is important because it shows that the addition of a noise model, advocated by many authors can improve the forecast at short lags, but the extension of the lead time can only be reached by using other techniques, such as the ones here proposed, that will increase the length of the good quality flood forecasting beyond the catchment characteristic response time.

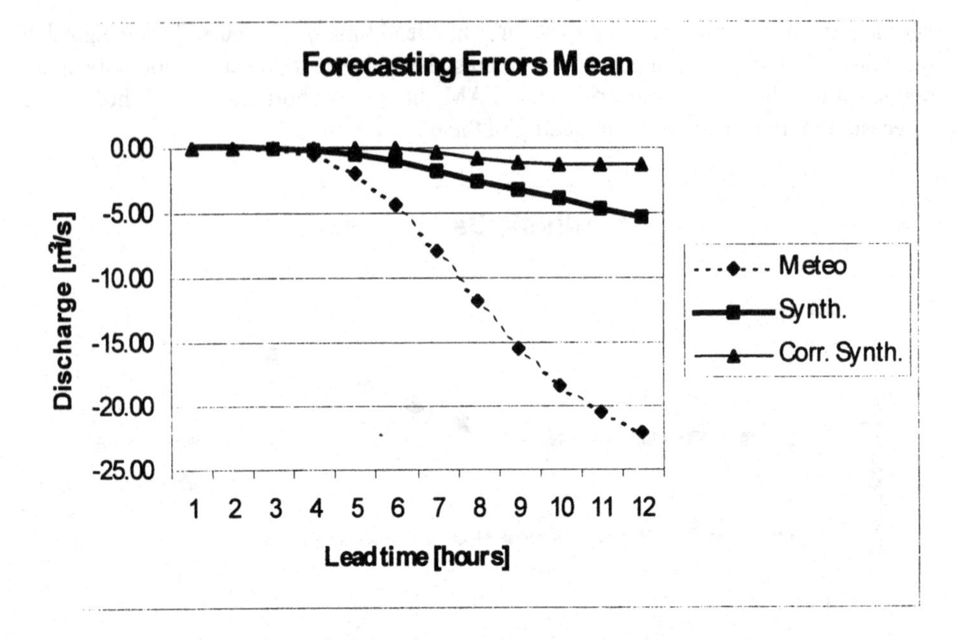

Figure 7. Bias of forecasting errors for 1 to 12 hours in advance flood forecasts using the LAM forecasted rainfall, the nearest neighbour scenarios and their Bayesian combination.

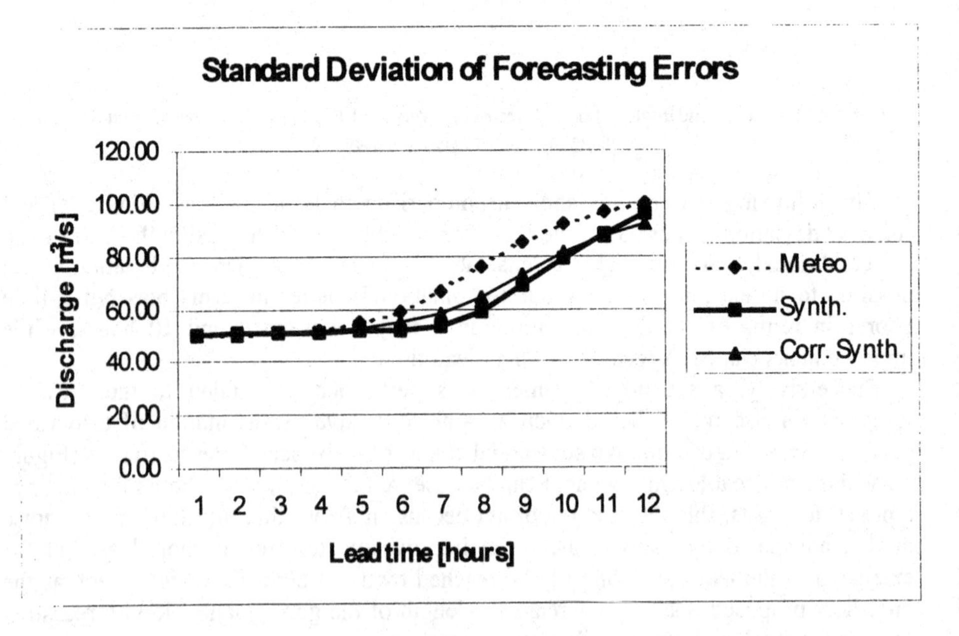

Figure 8. Standard deviation of forecasting errors for 1 to 12 hours in advance flood forecasts using the LAM forecasted rainfall, the nearest neighbour scenarios and their Bayesian combination.

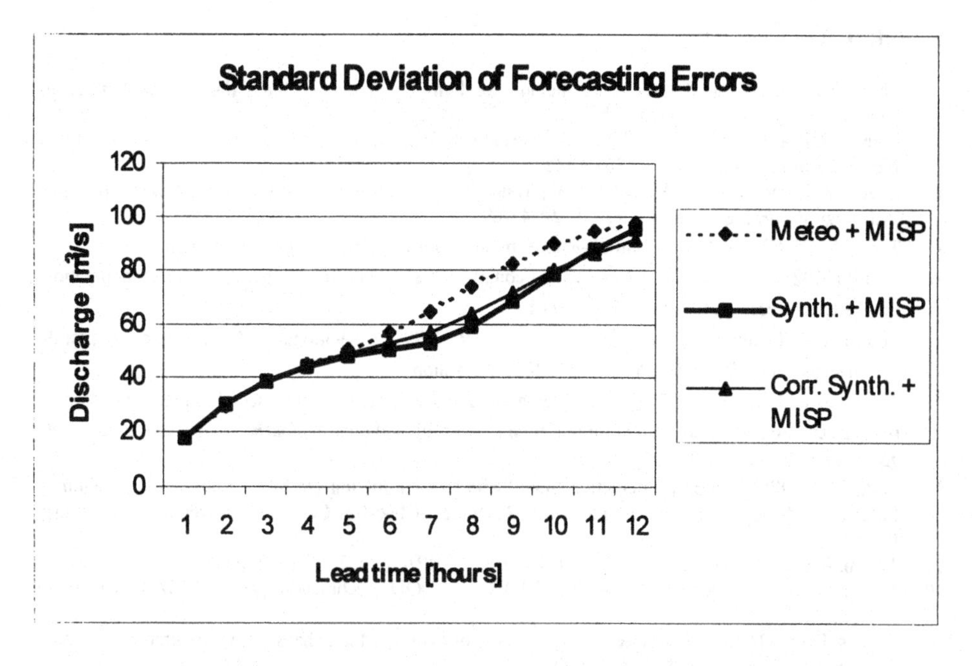

Figure 9. Standard deviation of forecasting errors for 1 to 12 hours in advance flood forecasts using the LAM forecasted rainfall, the nearest neighbour scenarios and their Bayesian combination, using MISP algorithm.

4. Conclusions

Experience gained from operational flood forecasting systems shows that real-time flood forecasts can be reliably issued up to a time horizon conditional upon the concentration time of catchment.

At the present time, the extension by a few hours (2-3) of this time horizon can be obtained using stochastic rainfall series (possibly using the Nearest Neighbour technique) generated on the basis of the latest rainfall measures.

The effect of a "noise model", although essential to reduce the forecasting errors at very short time horizons (2-3 hours), does not affect this increase in forecasting capability by extending the lead time.

The LAM forecasts (note that these considerations have been drawn on the basis of the used LAM, namely LAMBO the operational model available inn Bologna, which is based upon the), given their present performances can be used to improve the reliability of flood forecasts only at lead times longer than few hours (> 10 hours) when the uncertainty of the conditional rainfall scenarios becomes larger and comparable to that of the precipitation forecasts produced by the LAM.

Improvements in the flood forecasting performances must be sought in the improvement of LAM forecasting capabilities, possibly using non-hydrostatic models with better rainfall parameterisations.

Last but not least the Bayesian combination of synthetic and LAM generated series should be repeated in terms of flood discharges instead of rainfall, and this could be the objective of new investigations to be carried out in the nearby future.

5. References

1. Zhao, R.J. (1977) Flood forecasting method for humid regions in China, East China College of Hydraulic Engineering, Nanjing, China.
2. Moore, R.J. and Clarke, R.T. (1981) A distribution function approach to rainfall-runoff modelling, *Water Resources Research*, **17**, 1367-1382.
3. Beven, K.J. and Kirkby, M.J. (1979) A physically-based variable contributing area model of basin hydrology, *Hydrological Sciences Bull.*, **24**, 43-69.
4. Todini E. (1996) The ARNO Rainfall-Runoff model, *Journal of Hydrology*, **175**, 339-382.
5. ET&P (1992) The Fuchun River project - a computer based real-time system. EC-China Cooperation. Final Report Research Contract CI13-0004-I.
6. Todini, E., Marsigli, M., Pani, G. and Vignoli, R. (1997) Operational real-time flood forecasting systems based on EFFORTS. Proc. RIBAMOD Workshop.
7. Todini E. and Bossi A. (1986) PAB (Parabolic and Backwater) - a0n unconditionally stable flood routing scheme particularly suited for real time forecasting and control, *Journal of Hydraulic Research*, **24**, 405-424.
8. Todini E. (1978) Mutually Interactive State/Parameter Estimation (MISP) - Application of Kalman Filter to Hydrology, Hydraulics and Water Resources. Chao-Lin Chiu Ed., University of Pittsburg, Penn.
9. Todini, E. (1998) Un progetto che fa PRIMAVERA. ARPA, n. 0, 23-25 (in Italian)
10. Todini, E. and Bongioannini Cerlini, P. (1999) TELFLOOD: Technical Report DISTGA Università di Bologna.
11. Takens, F. (1981) Non-linear forecasting as a way of distinguishing chaos from measurement errors in time series. *Nature* vol. 344,6268, 734-741.
12. Casdagli, M. (1991) - Chaos and deterministic versus stochastic non-linear modelling , *Journal Royal Statistics Society* B, **54**, 303-328.
13. Yakowitz S. (1987) Nearest neighbour methods for time series analysis. J. Time Ser. Ansl., **8**, 235-247.
14. Yakowitz S. and Karlsson M. (1987) Nearest neighbour methods for time serie with application to rainfall-runoff prediction, in J.B. Mac Neil and G.H. Humphries (eds.) *Stochastic Hydrology*, Reidel, Hingham, Mass., pp.149-160.

ANALYTICAL REPRODUCTION OF RADAR IMAGES OF RAINFALL SPATIAL (2D) DISTRIBUTION AS AN INPUT TO GIS BASED FLOOD MODELLING

B. DODOV[1], R. ARSOV[1] and P. MARINOV[2]

[1] *University of Architecture, Civil Engineering and Geodesy; Faculty of Hydrotechnics; 1 Chr.Smirnensky Blvd., 1421 Sofia, Bulgaria; Tel./ Fax: +359 2 668 995;*
[2] *Bulgarian Academy of Science; Central Laboratory for Parallel Processing; Acad. G. Bonchev Str., Bl. 25 A; 1113 Sofia, Bulgaria;*

1. Objectives

Increasingly, implementation of remote sensing technologies for atmosphere observation and availability of automated radar, satellite and airborne rainfall images in particular, is a good base for improving the accuracy of the rain fields assessment as an input to the adequate floods and pollution transfer modelling. Although the information from these observations is enough detail to support directly sophisticated hydrological models, it is huge in quantity and is organised in a manner which makes very difficult extraction of general trends with respect to process development. Regardless of the fact that this information is much more detail than the conventional one, the resolution of the most popular and accessible radar images is usually not less than *2x2 km* per pixel, with time interval of about 4-5 minutes. Such a resolution is still too "coarse" for adequate flood modelling, particularly for urban areas. It is reasonable question to be asked if there is such a technique, which could allow extraction of the most important features of any sequence of images, describing a rainfall event with minimum information resources in a manner, that would allow subsequent restoration of the process to prescribed resolution, preserving the visual resemblance and statistical properties of the original data. The importance of this question motivated the reported research and fortunately now we can answer it positively.

In order to explain better the idea, we find it reasonable to introduce some important definitions and explanations, concerning the terms and analytical procedures, used in the paper.

There is not a unique definition of *fractals*, but they could be described as objects of a complex structure, consisting of parts, which are at any scale similar to the whole in some way. An important feature of the most of the fractals is the *scale invariance* (or *scaling*). A fractal function $f(\lambda t)$ is considered to be scaling with factor $\alpha > 0$ if the homogeneity relation $f(\lambda t) = \lambda^{\alpha} f(t)$ holds for all $\lambda > 0$ [1].

While the fractal is a set, the *multifractal* could be considered as a measure or distribution (of mass, energy, probability, feature, etc.). The *Hausdorf dimensions* of the sets, represented in a relevance to their *singularity exponents* [1], define the

J. Marsalek et al. (eds.), Flood Issues in Contemporary Water Management, 271–281.

multifractal spectrum of the relevant density (precipitation) distribution. The singularity exponent measures the "strength" of singularity of the relevant distribution at particular position x_0 in the domain under consideration. The estimated spectrum is known also as Legendre's one, because it is calculated trough Legendre transform of the so-called *mass exponents* – $\tau(q)$, accounting for the scaling lows of the q-th order moments of the data. For the distribution μ this parameter could be defined as follows [2]:

$$\tau(q) = \lim_{\delta \to 0} \frac{\log S_\delta(q)}{\log \delta}, S_\delta(q) = \sum_{C \in G_\delta} \mu(C)^q, \qquad \text{(1a, 1b)}$$

where $\mu(C)$ is the total mass of μ over the (hyper)cube C with side δ;
G_δ - the set of all the non-overlapping cubes, covering the support of μ.

If the distribution μ is scale invariant, the values of S_δ for particular q form a straight line in a log-log plot (1a), which slope is the corresponding value of τ. Then the Legendre spectrum could be estimated through the transform:

$$\alpha = d\tau / dq, f(\alpha) = q\alpha - \tau(q), \qquad (2)$$

where α is the singularity exponent, $f(\alpha)$ - corresponding value of the singularity spectrum.

The multifractal theory implementation with respect to the atmospheric fields modelling is founded on the assumption that fluxes of mass (water) and energy in the atmosphere could be represented by the so called *multiplicative cascade processes*, successively transferring these quantities from larger to smaller scales [3,4,5]. The cascade processes could be considered as eddies of fluxes, broken down into smaller sub-eddies, receiving a part of the flux of its parent eddy and the amount of which is being determined by a random factor.

Applying different strategies and "levels" of randomness, as well as some flux dissipation, an imitation of very complex structures such as rainfall patterns or clouds could be simulated [5]. The artificial fields produced by cascade processes exhibit usually good multifractal properties, which could be changed at different stages of resolution [6]. Regardless of this, the fact that they are stochastically based makes the chance to fit some real data insignificant. Maybe the most promising deterministic technique to produce "naturally looking" fields is the so-called *"Fractal/Multifractal" (FM) approach*, which could be considered as a deterministic cascade process "refracted" through a *Fractal Interpolation Functions* (FIF) [8,9,10,11,12,13,14].

Morlet [7] introduced the *wavelet transform (WT)* in connection with the analyses of geophysical signals with very different scales. This transformation basically leads to decomposition of the signal in terms of some elementary functions $\psi_{b,a}$, obtained from a "mother" (basis) function ψ by dilations and translations, such as:

$$\psi_{b,a}(x) = a^{-1/2} \psi_{b,a}((x-b)/a), \qquad (3)$$

where a and b are real numbers.

The wavelet transform applied continuously with real values of a and b is named *continuous wavelet transform* (CWT) since it produces continuous function in scale-space domain (called often *wavelet space*). Later on, the *multiresolution analysis* discovered by Mallat [15,16] and describing any orthogonal wavelet basis as common multiscale structure, was combined with the compactly supported wavelet basis [17,18] to give a fast transform, working on discrete scale-space domain. This transform was respectively called *discrete wavelet transform* (DWT). The inherent scaling structure of the wavelet transform makes it a perfect tool for multifractal analyses, including continuous [19] and recently - discrete wavelet transform techniques [20,22].

The so-called *Haar scaling and wavelet functions* provide the simplest example of an ortho-normal wavelet basis. The Haar *scaling* and *wavelet coeficients* could be represented as:

$$S_{j,k} = 2^{-1/2} (S_{j+1,2k} + S_{j+1,2k+1}),\ W_{j,k} = 2^{-1/2} (S_{j+1,2k} - S_{j+1,2k+1}). \tag{4}$$

In the notion of the multiresolution analyses the same expression could be written as

$$\begin{pmatrix} S_{j,k} \\ W_{j,k} \end{pmatrix} = \frac{1}{\sqrt{2}} \begin{pmatrix} \lambda_{j+1,1} & \lambda_{j+1,2} \\ \lambda_{j+1,2} & -\lambda_{j+1,1} \end{pmatrix} \begin{pmatrix} S_{j+1,2k} \\ S_{j+1,2k+1} \end{pmatrix}, \tag{5}$$

where $\lambda_{j+1,1} = \lambda_{j+1,2} = 1$.

Wavelet decompositions contain considerable information about the oscillations (singularities) of the recorded signal. The greater the oscillations at particular scale are, the greater are the WT coefficients at positions at which the basis "fits" best the data. As has been observed in many contexts, most wavelet coefficients are very small and the large ones draw the so-called *"maxima" lines* in the wavelet space. These features are the reasons for the image compression and detecting of the singularities (edge detection in 2D) to be among the main applications of the WT.

In his recent work Riedi [2] introduces the term *asymptotically wavelet self-similar* with parameter H_{wave} for the increments X of process Y if thc following limit exist:

$$H_{wave} = \frac{1}{2} - \frac{1}{2} \lim_{j \to \infty} \frac{1}{2} \log_2 var(W_{j,k}) \tag{6}$$

where $W_{j,k}$ are the wavelet coefficients at the j-th level of transformation.

The parameter H_{wave} could be used as equivalent to the H parameter of the so-called *long-range dependence* (LRD), governing the correlation at large lags. This means that changing the trend of the wavelet coefficients' decay, we could control the *power low* behavior at prescribed scale intervals and therefore model long-range dependence observed in real data.

The values of the parameters $\lambda_{j,1}$ and $\lambda_{j,2}$ could be optimized in a way that provides the lowest possible values of the squares of the wavelet coefficients, generated at the j-th level of transformation. Such optimization, called *adaptive wavelet approximation (AWA)*, has a unique and relatively simple solution, which is discussed elsewhere [5].

The support of the wavelet basis could be extended to work over four or eight values of the approximated function using the so-called *Rademacher ε-matrices.*

This extension provides possibilities to apply the Haar WT and the adaptive wavelet approximation with respect to the surfaces and volumes. More information could be found in [21-27].

The subject of this paper is a new approach for the radar rainfall images compression, recovering and interpolation, based on the Haar WT and the AWA. Since the AWA's basis (mother) function is generically multiplicative, it is a very suitable tool for representation and analyses of multiplicative (cascade) processes, such as rainfall distributions are expected to be, because of the turbulent environment of their genesis.

2. Results and Discussion

Since the wavelet transform (WT) is a linear operation, its product – the wavelet transform coefficients (WTC), their absolute values in particular, should not have a different type of distribution than the original signal. Instead, the parameters of the distribution of the WTC will change at any successive level of the WT. As the processes under consideration are expected to be multiplicative cascades, a log-normal distribution of the wavelet coefficients could be observed. This means that exploiting (6), a linear correlation could be expected in regard to the parameters of the normalized distribution of the WTC, produced at any successive level of the WT. If confirmed, this would allow adequate interpolation of data, resulting from multiplicative cascade process.

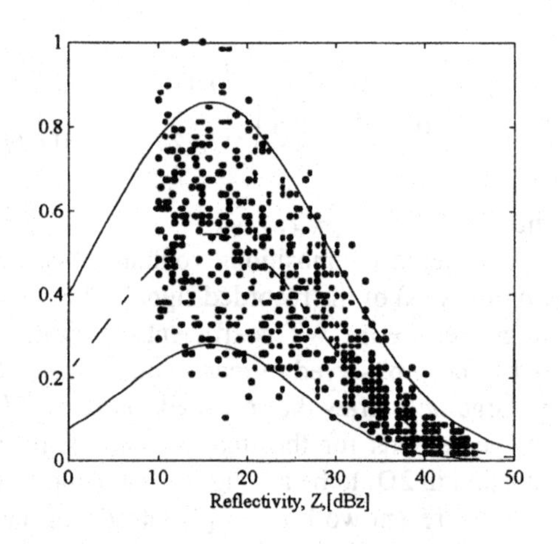

Figure 1. Z – reflectivity distribution.

To confirm the above statement we processed 20 images, representing values of the radar reflectivity (so-called *dBz*-reflectivity - *Z*), related to the physical reflectivity - z, [mm^6/m^3] according to the expression $10\log_{10}z = Z$. As shown in Figure 1, a normal statistical distribution of the relevant data *Z* has been obtained. Therefore, the statistical distribution of the radar reflectivity (*z*) itself is obviously log-normal.

In the term of the multiplicative cascades, application of the adaptive approximation means that for a deterministic binomial process the mean and the variance of the wavelet coefficients at any successive level of transformation will be practically zero. Because of the randomness of the process, the mean and the variance of the coefficients rise up to the level, where "absence of adaptivity" is observed and measured by the corresponding Haar transform. Figures 2a and 2b present a comparison between Haar WT and AWA with regard to the mean and standard deviation of the first level wavelet coefficients of binomial cascades with base 0.1 and 0.4, respectively for a tolerance of randomness from 0 to 100 % (from 0 to 0.1 and from 0 to 0.4, respectively). Obviously the AWA ensures higher compression of the signal than the conventional Haar WT . The values of the ratios $\lambda_{j1}/(\lambda_{j1}+\lambda_{j2})$ for cascade with base 0.4 are shown in Figure 3 where it can be seen that, irrespective of the degree of randomness, the adaptive wavelet approximation detects the base of the underlying cascade. The base at any level can be calculated for data with length $2^{S\times N}$ through the following expression:

$$b_{j,i} = \lambda_{j,i}/(\lambda_{j,1}+ \ldots +\lambda_{j,2^S}), \tag{7}$$

where $j = 1,\ldots,N$; $S = 1,2,3$; $i = 1,\ldots,2^S$ and N represents the index of the highest level of resolution. It can be seen from Figure 3, that the higher the randomness is, the higher is the variance of the λ-s. Figure 4 represents the ratios $\lambda_{j1}/(\lambda_{j1}+\lambda_{j2}) = b_{j1}$ of a

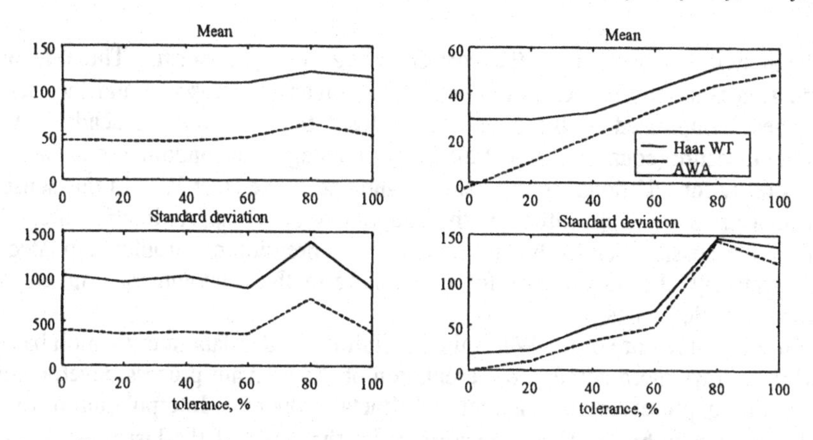

Figure 2. Comparison between Haar WT and AWA - mean and standard deviation of the first level wavelet coefficients of binomial cascades with base 0.1 (a) and 0.4(b).

binomial cascade with base 0.4 and length of 64, 128, 256, 512, 1024 and 2048, respectively with 100 % tolerance of randomness. It can be seen that the values b_{i1} (and respectively, the base vectors) alternatively change their directions, decaying to b_{N1}. This corresponds to the eddies' direction and size changing at the successive scales of the turbulence, which at the finest resolution must disappear. Therefore, the AWA application allows revealing of the availability of cascade behavior of given process (signal) and could be denominated also as *inverse cascade approximation (ICA)*.

Some of our preliminary analyses of real data indicate that the values of $\lambda_{N,i}$ tend to one. This could be expected, since any other value means that the energy and the mass will never dissipate in the turbulent processes. In this sense the AWA is a good tool for verification of different types of turbulent processes' models. The first model which we tested and which exhibited good properties in this sense is the FM one.

The analyses carried out so far show that the adaptive wavelet approximation exhibits high sensitivity with respect to the existence of a "cascadeness" in the analyzed signal. This sensitivity reflects the possibility to extract the main features of the signal with a minimum amount of information, or in other words – with higher

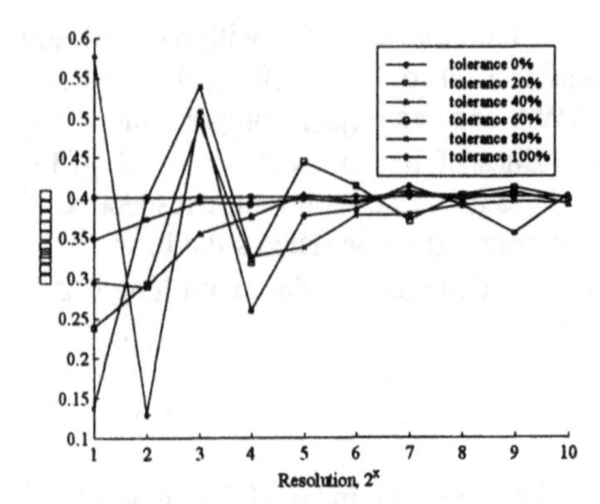

Figure 3. Values of the ratios $\lambda_{j1}/(\lambda_{j1}+\lambda_{j2})$ for cascades with base 0.4 and different tolerances of randomness.

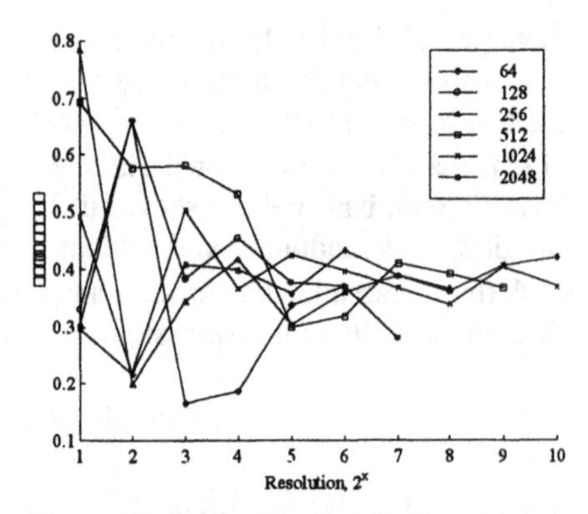

Figure 4. Ratios $\lambda_{j1}/(\lambda_{j1}+\lambda_{j2})$ of a binomial cascade with base 0.4 and different lengths with 100 % tolerance of randomness.

compression in comparison to the corresponding Haar transform. The rate of the compression is strictly related to the base of the underlying cascade and the "level of randomness" implanted in the signal. The closer the underlying cascade is to the Lebesque measure (equal bases) and the higher the degree of randomness is, the less is the possibility for additional compression, compared to the Haar WT. In this sense, the practical aspects of application of the adaptive wavelet approximation are related mainly to the possibilities for higher degree of compression of turbulence-related data and respectively, the possibilities for minimizing of the randomness, implied in the restored (modelled) ones.

An important feature of the WT is the possibility for the data interpolation based on (6), which allows more detailed representation of the relevant process development in space and time, preserving its inherent multifractal properties. Interpolation of the data must be performed by the Haar transform or by the AWA if the last three λ-sets are relatively equal and the trend of the WTC is captured. If the last condition is not fulfilled, the AWA is not applicable, because the λ-s distortions in the WTC distribution and (6) will not be valid.

The practical conclusions from the above notions in the context of the radar data analyses are related to the possibilities for:

- storage and analyses of the λ-coefficients, the core wavelet coefficients marking the general maxima lines and the trends of decay (or distributions at any level of transformation) of these coefficients;
- analyses of the obtained information in multiparametric probability space to obtain the trend in a particular process, or to find the position of this process in probability space;

- restoration, interpolation and modelling of radar images, based on the derived coefficients, filling in the gaps of the missing coefficients with random ones (but derived from distribution, inherent for the data) and finally - permutation of the random coefficients to provide positive values for the recovered data;
- governing the scaling behavior of the model through control of the rates of decay of the wavelet coefficients at any level of the transform and therefore match the bifurcation points (or zones) of the natural turbulent process; this is essential for the successful interpolation of this process in space and time.

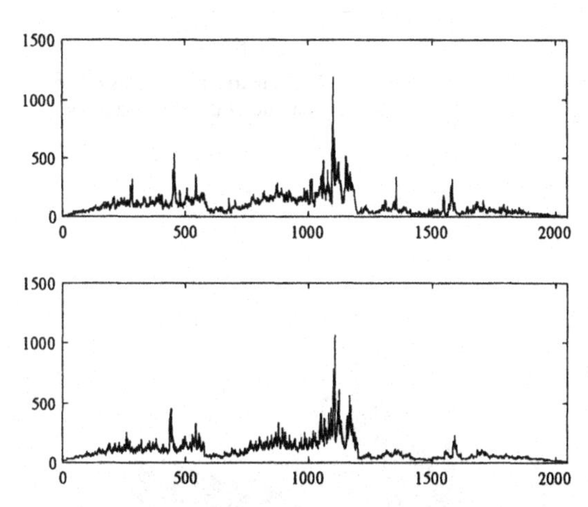

Figure 5. Comparison between signals produced by FM procedure with length 2048 (up) and the same signal with length 128 interpolated through WT to 2048 (down).

An example of 1D interpolation of data, produced by means of the FM approach is shown in Figure 5. The corresponding $\tau(q)$ and $f(\alpha)$ are given in Figure 6 and the power spectrum is given in Figure 7. A similar example but in 2D is represented in Figures 8a and 8b.

Interpolation of real data of a radar image is illustrated in Figure 9. The interpolated image is not an exact copy of the natural one in the visual aspect (as it is at the examples, illustrated in Figure 5 and Figures 8 a, b), because of missing detailed data of the registered radar signal. The low intensity signals are cut automatically by the radar being used (as could be seen in Figure 1), since they are currently considered to be of little practical importance. Taking into account as many details as possible, however, is of crucial importance for the adequate modelling of the multiplicative cascade processes [14]. Missing the low intensity signal data changes the pattern of "cascades" of the turbulent process (and respectively, the distribution of the wavelet coefficients) and does not allow an adequate interpolation. This reflects the accuracy of the modelling, as well.

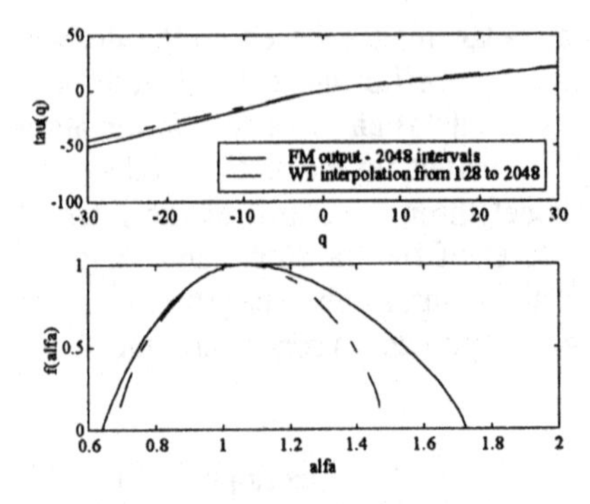

Figure 6. Comparison between $\tau(q)$ and singularity spectrums $f(\alpha)$ of the above signals.

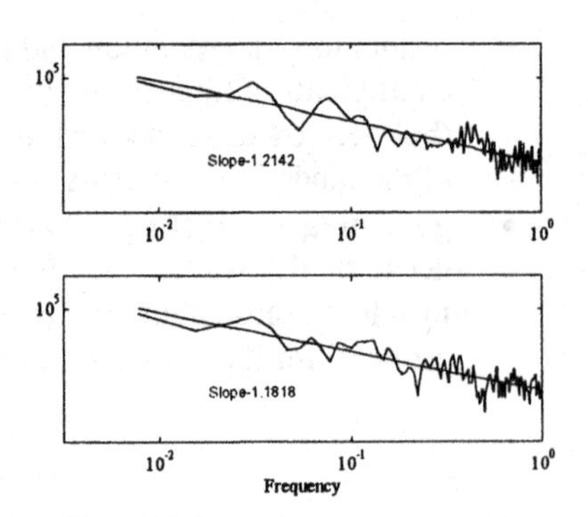

Figure 7. Comparison between power spectrums of the of the above signals.

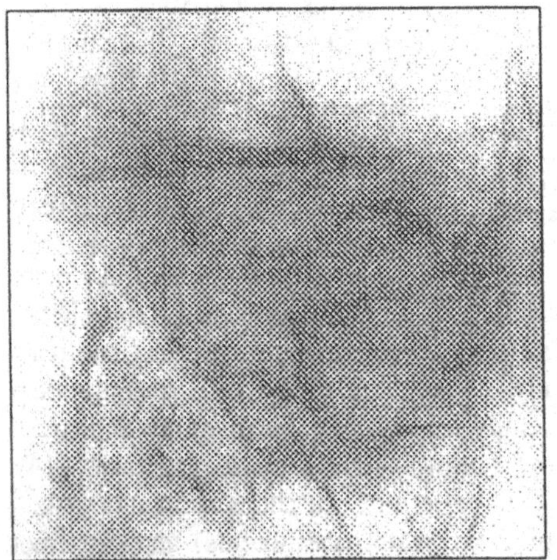

Figure 8 a. FM-output image 128x128 (left) interpolated to 512x512 by WT (up).

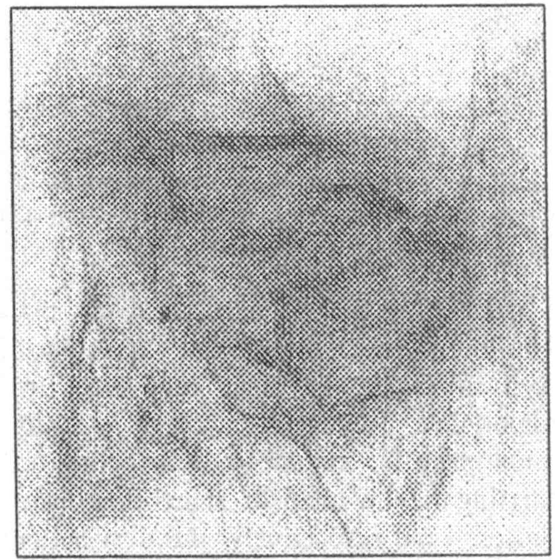

Figure 8 b. FM image 512x512 produced by means of the same FM parameters as for the one, given in Fig. 8a (left).

3. Conclusions

Adequate assessment and modelling of the rainfall processes have been recognized as a key issue in the successive flood prediction and mitigation. The main emphasis in rainfall/runoff processes modelling in current practice is on developments on the ground. Even in these cases, the possibilities offered by the GIS have not been used in the advanced models in a way which allows an accurate accounting for the topology and land use on the runoff development. Higher space and time resolution can be implemented for more adequate modelling both of rainfall and runoff processes in a GIS environment, than the currently used approaches offer.

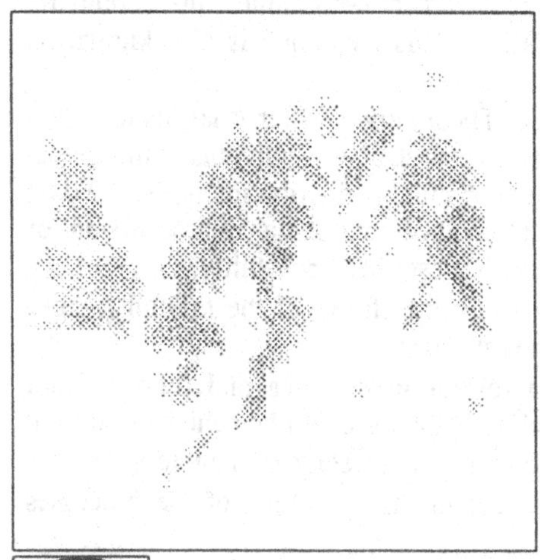

Figure 9. Radar image in it's original size 100x100 pixels or 400x400 km (left) and its interpolation (zoom) of 16 times. (up)

Since many of the existing representations are limited by their analytical tractability and because there has been a recognition of chaotic effects in atmospheric precipitation, a new class of models, based on Fractal/Multifractal (FM) approach, wavelet transform (WT) and adaptive wavelet approximation (AWA) have been considered in the rainfall pattern description.

The multifractal theory implementation with respect to the atmospheric fields is founded on the assumption that fluxes of water and energy in the atmosphere could be represented by multiplicative cascade processes, successively transferring these quantities from larger to smaller scales. For this considering of the smallest details of the turbulent process is of crucial importance.

The new technology discussed in this paper is based on the advantages of the WT and AWA approaches and allows analyses, modelling, and detailed interpolation of turbulence-related processes, such as rainfall. It offers a possibility for semi-deterministic reproduction of radar rainfall images by means of a restricted number of parameters and allows statistical analyses to be performed on them. This technology promises to be a powerful tool for reliable prediction of storms detailed spatial and temporal distribution with prescribed probability.

4. Recommendations

Steps should be taken to create a catalogue, with optimal number of the coefficients, which for prescribed size of radar images, could allow detailed restoration of the data in statistical and visual sense. Fortunately the numerical aspects of the notion "similarity in statistical and visual sense" has its mathematical analogue, such as conventional statistics and multifractal analyses for statistical similarity, as well as *integrated Hausdorf distance* [27] for visual similarity.

In conclusion, the stages of implementation of the Haar WT and adaptive (extended Haar) wavelet transforms (AWA) in radar data analysis and modelling for a given region and resolution (image size) could be summarized as follows:

- determination of the optimal data quantity and quality to be stored in the catalogue;
- filling in the catalogue in a two-step procedure: (i) extracting the information per image for all the images, scanned during particular rainfall - 2D transform, and (ii) extracting the information for the same rainfall as 3D representation (images as slides in time step) – 3D transform. (If there are missing slides, they could be obtained through interpolation in time in 3D. This technique is also known as "morphing");
- analyzing of the data by means of Fuzzy Set Theory and Cluster Analysis to reveal the trend of the parameters development in respect of an individual rainfall 2D image. Another possibility is analyzing the probabilistic distribution of the 3D data in multiparametric space, to define the likelihood of a particular rain occurring, or to "construct" the parameters of a rainfall with prescribed probability;
- reproduction, through inverse adaptive wavelet transform, of the rainfall images from the chosen set of parameters in "standard" size;
- interpolation, if necessary, to prescribed resolution in space and time through inverse Haar transform. Anisotropy in scaling behavior could be achieved through control over the decay of the wavelet coefficients at any stage of the interpolation;
- implementation of the produced grid (image) in the modelling of the processes through the conventional GIS tools.

5. References

1. Feder, J. (1988) Fractals, *Plenum Press, New York.*
2. Riedi, R. H., Multifractal Processes, (submitted) *Stoch. Proc. and Appl.*
3. Olsson, J., Niemczynowicz, J. & Berndtsson, R. (1993) Fractal of high-resolution rainfall time series," *J. Geophys. Res.*, 98 (D12), 23265-23274.
4. Olsson, J. & Niemczynowicz, J. (1994) Multifractal relations in rainfall data, *Nordic Seminar on Spatial and Temporal Variability and Interdependancies Among Hydrological Processes*, Kirkkonummi, Finland, 14-16 Sept.
5. Tessier, Y., Lovejoy & Schertzer, D. (1993) Universal multifractals: Theory and observations for rain and clouds, *J. Appl. Meteorol.*, 32 (2), 223-250.
6. Marshak, A., Davis, A., Cahalan, R. F. and Wiscombe, W. J. (1994) Bounded Cascade Models as Nonstationary Multifractals, *Physical Review E*, 49, 55-69.
7. Morlet, J. (1983) Sampling theory and wave propagations, *NATO ASI series FI* (Springer, Berlin), 233-261.
8. Barnsley, M.F., (1988) Fractals Everywhere, *Academic Press*, New York.
9. Lutton, E., Levy-Vehel, J., Cretin, C., Glevarec, P. and Roll, C., (1995) Mixed IFS: resolution of the inverse problem using Genetic Programming, *Complex Systems*, 9, (5), October and *Inria research report No. 2631*, August.
10. Puente, C. & Klebanoff, A. (1994) Gaussians everywhere, *Fractals*, 2 (1), 65-79.
11. Puente, C. (1994) Deterministic fractal geometry and probability, *International Journal of Bifurcation and Chaos*, 4 (6), 1613-1629.
12. Puente, C. & Obregon, N. (1996) A deterministic geometric representation of temporal rainfall: Results for a storm in Boston, *Water Resources Research*, 32 (9), 2825-2839.
13. Puente, C., Lopez, M., Pinzon, J. & Angulo, J. (1996) The gaussian distribution revisited, *Adv. Appl. Prob.*, 28 (2), 500-524.
14. Puente, C. (1996) A new approach to hydrologic modelling: derived distributions revisited, *Journal of Hydrology*, 187, 65-80.
15. Mallat, S.G. (1989) A Theory for multiresolution signal decomposition: The wavelet representation, *IEEE Trans. On Pattern Analysis and Machine Intelligence*, 11, 674-693.

16. Mallat, S.G. (1989) Multiresolution approximations and wavelet orthogonal bases of $L^2(R)$, *Trans. of the Am. Math. Soc.*, 315, 69-87.
17. Daubechies, I., Grossmann, A. and Meyer, Y. (1986) Painless nonorthogonal expressions, *J. Math. Phys.*, 27, 1271-1283.
18. Daubechies, I. (1992) Ten Lectures on Wavelets (S.I.A.M., Philadelphie).
19. Muzy, J. F., Bacry, E. and Arneodo, A. (1994) The multifractal formalism revisited with wavelets, *International Journal of Bifurcation and Chaos*, 4(2), 245-302.
20. Goncalves, P., Riedi, R. H. and Baraniuk, R. G., (1998) Simple Statistical Analysis of Wavelet-based Multifractal Spectrum Estimation, *Proceedings of the 32nd Conference on 'Signals, Systems and Computers'*, Asilomar, 7.
21. Riedi, R. H., Crouse, M. S., Ribeiro, V. J., and Baraniuk, R. G., A Multifractal Wavelet Model with Application to Network Traffic, *IEEE Special Issue on Information Theory*, 45 (4), 992-1018.
22. Sendov, Bl. (1997) Multiresolution Analysis of Functions Defined on the Dyadic Topological group, *East Journal of Approximations*, 3 (2) 225-239.
23. Sendov, Bl and Marinov P. (1997) Orthonormal Systems of Fractal Functions, *Mathematica Balkanica, New Series*, 11 (1-2), 169-200.
24. Sendov, B. & Marinov, P. (1998) Binary exponential fractal functions, *Fractal Calculas & Applied Analysis*, 1, 23-48.
25. Sendov, Bl. (1998) Adaptive approximation and compression, *Approximation Theory IX*, (eds. Charles K. Chui and Larry L. Schumaker), 1-8.
26. *Approximation Theory*, Vanderbild Univ. Nashville, Tennesse, USA, January 3-6, p. 8.
27. Sendov, Bl. (1999) Adaptive Multiresolution Analysis on the Dyadic Topological Group, *Journal of Approximation Theory* 96, 258-280.
28. Sendov, Bl. (1999) Walsh-Similar Functions, *East Journal of Approximations*, 5 (1),1-65.

LESSONS LEARNED FROM RECENT FLOODS

PETR VOZNICA

Territorial Defense Forces HQ
Army of the Czech Republic
Tabor, Czech Republic

1. Introduction

The Territorial Defence Forces of the Army of the Czech Republic organise and provide military relief work during disasters at an operational level with activities throughout the whole territory of the Czech Republic. An overview of such work is provided with respect to three aspects:

- the current organisation of the protection of the Czech population against various disasters and also the place and role played by the Army of the Czech Republic,
- the experience from directing the operation to rescue persons and property and remediation of the effects of floods in the Czech Republic during recent years, and
- the measures planned for improving the system of the protection of the Czech population.

2. The Organisation of Protection of the Czech Population against Various Disasters: the Place and Role of the Army of the Czech Republic

From analysis of the current state and prognosis of a probable development of security situation follows a set of security risks and threats that mankind faces in Europe and in the Czech Republic, not only at present but also in the near future. During last decades, in the period of cold war, the influence of military danger prevailed. Military security was considered as a main aspect of security and other security factors were neglected. However, it is the duty of every democratic state, the Czech Republic (CR) not excepted, to ensure the safety of its inhabitants against all threats.

The National Defence Strategy of the Czech Republic determines, on the basis of evaluation of political, military, economic, social, ecological and other factors, possible military, non-military, and combined threats. Currently, the CR does not face any direct military threats, so non-military threats are considered to be more probable. Such threats are primarily industrial, ecological and natural disasters, and catastrophes. From the point of view of causal factors, these are primarily disasters or social dangers.

J. Marsalek et al. (eds.), Flood Issues in Contemporary Water Management, 283–292.

Disasters may be of the following nature:

- natural disasters: fires, floods, gales, large landslides, earthquakes, blizzards, freshets, snow calamity and extensive environmental damage caused by pests;
- industrial accidents: explosions and fires, leakage of dangerous substances, radiation accidents, destruction of buildings, extensive failures of engineering networks, and large traffic accidents;
- epidemics: the serious illness of a large number of inhabitants or animals.

To ensure the protection of the population and tangible property in the case of disasters, the Czech Republic has established the so-called Integrated Rescue System.

The Integrated Rescue System (IRS) is constituted by government decree No. 246, dated 19.5.1993, as a system of links ensuring a co-ordinated approach of the rescue, emergency, special and other bodies, state authorities and local administrations, private and legal persons in dealing with accidents. The Integrated Rescue System is directed towards the protection and rescue of persons, and the protection of the environment, the remediation of impacts and the restoration of values during disasters and accidents. The IRS consists of basic components that are in the state of continual readiness, and of other components, as presented below.

The IRS also functions in wartime by carrying out rescue work for the benefit of the civilian population. The authority responsible for the IRS organisation is the Ministry of Interior of the Czech Republic. This ministry co-ordinates work of subsidiary regional state authorities and local administrations. It may assume control of remedying the effects of a pertinent threat in cases where this threat exceeds the authority of one single regional administration or extends beyond a state border. The creation and operation of the Integrated Rescue System is by law the responsibility of the Heads of District Councils.

Rescue work is directed by the relevant crisis staff, accident or flood commissions, dealing primarily with the issues concerning the exchange of information, co-ordination, readiness of the IRS elements, and their action during a joint intervention. The members of this staff or commissions are representatives of state administration and commanding members of the IRS. Their membership may also include legal and private persons and representatives of civilian organisations. A representative of the Czech Army may also be a member, as long as the Army contributes the appropriate forces and equipment.

For dealing with the most common threat to the Czech Republic, flooding, the 'Flood Plan for the Czech Republic' has been prepared under the supervision of the Ministry of Defence. In the case of occurrence of a large flood, the Central Flood Commission and its staff is established. According to the disaster character, the Government Commission for Radiation Accidents, the Central Epidemic Control Commission and others are created under the control of particular departments.

The Army of the Czech Republic is not a permanent member of the Integrated Rescue System, except for the Civil Defence units, according to the principle that army forces and equipment may only be employed in cases where other forces, equipment and resources of competent departments of state authorities are insufficient, or the extent and intensity of distress exceeded their abilities from the very beginning of the disaster.

The role of the Army of the Czech Republic (ACR) within the IRS is very significant, even though its abilities to provide humanitarian relief work during disasters are limited. Military forces and equipment are the best organised and flexibly controlled to be able to provide such services as rescue operations, communications, transport, medical and first aid services, and search and supporting activities. The Army can respond promptly and operate quickly, actively and independently, without outside help, and it is highly mobile. Soldiers are trained in particular professions that are needed for fulfillment of function tasks and they can be also used under very difficult and quickly changing conditions (adverse weather, etc.). From that point of view, it is obvious that military resources have considerable potential to be another tool for providing relief work during disasters. The fulfillment of tasks is limited by the need to replace units, material and by other factors.

The Army may act:

- by deploying small units to reinforce permanent elements of the IRS;
- independently as part of the intervention in carrying out tasks for which the other elements do not have sufficient equipment or scope, or it may contribute to carrying out special tasks.

At present, the ACR does not maintain any special forces or equipment in readiness for immediate deployment in the case of natural disasters and catastrophes. Depending on the development and character of a crisis situation, selected forces and equipment are brought to a level of intervention readiness and they are deployed progressively or en masse to the crisis area. The ACR units then carry out the tasks depending on their predetermination or special equipment.

The document 'The Guidance of the Chief of the General Staff of the ACR for Providing Military Relief Work During Disasters', determines principles and measures for providing military relief work during disasters. Above all, this work is directed to rescue human lives and important tangible property, and to secure a consistent approach within the ACR and with respect to state authorities. The 'Territorial Defence Forces (TDF) Commander's Guidance for Providing Military Relief Work' follows from the previous document and applies to the whole army. It was promulgated on April 1,1999.

The main principles of the TDF Commander's Guidance are:

- the plan and control system of relief work provided by units of the ACR is built on a basis of military regional defence authorities (TDF Headquarters (HQ), Regional Defence Forces (RDF) HQs, Military District Authorities), using formations of all kind of units and the Military Police,
- system of relief work of the ACR conforms to the structure and activity of the IRS,
- soldiers in active service and formations of the ACR are used during disasters to provide military relief work,
- direction and co-ordination of activities of deployed forces is provided at all levels of command through operational centres,
- responsibilities of particular command levels for organising, planning and providing relief work and also the rights and duties to communicate with the corresponding state authorities and regional state authorities are firmly determined,

- the system of authorities for providing military relief work is set. It is the garrison commander, the TDF Commander and the Minister of Defence or the Government of the CR who decide about using the forces,
- the content and scope of crisis management documentation is strictly defined; the Working Plan for Providing Military Relief Work serves as a basic document,
- for providing this military relief work, the appropriate forces will be permanently designated on a regional basis.

Some more details about the military part of Civil Defence (CD), the rescue and training bases of the CD that are parts of the IRS. These formations work according to special guidance and maintain intervention forces and equipment in graduated time standards (a reconnaissance rescue squad, a rescue and recovery detachment). Each of these unit has a binding responsibility for certain parts of the territory of the Czech Republic. The deployment of the equipment may be required by, in succession, by the relevant Head of the District Committee, the Head of the Regional Civil Defence Office, the Civil Defence Unit located in the area, and finally the Chairman of the Local Council.

3. Experience Gained from Directing Rescue Operations and Remediation of Flood Effects in the Czech Republic in 1997 and 1998

3.1. 1997 FLOODS

On July 4, 1997, flood conditions arose following substantial rainfall in the upper parts of the Morava, Oder and Elbe river basins. The flooding affected approximately one quarter of the territory of the Czech Republic, with serious effects in more than 20 districts of the CR and less serious effects in many others. It was undoubtedly the largest flood in Moravia and Silesia during this century. This flood surprised all by its magnitude and intensity, and continuation of extreme flows. The damages caused to properties of the state, towns, villages, and inhabitants exceeded 63 billion crowns; fifty people died, 2,152 apartments were completely destroyed and 18,526 houses were damaged.

As the flood situation was developing in the particularly afflicted areas, the specific Flood Commissions were activated and particular components of the IRS started to work. The deployment of the ACR forces and equipment was carried out mainly in the first days of the flooding on the basis of decisions made by garrison commanders and unit commanders in the affected areas. In the following days, the co-ordination of military relief work was carried out in accordance with the requirements of individual District Crisis Staff by the Crisis Staff of the ACR from the Olomouc area under the command of Maj.-Gen. Voznica. The mass deployment of the ACR forces and equipment was carried out on the basis of the Chief of the General Staff directive from July 10, 1997.

In this action, attention concentrated primarily on:

- providing assistance during rescue work in the affected areas and evacuation of inhabitants,
- assistance in the remediation of flood impacts,
- providing assistance during the distribution of humanitarian aid, and
- carrying out tasks in support of the Police of the Czech Republic.

During the most critical period of the flooding (July 12), the number of persons engaged from all sections of the Czech Army was 5,472 in total, and up to 819 pieces of equipment were employed. The number of helicopter sorties carried out was 2,332, with a total of 738 flying hours. In total, 36,920 persons were evacuated, and 605 tonnes of bottled drinking water, 330 tonnes of water in cisterns, 175 tonnes of bread and 384 tonnes of other materials were transported to the affected population. The total costs to the Army of the Czech Republic for the whole duration of the floods reached 807.8 million crowns. The army performed tasks in afflicted areas until mid November 1997.

The forces and equipment of the CD rescue and training bases were used to carry out rescue work and the provision of humanitarian aid to the population. During the flood culmination, the numbers of soldiers and equipment deployed were as high as 1,300 persons and 290 pieces of equipment.

The extent of the floods, the loss of life and property and the extent to which ACR forces and equipment were used, totally exceeded the existing policies. For this reason, all the benefits and shortcomings of the activities of the ACR during crisis situations manifested themselves to their full extent.

In this connection, one can name the following list of positive factors:

- the ability of commanders and staff to react operatively and with initiative to the situation which had arisen;
- the responsible approach to carrying out tasks of the overall majority of Czech Army members who contributed to the implementation of measures during the crisis situation.
- the army proved great ability to prepare and perform any non-standard operation at any location in the CR, whenever necessary.

Among negative factors, one could name the following:

- the floods showed that with respect to the organisation of the ACR, force numbers and local distributions, the present ACR is only capable of carrying out crisis tasks to a limited extent,
- insufficient equipment of troops, in particular with portable communication equipment (cell phones and so on),
- apart from this, there were some inadequacies in legislation (not only relating to the use of the ACR for remediation of flood damages), which led to problems in the co-ordination of rescue work. There is no law which would clearly state which supreme body should direct and co-ordinate all rescue work. In certain cases, therefore, the military crisis staff were established before the civilian component crisis staff whose decisions they were supposed to implement,
- some organs of state administration failed to react in time to the situation which arose, predict how the situation would develop, and direct all the necessary

action operatively. Documentation provisions were not based on material provisions.

- some civilian bodies were not capable of creating the necessary directing organs even after several days and weeks, and were completely dependent on the ACR directing organs. The army that should have been only one of the components involved in rescue of persons and property became the entity that organised work and engaged its dispensable forces and equipment during the whole period of crisis and also during substantial time of the remediation of flood consequences.

The ACR (as one of the components of the IRS) is obliged to carry out tasks according to the requirements of civilian directing organisations. In no case should it be used to substitute for civilian organisations and institutions which should be responsible to a decisive extent for dealing with the crisis situation.

3.2. 1998 FLOODS

The 1998 flood afflicted a part of Northeast Bohemia, especially the Rychnov and Kneznou District territory, and also partly the City of Hradec Králové. On a relatively small area of approximately 500 km^2, that was afflicted by an extreme rainstorm, arose a devastating catastrophic flood that resulted in 6 deaths, 40 completely destroyed houses, hundreds of flooded houses with big damage to equipment and property, and dozens of destroyed bridges. The total flood damage was estimated at approximately 1,8 billion crowns. During the culmination period of the flood, the number of soldiers and equipment of the ACR employed reached 591 persons and 67 pieces of equipment (not counting the equipment of rescue and training bases of the CD).

The experience with organising military relief work in the Rychnov nad Kneznou district definitely proved essential improvement in comparison to coping with a similar situation in Moravia in 1997, especially with respect to:

- the components of the flood service at all levels were activated immediately after the flood situation arose,
- the district crisis staff worked highly professionally and mastered organising rescue work and clearance of flood damage. Of course there was a great influence of experience gained from coping with previous flood situations,
- co-ordination of all components that participated in relief work and in remediation of flood damage was organised at a high level. Summaries of information about flood situation were provided continuously and all available forces and equipment were used effectively to fulfill emergency tasks,
- parts of the ACR, including the CD forces, were employed since the first day of the flood, July 23, 1998. The deployment of forces was organised professionally and promptly. Directly in the Town of Rychnov nad Kneznou, a control group of military relief work was established and consisted of members of the TDF HQs that was the only body commanding the employed forces and equipment of the ACR and also providing their direct control and co-ordination of activities,
- assistance of parts of the ACR including the rescue and training bases of CD was well-timed and effective and was very positively accepted by the state

authorities and civilian population. This appreciation was documented by the recognition of the best members of the ACR who participated in work and also by a festive farewell to the deployed forces at the town square in Rychnov nad Kneznou, after completion of work.

Of course, during the deployment of the army, some problems appeared again, for example:

- there was still an absence of the necessary legislative measures for prevention of the occurrence of, and solutions to, crisis situations and specification of obligations of particular components and institutions. Such measures should define the place and role of the ACR,
- the principle of using task forces only if the abilities of the state authorities are not sufficient has not been always followed,
- the ACR does not posses enough special equipment or its operators when there is an emergency need (especially engineering equipment),
- at the higher levels of command, there is an absence of special permanent positions for dealing with crisis situations.

With respect to the 1997 and 1998 floods, it can be stated that in spite of the earlier presented problems, the ACR fully accomplished the tasks of military relief work during its deployment in 1997 and 1998, and its activities significantly strengthened its relationship with the general population.

4. Planned Measures to Improve the System of Protection of the Population of the Czech Republic

The crisis readiness of state has not been yet solved with respect to legislation and organisation. However, last year and also this year, significant steps have been taken to build a system of crisis management. Based on the decree of the CR government, dated June 10, 1998, and on the basis of authorisation of the Constitutional Law of Security of the CR, the State Security Committee has been established as an advisory body of the government for ensuring security of the CR. Furthermore, the Defence Planning Committees and the Civil Emergency Planning Committees were also established.

The Civil Emergency Planning (CEP) is considered as reaching and maintaining the readiness of state offices to solve non-military crisis situations and to secure civil resources for solving both military and non-military crisis situations. The CEP is under the Ministry of Interior and creates narrow concurrence on defence planning, whose role is, among others, to plan support activities of the ACR during solving serious non-military crisis situations causing distress of population. The Ministry of Interior is the central administrative authority for co-ordination of preparations and solving crisis situations, in the sphere of domestic security and public order, through a system of protection of population.

The Ministry of Interior leads the preparation of a draft of the crisis management law that determines concrete tasks for state authorities, local administration, and legal and private persons for building up the crisis readiness of the state and while managing crisis situations. The Fire Brigade Management of the CR is preparing a draft of the Integrated Rescue System Law. Based on recommendations of the Legislative

Committee of the CR government, both drafts will be fused into one common draft of the Crisis Management and Integrated Rescue System Law.

The draft law is constructed on the principles of full consideration of directing crisis situations and on the principle of prime control role of the state in crisis situations. It is a responsibility of state authorities and regional state authorities to secure directing, organising and co-ordination roles to prevent occurrences of crisis situations and to solve them.

State authorities and regional state authorities will have an extra important role during the process of creating conditions for asserting control in crisis situations. The main reason for that is that they are the bodies with the largest burden of organisation, coordination and implementation of specific measures during preparation for crisis situations, but mainly during defusing such situations.

The draft also assumes to establish Security Committees of superior regional state authorities, District Security Committees, and Local Administration Security Committees that should replace the present committees which were established to solve specific extraordinary events.

The specific sphere of the proposed legislative adjustment is the legislative foundation of the IRS. The necessity of its legislative adjustment is due to its social need. The IRS is constructed for liquidation of extraordinary events when human lives, property or environment are in danger and when solving extraordinary events needs to determine specific authorities and obligations. The system will enable coordination of state authorities and security, rescue, emergency and expert services, state authorities and regional state authorities that participate in remediation of accidents.

The basic components of the IRS will be the Fire Brigade Corps of the CR, other Fire Brigade units, Medical Rescue Service and the Police of the CR that will be able quickly and continuously intervene and operate over the whole territory of the CR. 'Other components' (among them also the ACR) of the IRS are called on to intervene according to the character of disaster. Their abilities to intervene and their jurisdictions are given by the legislative regulations.

The new law, that should be proclaimed approximately on Jan. 1, 2000, will, of course, specify a definite role and place of the Ministry of Defence and the ACR. The Ministry of Defence is, according to the draft law, the central state authority for co-ordination of preparations and solving crisis situations in the field of external military threats to the CR, for co-ordination of fulfilling military alliance obligations abroad and accession of the CR into international humanitarian operations and international peace support or peace keeping operations carried out by the armed forces.

The ACR is then also involved in a system of domestic security. Together with other armed security and rescue, accidents, emergency, expert and other services, forces and public associations, it will create the so-called other components of the IRS that will provide assistance on call.

The constitutions of the crisis management law and the IRS will be, after its endorsement, put in details in particular departments' guidance documents and standards. It is expected that the system of providing military rescue relief work will not only respect the constitutions of the new law but also that the experience gained from remediation of flood damages in previous years will be considered there. Those are especially the following measures.

- A crisis management system will be created in the ACR to deal with cases of non-military distress, determination of crisis management authorities and set up of their responsibilities. The creation of crisis management working positions will be a great contribution as well.
- Jurisdiction of the governing authorities will be specified and determined with respect to the levels of coordination with state authorities and regional state authorities.
- Regional principle of control authorities of the ACR for planning and directing military relief work will be preserved.
- Military relief work will be based on a permanent predetermination of forces.
- A system of preparedness for coping with crisis situations will be established.

In practice, the main principles of military relief work will be based on planning and control processes of military regional authorities, i.e. the Territorial Defence Forces Headquarters, the Regional Defence Forces Headquarters, the Military District Authorities that will use all kind of forces, the Military Police and also, after reorganisation, the training and rescue bases of the CD, which are subordinate to the TDF HQs, according to a permanent predetermination. Co-ordination and control of the activity of the deployed forces will be realised at all levels of command through operational centres. For providing military relief work, a system of jurisdiction will be specified, including when a garrison commander and the TDF Commander can decide about using forces (it is proposed as 50 persons only for up to 24 hours) and similar decisions need to be made by the Minister of Defence of the CR or the CR government.

The responsibility for organisation, planning and provision of military relief work will rest with:

- the Chief of the General Staff of the ACR, at the central level and in Prague garrison,
- the TDF Commander at an operational level,
- the RDF Commanders, at the regional level,
- Chiefs of the District Military Authorities and of Local Military Administrations in Brno, Plzeň, Ostrava, at a district level, and
- Garrison Commanders in towns with garrisons.

Improved military relief work will bring the expected accession of rescue and training bases of CD from subordination to the Ministry of Interior into the organisation of the territorial defence forces. After their reorganisation, they will fulfill tasks under the command of RDF Commanders in particular regions. There will also be detached reserves from these forces and equipment. By these measures, the readiness of the ACR to respond to requests for remediation of consequences of possible disasters within the whole IRS will increase as a whole.

5. Conclusions

The Army of the Czech Republic demonstrated during the deployment in 1997 and 1998 its ability to carry out non-standard intervention (providing of military relief work) wherever and whenever necessary. To its credit, it also saved many human lives and a large amount of material property. According to public opinion, the ACR provided one of the most significant and effective forms of assistance during the flood period. As a result, the Army's public prestige has significantly increased. At present, in concurrence with the accession of the CR into NATO, it is necessary to review the current legislative measures and the state structure, to learn lessons from the inadequacies and to universally improve the system of protection of the population and its property against crisis situations.

EMERGENCY FLOOD CONTROL OPERATION DURING THE 1997 RED RIVER FLOOD

R. J. BOWERING, P. ENG.

Manager, Surface Water Management Section
Manitoba Water Resources
Box 14, 200 Saulteaux Crescent
Winnipeg Manitoba CANADA R3J 3W3

1. Introduction

In the spring of 1997 the largest flood in over 100 years inundated much of the Red River Valley. In the Canadian portion of the basin 1800 sq. km. (700 sq. mi.) of valley lands were inundated, creating a lake 80 km (50 miles) long and up to 40 km (25 miles) wide. Approximately 20 million sandbags were used to protect homes and property in the valley, and 28,000 residents were evacuated from their homes. The flood protection works that had been constructed over the past 40 years were pushed beyond their design capacity, but were successful in preventing serious damage in any of the protected communities. This paper will describe the flood protection works, and operation challenges faced during the 1997 flood.

2. The Red River Basin

The Red River rises near the Town of Whapeton, North Dakota, approximately 400 kilometres south of Winnipeg and flows northerly into Canada. The Assiniboine River joins the Red River in the City of Winnipeg approximately 97 kilometres upstream of the mouth of the Red River at Lake Winnipeg (Figure 1). The total drainage area of the Red River at Winnipeg is 287,500 square kilometres (111,000 square miles).

The basin is remarkably flat, dropping only 71 metres (233 feet) over a distance of 870 river kilometres (550 miles). The slope of the river varies from 25 cm per kilometre in the headwaters to four cm per kilometre at the Manitoba border.

Major floods on the Red River have all been associated with snowmelt. Most large floods have been caused by a combination of heavy fall rains which saturate the clay soils, a large snow pack, and a rapid melt. Heavy rainfall during the melt period often exacerbates the flooding.

J. Marsalek et al. (eds.), Flood Issues in Contemporary Water Management, 293–301.

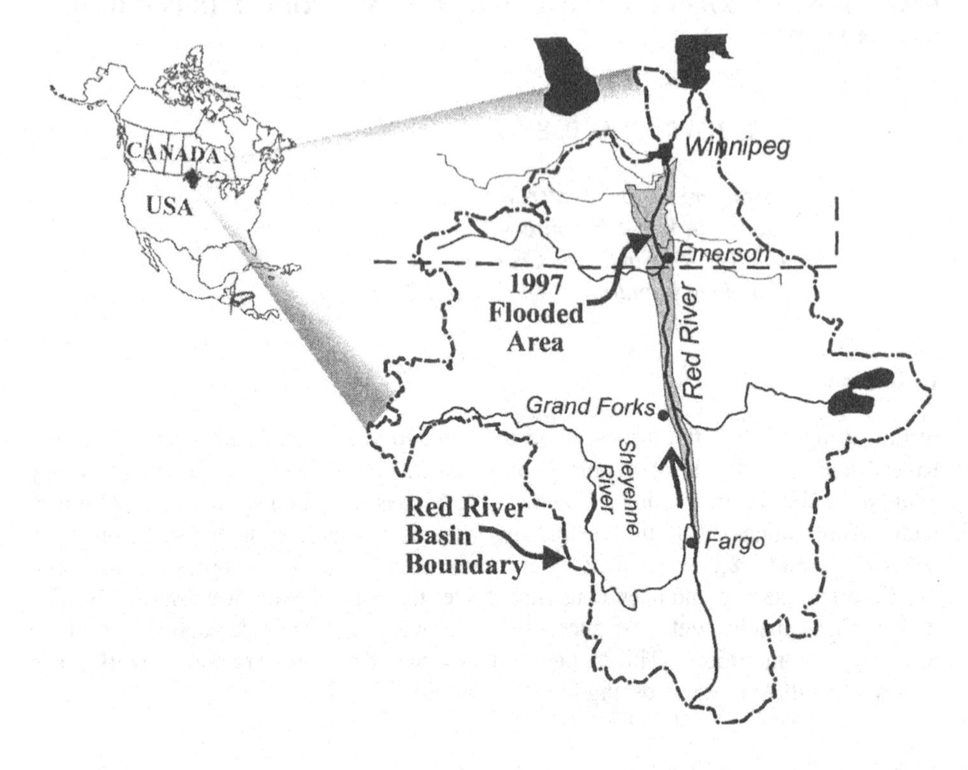

Figure 1. Red River Valley Location Plan.

The very flat valley slopes result in a slow progression of the flood wave. In most years this enables flood forecasters to provide considerable warning to valley residents, thereby enabling them to take emergency actions to minimize their vulnerability to flooding. Such actions would include moving of farm animals, relocation of equipment, and raising of temporary dikes. However it also increases the period when rainfall can add to the flooding.

3. History of Flooding

Although the 1997 Red River flood was denoted the "Flood of the Century", over the past 170 years, major floods also occurred in 1852, 1861, 1882, 1950, 1966, 1974, 1979, and 1996. The most severe documented flood occurred in 1826. It essentially destroyed the Red River Settlement, now the City of Winnipeg. The magnitude of the flood, some 1.9 metres (6.2 ft.) higher than the benchmark flood of 1950, forced the local residents to retreat to the only areas of high ground, Stony Mountain and Bird's Hill, in the vicinity of Winnipeg.

The long gap between 1861 and 1950 is of particular significance in that this was a period of major settlement and development in the valley. The valley residents experienced some local flooding but they appeared to have regarded high water as a part of life in the region. Communities such as Morris, Emerson, Grand Forks, and Fargo simply evacuated low-lying lands and rebuilt after each inundation.

The first major flood of the 20th century occurred in 1950. Flooding of communities from Fargo to Winnipeg caused widespread evacuation of residents and extensive damages. In Winnipeg 100,000 people were evacuated from their homes, and in the valley south of Winnipeg the total population was evacuated. Damages were extensive, particularly in the City of Winnipeg where most of the downtown area was under water.

4. Development of Flood Protection Works

Immediately following the 1950 flood on the Red River, the federal government, together with the Province of Manitoba, set up a fact-finding commission to appraise the damages and make recommendations. The commission recommended in 1958 the construction of the Red River Floodway, the Portage Diversion and the Shellmouth Reservoir. Figure 2 shows the flood protection works now in place.

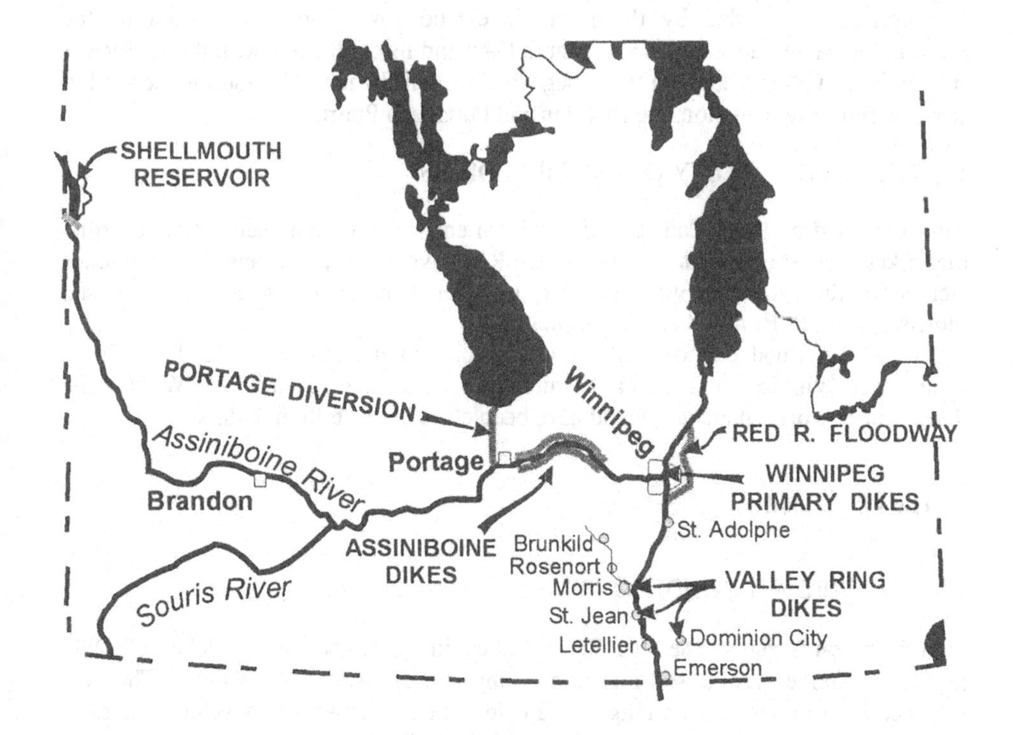

Figure 2. Flood Control Works in Manitoba.

4.1. THE RED RIVER FLOODWAY

The Red River Floodway is a 46.7 km (29 mile) channel that diverts floodwaters around the City of Winnipeg. It has a design capacity of 1,700 m^3/s (60,000 cfs) and an emergency capacity of 2,830 m^3/s (100,000 cfs). Flows into the Floodway are controlled by two submerged gates in the control structure in the river just north of the Floodway entrance. Construction was completed in 1968.

4.2. PORTAGE DIVERSION

The Portage Diversion is a 29 km (18 mile) channel designed to carry flows up to 700 m^3/s (25,000 cfs) away from the Assiniboine River and into Lake Manitoba. Construction was completed in 1970.

4.3. SHELLMOUTH DAM AND RESERVOIR

The Shellmouth Dam is an earth structure on the Assiniboine River in western Manitoba. It is 21m (70 feet) high and 1,270 m (4200 feet) long. Construction was completed in 1972.

The reservoir created by the Shellmouth Dam is approximately 56 km (35 miles) long and is capable of storing 480 million cubic metres (387,000 acre feet) of water. The protection afforded by the reservoir extends over the entire reach of the Assiniboine River between the Shellmouth Dam and its confluence with the Red River at Winnipeg. Cities other than Winnipeg which benefit by not only flood reduction but also low flow augmentation, are Brandon and Portage la Prairie.

4.4. RED RIVER VALLEY COMMUNITY DIKES

After the flood of 1966, Canada and Manitoba entered into an agreement to construct ring dikes around eight communities in the Red River Valley. The eight communities included in the agreement were Emerson, Letellier, Dominion City, St. Jean Baptiste, Morris, Rosenort, Brunkild and St. Adolphe.

In 1979, a flood occurred which was similar in magnitude to the 1950 flood. Damages in Winnipeg and in rural Manitoba were less than $10 million. Without the flood control works, damages would have been close to one billion dollars.

5. The 1997 Flood

5.1. ANTECEDENT CONDITIONS

Significant rains across the Red River Valley in September and October of 1996 resulted in above average soil moistures going into the winter of 1996/97. This was followed by an unusually prolonged and cold winter. Snowcover developed in early November and remained until mid April, 1997. Over the winter there were four blizzards producing record snow packs throughout much of the Valley.

On February 24, 1997 the first Manitoba flood forecast was released. This forecast stated that with adverse weather over the remainder of the winter, flood levels on the Red River could "....surpass all previous floods this century". An update on March 21 reiterated the potential for a major flood to develop.

A two-week period of gradual melting from March 21 to April 4 reduced the snow pack, and the flood situation appeared to be improving. However, a major storm on April 4 dumped from 50 to 90 mm of precipitation over the entire basin. This storm virtually guaranteed that a major flood was imminent.

5.2. FLOOD PREPARATION

5.2.1. Community Ring Dikes

The forecast issued after the April 4 blizzard showed that there was a potential that the flood could overtop some of the ring dikes around Manitoba communities. Therefore, the ring dikes were raised by from half to one metre, and during the flood period the ring dikes were constantly monitored to guard against failure.

5.2.2. Flood Forecast Communication

As the flood progressed the forecast centre distributed daily flood information sheets showing current levels and flows along the river and forecast peak levels and dates. These were put on the internet and were distributed widely by e-mail and fax. Also, a full time flood media contact was appointed. He, along with an emergency measures spokesman, briefed the media through interviews and daily news conferences.

5.2.3. Flood Liaison Offices

To assist Valley residents to protect their property, staff were located in Flood Liaison Offices in the Valley. They provided site specific water level forecasts to the rural residents so that they could construct temporary dikes. Many of these local sandbag dikes successfully protected the homes, but many others failed due to inadequate construction, or simply too much water pressure caused by the depth of water and by wave action.

5.2.4. Extension of the Floodway West Dike

The West Dike of the floodway was constructed to prevent Red River water overflowing into the La Salle River, thereby bypassing the Floodway Control Structure and increasing flood levels in Winnipeg. It continues for 34 kilometres (21 miles) south and west of the control structure, tying into high ground south of Domaine, Manitoba.

The April 18 forecast predicted that levels between Morris and Brunkild Manitoba could rise to 784 feet. The land west of the West Dike is at around 780 feet. Therefore the province quickly initiated an emergency raising and extension of the dike. Between April 24 and 30 the West Dike was raised where necessary, and a 34 kilometre (21 mile) extension was constructed. A task that would normally take months to complete was accomplished in just six days. At the peak of construction more than 400 pieces of heavy equipment and 450 civilians worked on the dike. Military equipment and skilled personnel contributed greatly to the effort.

At the same time that the dike extension was under construction, a flood task force was established with significant support from virtually all sectors of the water resources community in Manitoba. Under the direction of the Provincial Water Resources Branch, this group studied the West Dike extension to recommend on additional protection measures, and to assess the downstream impact of a dike failure. A dam break analysis indicated that a major failure of the dike would inundate the La Salle River valley from Starbuck to St. Norbert, and would permit a flow of 1100 m^3/s (40,000 cfs) to bypass the control structure.

Based on this analysis a decision was made on April 28 to evacuate the La Salle River Valley and the Village of St. Norbert. Residents were given 36 hours warning. Approximately 3,000 people were evacuated. At the same time the City informed residents along the Red River in Winnipeg that they must be prepared to evacuate on short notice should the extended West Dike fail.

The Task Force then developed an emergency operating plan for the Floodway Control Structure. The purpose of the plan was to enable the operators to minimize flooding in Winnipeg under a dike breach scenario. Figure 3 was developed by the Task Force to show flows and levels that would occur under the scenario of a multiple breach of the West Dike and emergency operation of the control structure. The analysis showed that with careful operation flooding in Winnipeg could be controlled, but levels south of the structure would have to be raised an additional 0.9 metres (3 feet), resulting in a floodway flow of 2240 m^3/s (79,000cfs). This flow would be 30% higher than the design discharge for the floodway.

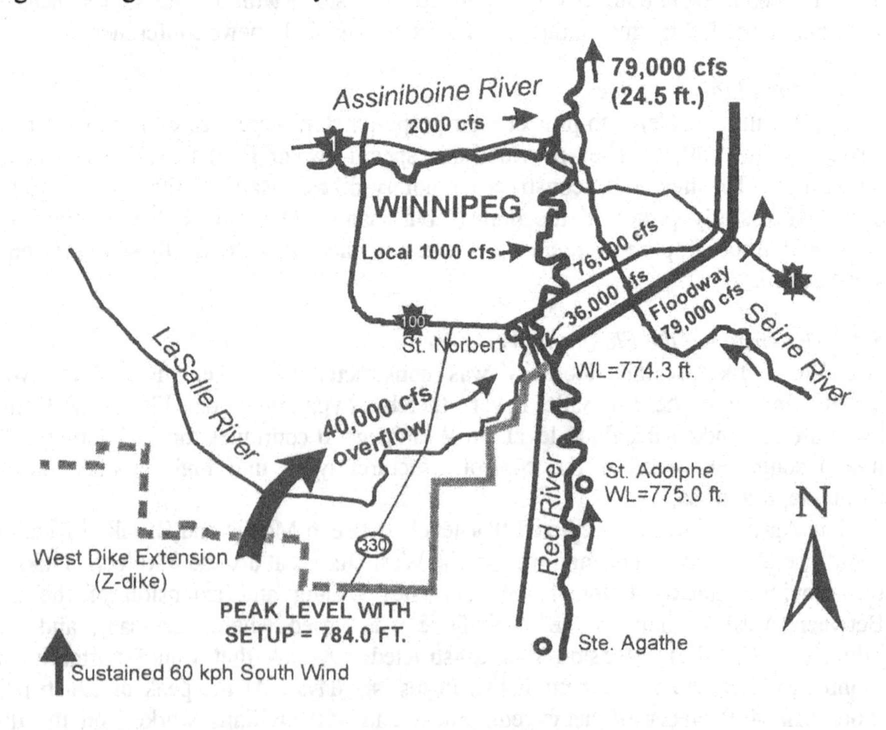

Figure 3. Worst case scenario of a West Dike breach.

5.3. FLOOD CONTROL STRUCTURE OPERATION

As the peak flows approached Winnipeg, the two structures on the Assiniboine River were operated so that flows joining the Red River at Winnipeg were as low as possible at the time of the crest on the Red River. The Shellmouth Reservoir had been drawn down over the winter 5.1 metres (16.7 feet) below the summer target level to provide flood storage. This is the lowest the reservoir has been drawn since it was built. Over the next six weeks, levels rose 6.5 metres (21.3 feet) storing 320 million cubic metres (259,000 acre feet) of runoff. The Portage Diversion was operated so as to divert virtually all of the Assiniboine River flows north to Lake Manitoba at the time of the peak on the Red River.

On the Red River the normal operating strategy for the Floodway Control Structure is to hold levels south of the structure no higher than they would be with no flood control works in place. The structure had been operated in this manner in every year of operation up to 1997.

Thursday, May 1, 1997 marked a historic day in the operation of the floodway, and proved yet again that we were truly in "uncharted waters". Levels in Winnipeg were approaching the top of the primary dikes, particularly in the south end of the City. If levels in the City rose any higher there would be considerable risk that the primary dikes may fail, thus resulting in extensive flooding in the City. Therefore, the floodway was operated to hold levels in the City constant, and to pond levels south of the City above the levels that would have occurred without flood control works in place. The result was that peak levels south of the structure were approximately 0.6 metres (two feet) higher than they would have been. This, of course, caused considerable consternation in the valley. The May 3rd, 1997 headline in the Winnipeg Free Press read "Suburb sacrificed to spare Winnipeg".

5.4. 1997 FLOOD DAMAGES

The only community in Manitoba that was flooded was Ste. Agathe. This community was not protected by a ring dike because it had never experienced flooding in the past. However, many homes in the valley were flooded, and many of these had to be torn down and rebuilt. The overall cost of flood preparation, flood fighting, and damages in the Manitoba portion of the basin has been estimated to be $500 million. Had the flood control works not been in place the damage, particularly to the City of Winnipeg, would have been catastrophic. The waters would have inundated thousands of homes and most of the downtown area. The famous intersection of Portage and Main would have been under 1.3 metres (4.5 feet) of water.

5.5. EFFECTIVENESS OF THE WEST DIKE EXTENSION

The raising of the West Dike proved to be critical in preventing flood waters from overflowing into the La Salle River. However, levels against the westward extension of the dike never reached the heights that had been forecast. In fact, the land remained dry at the location where most concern was focused. What had not been recognized was the effectiveness of the road grids in the valley in retarding the progression of the flow towards the dike. More importantly, wind conditions were favourable during the high-water period. With a strong and sustained south wind significantly more water would have reached the West Dike extension.

6. Post Flood Activities

6.1. MANITOBA WATER COMMISSION STUDY

Although major damages had been averted in the Canadian portion of the basin in 1997, many questions were raised, particularly with respect to the decision to raise levels in the valley south of the control structure to prevent flooding in Winnipeg. Therefore, Manitoba re-constituted the Manitoba Water Commission to conduct an independent review of action taken during the flood, and to provide recommendations to the Minister. The Commission reported in June, 1998 and presented 58 recommendations to the government. These recommendations are currently being addressed.

6.2. INTERNATIONAL JOINT COMMISSION STUDY

Since the Red River is an international river, there are many inter-jurisdictional issues that must be managed and coordinated during a flood. Therefore, the governments of Canada and the United States asked the International Joint Commission to study the flood management in the basin from an international perspective. The focus of this study is to develop tools to aid in flood management, and to examine actions that could be taken to reduce the risk of future flood damage. The report of the IJC is expected late in 1999.

6.3. CANADA-MANITOBA PARTNERSHIP AGREEMENT ON RED RIVER VALLEY FLOOD PROTECTION

Shortly after the Flood, all levels of government recognized that there was a need to improve the level of flood protection in the Red River Valley. A Phase I Agreement was signed between Canada and Manitoba providing a total of $34 million for flood proofing individual homes and businesses in the valley to a level 0.6 metres (two feet) above the 1997 peak level.

Then in March of 1999, a Phase II Agreement was signed providing an additional $100 million to continue individual flood proofing, to construct additional community dikes, to upgrade flood works in the City of Winnipeg, and to expand data monitoring in the valley. This work will be done over the next four years.

Ultimately, however, there remains the need to address the question of what level of protection is adequate. Is a 100-year level of protection acceptable? The 1997 Flood has a return period of just over 100 years, and pushed the flood protection system close to the limit. This issue will continue to be debated over the coming months.

7. Lessons Learned

Although the Red River flood protection works successfully prevented flooding in Winnipeg and the valley towns, the 1997 flood demonstrated that management of a large flood can be very stressful. Four principles were reinforced by the flood.

a. Keep it simple. During a major flood, demands for updated forecasts and operating decisions are relentless. Increased computer power, and developments in runoff and routing software tempt forecasters to use more and more complex forecasting procedures so as to improve accuracy. But, increased complexity creates more things that can go wrong. Manitoba's experience is that the procedures should remain the simplest that will produce reasonable results.

b. Plan for the emergency before the emergency. The towns in the Red River Valley that had a flood response plan in place, and that had tested it in a simulated flood event, were the most successful in managing the flood.

c. Good communication is critical. An excellent forecast is useless if it doesn't get to the people who need it.

d. Permanent works are much safer than emergency works. Although the west dike extension of 34 kilometres in six days was a remarkable feat, the resulting dike was much more vulnerable to wave erosion that a carefully constructed permanent dike would be.

INVESTIGATIONS OF THE EFFECTS OF ON-SITE STORMWATER MANAGEMENT MEASURES IN URBAN AND AGRICULTURAL AREAS ON FLOODS

F. SIEKER

University of Hannover
Institute of Water Resources
Hannover, Germany

1. Introduction

There is no doubt that extreme floods are firstly the result of heavy and widespread rainfall events. On the other hand, there are some impacts on floods due to human actions. For instance the process of urbanising consisting of "sealing the landscape" by buildings and traffic areas and the process of developing sewer systems to discharge stormwater runoff by the principle "as fast and effective as possible" changes the water balance of the catchment significantly. The effects on floods can be twofold: increased flood volume caused by an increase in surface runoff and increased peak values by reducing the duration of runoff. Another example of human impacts on floods is increased runoff from agricultural areas, especially arable lands. Because of the increasing weights of farming machines, the soil becomes compressed and the infiltration capacity decreases. The effects of human actions on floods may differ from event to event with the tendency, "the higher the flood the lower the human impacts", because of the growing runoff from natural parts of the catchments in cases of extreme floods. But, there is no doubt that human activities mentioned before have led to a certain intensification of all floods. So it is logical to ask if there are possibilities to replace these activities partly by alternatives with less impacts on floods and without significant disadvantages. This paper aims to investigate such alternatives related to urban drainage systems and arable lands.

2. On-site Water Management Measures in Urban Catchments

Against the principle of conventional urban hydrology to discharge rainfall runoff as fast and complete as possible to the next receiving water, the principle of on-site storm water management aims to retain water as much as possible on-site without disadvantages for the user of the sites. The measures, applied individually or in combination, consist of man-made infiltration, surface and subsurface storage measures and discharge controlled by throttles. The combination of all three components can be applied independent of the local circumstances. It is called (in German) the "Mulden-Rigolen-System", which may be translated as the "Swale-

J. Marsalek et al. (eds.), Flood Issues in Contemporary Water Management, 303–310.

Infiltration Trench System". Figures 1 and 2 show a cross section and a longitudinal section of the system as well [1].

The layout of the MR-System is rather flexible and can be easily adapted to various soil conditions. In soils with intermediate permeability (10^{-6} m/s), where no discharge from the trenches is expected, the sewer system can be eliminated and the MR-System just comprises unlinked swale-trench-elements. In soils with high permeability (10^{-5} m/s) even the trenches can be eliminated and the system is reduced to swales only. Both these variants are special cases of the overall MR-System. In practice, a certain urban area may be divided into parts of different soil conditions so that the application of different variants within an area may be required.

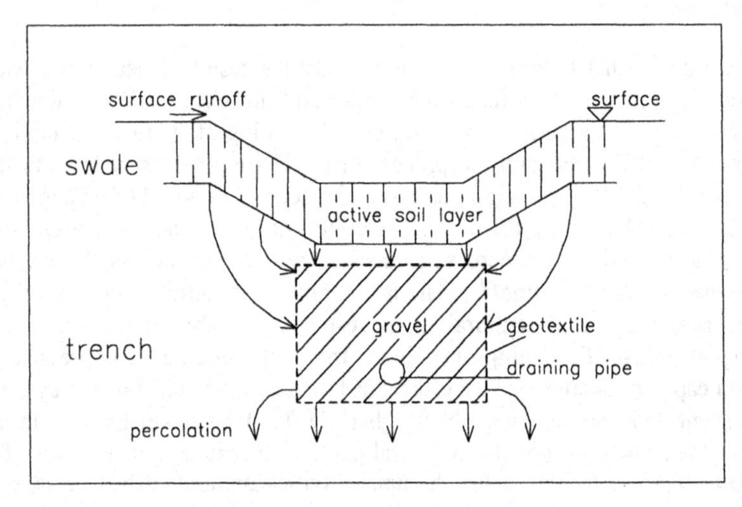

Figure 1. Cross section of the MR-element.

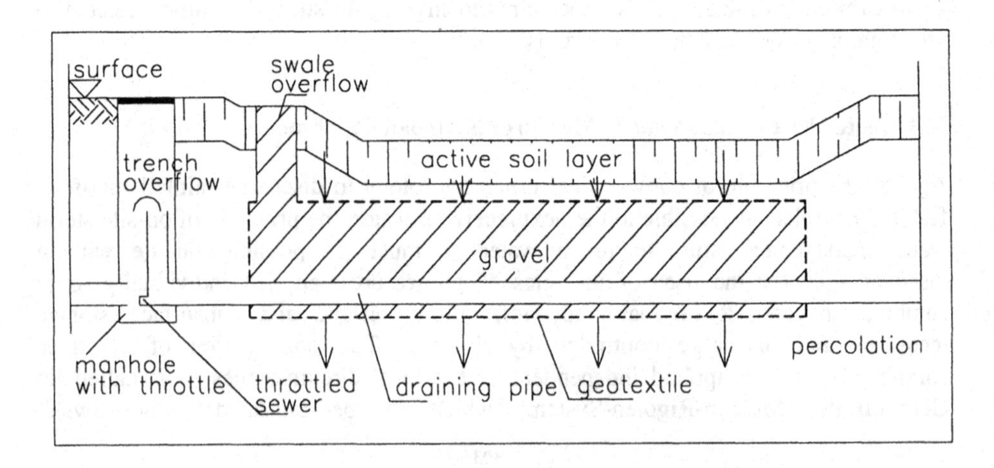

Figure 2. Longitudinal section of the MR-element.

3. Effects of the MR-System on Flood-causing Rainfall Events

The MR-System and its variants are normally constructed to manage short but intense storm events, which are typical for the dimensioning of urban drainage systems. To manage these storm events, the storage volume of MR-elements is usually chosen to have a capacity of about 50 mm rainfall runoff (from the connected impervious area). But extreme floods of rivers with large catchments (e.g. > 10.000 km^2) are caused in Europe by rainfall events of about 200-400 mm, although these rainfall events normally occur within a period of some days up to some weeks. The question is, can the MR-elements manage these events? "To manage" means that the throttled discharge of the elements is at least equal or (better) lower in comparison to the surface runoff of natural pervious areas of the given catchment. To investigate this question, the hydrological processes of various MR-elements were simulated with variation of different parameters like geometry and dimensions of the storage volume, the permeability of the surrounding natural soil, the amount of the throttled discharge, etc.

As rainfall input, a period was used which led to the extreme flood of January 1995 in the catchment of the river Rhine. Figure 3 shows the daily data which were divided for the simulation process in intervals of five minutes. Mainly responsible for the flood is the part of the period between January 20 and 29 with a total of 217 mm and peaks of 72 and 52 mm on January 22 and 25.

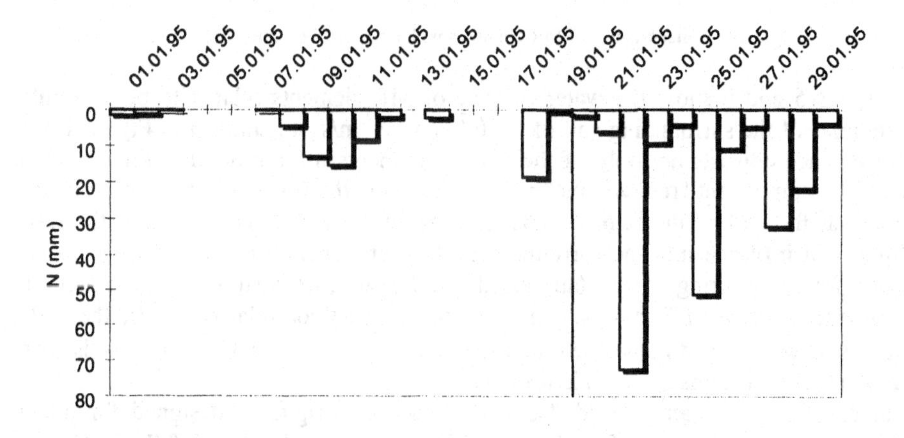

Figure 3. Rainfall period of January 95 at the river Rhine causing an extreme flood.

Figure 4 shows the simulated water balance of an MR-element with different storage volumes described by the parameter "n" (yearly overflow-frequency) and a permeability of the surrounding soil of 10^{-6} m/s. The latter represents a permeability which is typical in the catchment of the river Rhine. The figure shows that in case of a yearly overflow-frequency of n = 0,2 (recurrence interval of 5 years), which is the most commonly used value, the discharge of the MR-element amounts to about 7 % of the input flow. This ratio is significantly below the surface runoff of the permeable areas outside of the urban areas, because the total runoff coefficient of the flood caused by the rainfall event mentioned above amounted to 50%, related to the catchment of the river Mosel, a significant tributary of the river Rhine. In the case of smaller storage volumes smaller than those related to n = 0,2, the runoff coefficient of the MR-element increases up to about 70% in case of n = 1,0. For n = 0,5 the runoff coefficient amounts to 50%, the average of the flood. So in order to reduce flood-related runoff by MR-elements it is necessary to dimension them with n < 0.5 But this is consistent with the aims of urban hydrology.

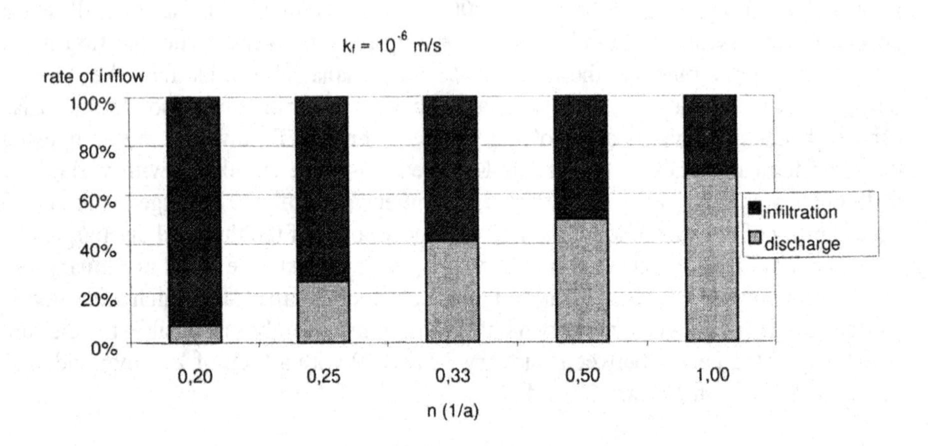

Figure 4. Water balance of MR-elements with a surrounding soil of 10^{-6} m/s.

Figures 5 and 6 show the water balance of MR-elements related to permeability coefficients of the surrounding soil of $5x10^{-7}$ and 10^{-7} m/s. In contrast to Figure 4, the water balance consists not only of the two components infiltration and discharge but also of the component "rest-volume" which represents the water which is stored in the element at the end of the rainfall. Because no other rainfall is following, the rest-volume will infiltrate into the surroundings afterwards and so it can be added to the direct infiltration during the rainfall period. In this way, the results of Figure 5 and 6 are similar to those of Figure 4. In order to reduce flood-related runoff, the MR-elements have to be dimensioned in these cases with n < 0,33. But, this is also consistent with the aims of urban hydrology.

In conclusion, it can be said that MR-elements - originally designed for urban hydrological targets and therefore dimensioned for short and heavy rainfall events - are also able to serve for the management of long rainfall events which lead to floods in river catchments. Surface runoff from connected areas can be significantly reduced. That means (in comparison to normal reservoirs or retention ponds) that not only the shape of the runoff curve will be damped but the volume of the curve will be also reduced. That is a significant difference between these two principles.

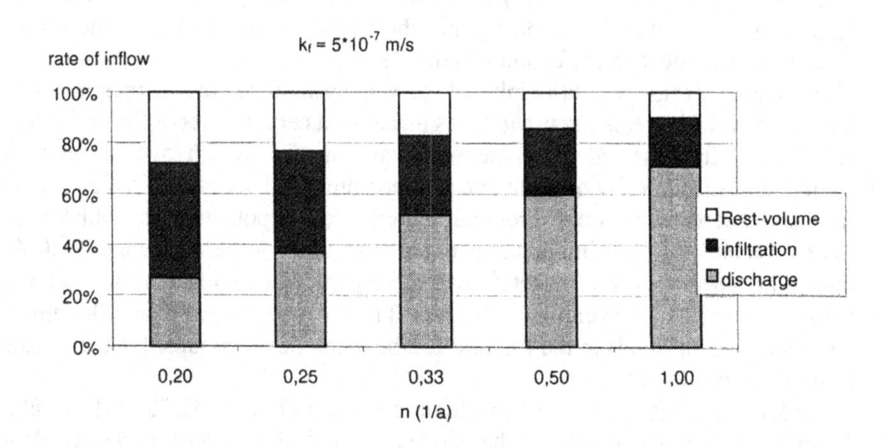

Figure 5. Water balance of MR-elements with a surrounding soil of $5*10^{-7}$ m/s.

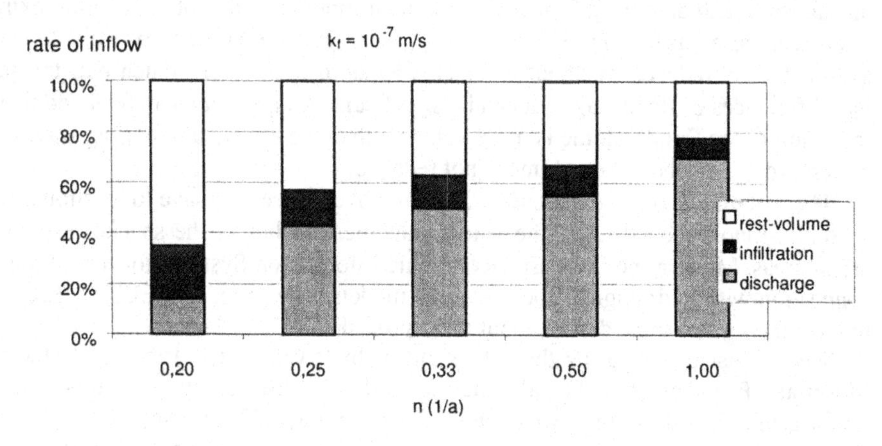

Figure 6. Water balance of MR-elements with a surrounding soil of 10^{-7} m/s.

4. Effects of On-site Measures on Floods in Case of Catchment-wide Application

There is a chance that MR-systems or similar measures will be widely applied in the future not only to new urban areas but also to existing urban areas with existing sewer systems in order to get rid of failings and disadvantages of those systems. Such failings and disadvantages are for instance hydraulic overloading of the systems, the problem of mitigation of combined sewer overflows, the pollution of the receiving waters in general, a deficiency of groundwater recharge and last but not least more or less the increase of floods. To mitigate all these failings and disadvantages without decreasing the standard of the hydrological aims in urban areas, the policy of urban hydrology is changing from the principle "discharge as fast and complete as possible" to the principle of "on-site management" respectively "Best Management Practices". A reason for this change of mind is the rising cost of hydrological infrastructure in urban areas. Therefore, there is not only the task to replace one principle by another but also to mitigate the costs without decreasing the hydrological standards. The principle of the MR-system meets these conditions in general.

Changing from the principle of sewer system to the principle of on-site management within existing urban areas means that certain parts of the connected areas have to be disconnected from the sewer system. To switch over a sewer system completely is technically difficult and not recommended because of the polluted runoff from some parts of the areas. From our experience, the potential area which is suitable to be disconnected in a simple manner can be assumed to be in a range of 20-40 % of the total connected area. To switch over these parts will also take a lot of time even if there is a special subvention program. But, in order to estimate the impacts of disconnecting on floods in the future it can be theoretically assumed that all these areas could be disconnected.

Urban areas with sewer systems cover a portion of about 12 % of the total area in Germany. This portion includes the permeable parts like gardens, parks etc. Assuming that a third of the urban area has been switched over from sewer system to on-site management and assuming further that the total runoff coefficient concerning extreme floods can be reduced from 50 % to 10 % (see Figures 4-6) the flood volume can be assumed to be reduced by about 3 %. This looks to be not very much but, translated into differences of flood stages, it can be significant. And, again it can be noted that the reduction of the flood volume is more valuable than the effects of damping waves only by reservoirs etc where the volume is not reduced.

The real effects of a catchment-wide switch over procedure have to be simulated by a rainfall-runoff modelling of the whole catchment including the special response of urban areas. Modern methods use Geographic Information Systems for this purpose in connection with hydrological and hydraulic models. The paper of HEIKO SIEKER (in this volume) gives some details about these procedures.

We are just investigating the catchment of the river Saar, a tributary to the river Mosel and Rhine as well. To calibrate the model for disconnecting scenarios, we are investigating four different towns concerning their potential for disconnecting and on-site management. Further, we are investigating and modelling two small natural subcatchments of the river Saar in order to get knowledge about the hydrological aspects of the natural parts of the whole catchment (see also paper of H.SIEKER). We hope to finish the research project on "Effects of on-site management measures on floods of the river Saar" at the end of the year 2000.

A preceding study, concerning the catchment of the river "Leine" in Lower Saxony [2], has shown that such investigations are worth performing.

5. On-site Storm Water Management Measures on Arable Lands

Agricultural areas contribute significantly to the development of extreme floods not only because of their extent in comparison with urban areas. Especially, arable lands tend to produce a great deal of surface runoff during the late autumn and in the winter when the surface is bare and uncovered. Soil compression by heavy farming machines, tracks of vehicles etc. led to an increase of runoff from arable lands during the last decades (see paper of VAN DER PLOEG). So it is necessary to think how to reduce these impacts on flood development.

Measures like the MR-system are not as suitable as a standard method in agricultural areas as they are in urban catchments because they are not compatible with the financial possibilities and practical aspects of farming. The best way to reduce surface runoff from arable lands is to renew the infiltration capacity of the soil. Deep loosening of the soil by machines is one possibility. Another more natural method is to do without conventional plough and apply another method called in German: "Mulch-Saat"-here translated to the term "compost-method". Remains of the harvest are left on the surface and just crushed so that the next sowing into the topsoil can be performed. The idea of this kind of "plough-less" farming is relatively old and has been applied especially to areas with wind and water erosion problems. The method has some disadvantages for farmers, because they have to apply more pesticides against weeds. They also have to fight against a growing population of snails and mice. So it is no wonder that this kind of farming is not so popular. But, because of the positive effects against wind and water erosion, the compost method is politically desired and is therefore supported by subvention from the state. In Saxony for instance, farmers get an amount of 50 DM per ha for the application of the compost method and another amount for growing an intermediate agricultural crop to avoid an uncovered surface of the fields during the autumn and winter time. A result of these subventions is that in Saxony, at present, about 50.000 ha are cultivated by the compost method.

But, in relation to the infiltration capacity, there is a significant advantage. The population of earthworms increases significantly if the upper soil layer is not disturbed again and again by ploughing. And earthworms produce positive hydrological characteristics by drilling deep into the soil (up to 1,5 m or more). These bore-holes have a vertical drainage effect and therefore increase the infiltration capacity not only in the upper soil layer but also in the deeper layers.

Some investigations at the institution "Landesanstalt für Landwirtschaft" in Leipzig have shown, that the permeability of arable lands (loess) increases by a factor of 10 by application of the compost method.

In order to bring the idea of water management by compost method closer to the farmers, the following calculations shall be made.

In Germany the water works have to pay about 0,10 to 0,20 DM per m^3 groundwater for extraction (different values from state to state). If, on the other hand, a farmer by application of the compost method increases the infiltration of his arable land by 100 mm per year, then he "produces" 1.000 m^3 per ha. Considering the disadvantages of the compost method it seems only right and proper that the farmer should be paid for the production as the water works have to pay for the extraction. So he can "earn" not only by production of wheat, corn etc., but also by groundwater production.

Another calculation is the following. Measures for flood protection like construction of higher dams, reservoirs etc. are expensive. For instance, the construction costs of a large reservoir lateral to the river Rhine amounts to 5 DM per m^3 storage volume. If a farmer, by application of the compost method, increases the infiltration in such a way that during an extreme flood 100 mm more can be infiltrated than into a conventionally cultivated field, he offers a storage volume of 1.000 m^3 per ha. This represents in comparison to the reservoir volume an amount of 5.000 DM per ha. This amount can be considered as an investment distributed over 20 years. The value of the farmer's preventive flood protection amounts to 250 DM per ha yearly.

The calculations show that farmers can be integrated into a water management program and even into a preventive flood protection program where they can be paid. So, they are motivated to accept the disadvantages which are unavoidable if they apply the compost method. To care for motivation (that means especially to care for a financial basic) is at least as important as the practical aspects of a preventive flood protection program.

In order to prove the effects of a widespread application of the compost method, we are just starting a research project concerning the German part of the catchment of the river "Lausitzer Neiße". It may be possible that later on the investigation can be an extended to the Polish and Czech parts of the catchment. The first step of the investigation is to identify the sizes and local positions of the arable areas which are suitable for the new method. The areas will be considered in different steps depending on soil conditions, motivation of owners etc. For each of the different scenarios, floods will be simulated along the river using historical flood-causing rainfall periods. The results will be compared with the present situation.

6. Conclusions

Flood protection measures may be divided in "reactive" and "preventive" kinds of measures. Reactive measures, for instance, are constructions of reservoirs. Preventive measures serve to reduce the runoff "at the source" by infiltration and on-site retention. There is a significant potential of runoff mitigation in urban areas and arable lands as well. In urban areas application of the so-called "Swale-Infiltration Trench-System", replacing conventional sewer systems, can contribute to a mitigation effect. In arable lands surface runoff can be significantly reduced by application of the so-called "Compost-method", replacing the conventional kind of ploughing. A significant effect of these measures on extreme floods can only be achieved if they are widely applied in a certain catchment. A real problem is to motivate the decision maker to apply these alternative methods. Motivation depends – firstly – on the costs of the measures in comparison with their financial benefits and – only secondarily – on the so-called ecological awareness. In Germany, at any rate, these financial benefits can be proved concerning both urban areas and arable lands.

7. References

1. Sieker, F. (1998) On-site stormwater management as an alternative to conventional sewer systems: a new concept spreading in Germany, *Water, Science and Technology*, **38**, 10, 65-71.
2. Huhn, V., B. Strauss (1999) Alternatives of Urban Rainfall Runoff Management and their Impact on Floods. Journal Wasser & Boden, Parey, Berlin, (in German).

GEOGRAPHICAL INFORMATION SYSTEMS USED AS A PLANNING-TOOL FOR ON-SITE STORMWATER MANAGEMENT MEASURES

H. SIEKER

Ingenieurgesellschaft Prof. Dr. Sieker mbH (IPS)
Berliner Str. 71, 15366 Dahlwitz-Hoppegarten, Germany

1. Introduction

It is undisputed, that a change of land use has an influence on the peak rate of a flood as well as on the flood volume. Sealed areas in urban regions along with a dense drainage system and drainage of farmland in rural areas increase flood waves especially in smaller catchments. Concerning the influence of urbanization on flood peaks in larger catchments, an intensive discussion about the extent of influence is taking place in Germany. But, regardless the extent, the influence of urbanization on floods is not positive.

Conventional technical flood protection measures try to minimize the consequences of floods - not only the 'man-made' part of it - mostly through storage, for example with detention reservoirs. Besides these 'reactive' measures, 'active' or preventive measures became more discussed in the last years. In urban areas the infiltration of stormwater became a real alternative to conventional sewer systems [1]. In agriculture, new (old) techniques of soil management were developed (sub-soiling, mulch) [2].

The legislation in Germany realized these effects and also the possible alternatives and reacted with new sections in the water law, for example in the German Water Act (Wasserhaushaltsgesetz, § la): *Everybody is obliged to take care of not polluting water bodies, using water economically, keeping up the capacity of the natural water balance and not increasing the runoff* (translated faithfully). In some states (North-Rhine Westphalia, Baden-Württemberg, Saarland), stormwater runoff of new buildings or other sealed areas has to be infiltrated by provincial water law. Almost every River Management Plan demands for a limitation of sealing and more retention in the catchment.

In the last few years, many demonstration projects for best stormwater management practices were established. These projects have shown, that alternatives for conventional drainage system are available and that they have many advantages in economical and ecological respect.

Nevertheless, from 1991 to 1996, the total length of public storm sewers (separate and combined) in Germany increased from 267,000 to 290,000 km (~9%) not including open channels and private sewers! With an estimated average width of the catchment of 100 m, this results in newly drained area of approximately 2300 km^2, which is almost the area of the Saarland or 1% of the area of Germany [3].

J. Marsalek et al. (eds.), Flood Issues in Contemporary Water Management, 311–321.

It has to be realized that there is a great difference between ideal and reality concerning stormwater management in Germany . What are the reasons for this difference? The most important reasons are - by opinion of the author - a false consciousness of tradition as well as a lack of information. Another major reason is the fuzzy formulation in the water law ('stormwater should be infiltrated, *wherever it is possible*') or in the River Management Plans. Of course, there are localities where it is not practicable to infiltrate stormwater. But there are also many places where it is possible and not carried out anyway. These cases can be reduced, if a River Management Plan contains a Stormwater Management Plan, which shows the possibilities for different stormwater management practices for every locality. With modern planning tools, like geographical information systems or simulation models, it is feasible to create such a detailed plan even for a large area.

2. Influence of Urbanization on Floods

To backup the opinion that urbanization has a significant influence on flood peak, flood volume and flood frequency, some simple mass balance calculations had been made for the Saarland, a federal state in former central Germany and its main river, the Saar. The Saarland has an area of approximately 2,500 km^2. As a former centre of heavy industry and coal mining, there is a high degree of urbanization in many parts of the Saarland. The mean imperviousness (the degree of sealing) is estimated to be about 13%.

The main river in the Saarland is the Saar, a branch of the Mosel, which is a tributary to the Rhine river. The catchment of the Saar is not identical with the Saarland - a major part of the catchment is in France - but most of the Saarland is in the catchment of the Saar. Frequent floods have been observed along the river, especially in the provincial capital Saarbrücken, where the main highway is flooded several times a year. The Rhine river is known by the two major floods in 1993/94 and 1995, which caused great damages downstream in Germany and in the Netherlands.

The flood in the winter of 1995 was caused by a 3-week rainfall period with a total precipitation depth of approximately 300 mm. The mean runoff coefficient calculated from a flood hydrograph measured at the point where the Saar flows into the Mosel is approx. 50%. The runoff coefficient of a sealed and drained area is almost 100% during this precipitation period. So, the runoff coefficient of the unsealed areas must be less than the mean runoff coefficient 50%, approx. 43%. Assumed that this runoff coefficient was valid for the sealed areas before they got sealed, the sealing of 13% of the area results in an increase of the runoff coefficient of 17%.

As another characteristic, the change of the mean drainage density illustrates the influence of urbanization. The total length of all rivers and creeks in the Saarland is approx. 1,400 km [4], so the natural drainage-density is 0.56 km/km^2. In consideration of the total length of the combined sewer system (approximately 6,000 km, [3]), the drainage density increases to approximately 3.0 km/km^2.

3. Planning of On-site Stormwater Management Measures

The variety of stormwater management alternatives is as large as the objectives of stormwater management [5]. Classical goals of stormwater management are, for example, the reduction of combined sewer overflows or the minimization of the sewer overflow frequency. Maintenance of natural groundwater recharge and flood mitigation in large catchments became objectives during recent years. Best Management Practices should consider all of these goals. Unlike conventional storage (e.g. retention ponds), which are not able to reduce the discharge volume, certain on-site stormwater management measures like infiltration facilities or trough-trench-systems turned out to be very effective in all of these aspects [6].

At the beginning of the planning process for an on-site stormwater-management system, the following questions have to be answered.

- *Are on-site storm water-management measures applicable under the local conditions?*
- *Which is the optimal technique for on-site storm water-management under the given conditions?*
- *Where are the optimal locations for storm water-management measures?*
- *Which degree of disconnection of presently connected areas can be achieved under the given conditions?*

The answers to these questions depend on a large variety of factors. They can be classified as

- *natural factors, i.e. the infiltration capacity of soils or the slope of the ground,*
- *quality factors, i.e. the pollution level of the connected impervious area or*
- *land use factors, i.e. the type of building, agricultural use or the space available for distributed measures.*

All of these factors have a spatial dimension and can be presented in maps. Besides that, non-spatial factors have to be considered as well, e.g. economical reasons, legal restrictions and acceptance by inhabitants, authorities or operating companies.

With a Geographical Information System (GIS) the factors can be collected, analyzed, transformed and presented. It is possible and necessary to

- *combine different maps with different scales*
- *create new classes of factors*
- *create DEMs (digital elevation models)*
- *combine different factors, so that new spatial contexts for on-site stormwater management measures can be visualized*
- *automatically create thematic maps*
- *update data and maps easily*

With the experience of many projects it was possible to develop a set of rules for a classification of the factors: For example,

- *distributed infiltration troughs at the side of streets are useful, if the slope is less than 4%,*
- *a trough-trench-system is applicable if the groundwater table is more than 1.50 m below ground level,*
- *a simple infiltration trough has an area demand of 20% of the connected impervious area, a trough-trench-system about 10%*

The following examples illustrate a technique that has been developed to plan on-site stormwater management systems in urban areas. It has been applied to several larger cities in Germany, for example Berlin (800 km^2) or Chemnitz (127 km^2). At the moment, a similar technique is used to investigate the potentials for on-site stormwater management measures in rural and agricultural areas. Of course, in this case other factors are important than in urban areas. But the technique of using GIS as a planning tool is the same.

4. Natural Factors

A Geographical Information System is able to change maps. The first step in using GIS as a planning tool for on-site stormwater management is to classify the different factors.

Normally, natural factors give answer to the question, which is the best technique for on-site stormwater management? Those natural factors can be divided in a few main factors. Most of them can be derived from soil maps, geology maps and digital elevation models (made by GIS). As an example, two typical natural factors and their possible classification are presented. The area of investigation is Chemnitz, a town in Saxony, East Germany. The total area is about 127 km^2.

4.1. INFILTRATION CAPACITY

Probably, infiltration capacity is the most important natural factor. The higher the infiltration capacity is, the easier and cheaper on-site stormwater management measures can be implemented. The following classification is applicable for climate conditions in Germany (and many other European countries).

TABLE 1. Application of different stormwater infiltration techniques depending on infiltration capacity

Infiltration technique	Infiltration capacity of soil (m/s)
Seepage in the area	$K_f > 2x10^{-5}$
Troughs	$K_f > 5x10^{-6}$
Troughs-Trenches-Elements	$K_f > 1x10^{-6}$
Troughs and Trench-Systems with throttled discharge	No limitation

The information about infiltration capacity of the examining soils can be obtained from soil maps and infiltration experiments ('open-end' tests). If no value for infiltration capacity is available, it can be derived from soil-types. The result is a modified soil map that is classified by infiltration capacity dependent on the technique for distributed stormwater management (Figure 1).

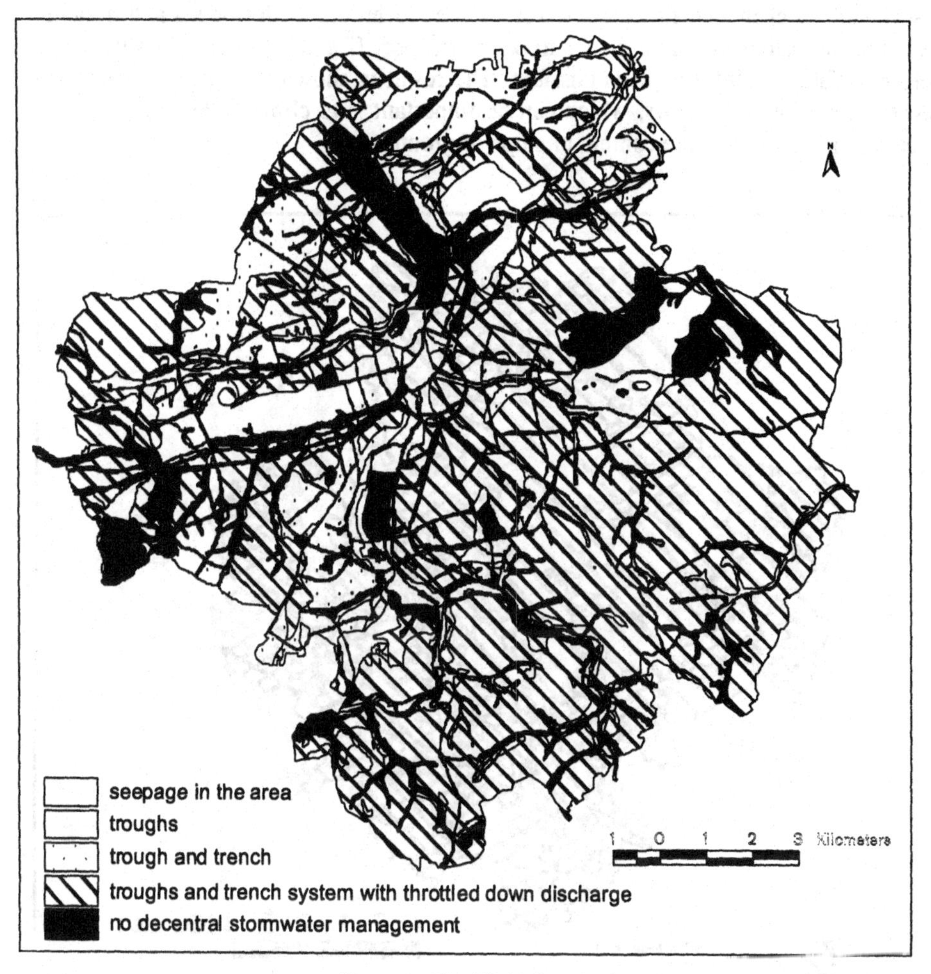

Figure 1. Classified infiltration capacity.

4.2. GROUNDWATER LEVEL

Groundwater level also influences the application of on-site stormwater management. To prevent water contamination, for example in areas of water reserve, there are often restrictions on stormwater infiltration seepage. In some states regulations demand a thickness of topsoil in stormwater infiltration measures of at least 1,00 m. Thus, the groundwater map will be classified in two classes: Groundwater level more or less than 1 m below surface. The result is shown in Figure 2.

Even if there are no other restrictions on stormwater infiltration, the combination of infiltration areas and groundwater level already results in "negative areas" (no stormwater infiltration recommended). This process of combination is called overlaying and is usually the second step after classification.

According to the classification of the infiltration capacity, different techniques for distributed water management are necessary. The technique with troughs and trenches demands an infiltration more than 1 m below surface. If groundwater is less than 1 m below surface, no infiltration is possible. The result is shown in Figure 3. There are several other natural factors, which can be important, e.g. climate, thickness of soils (weathering), soil moisture content or slope of ground.

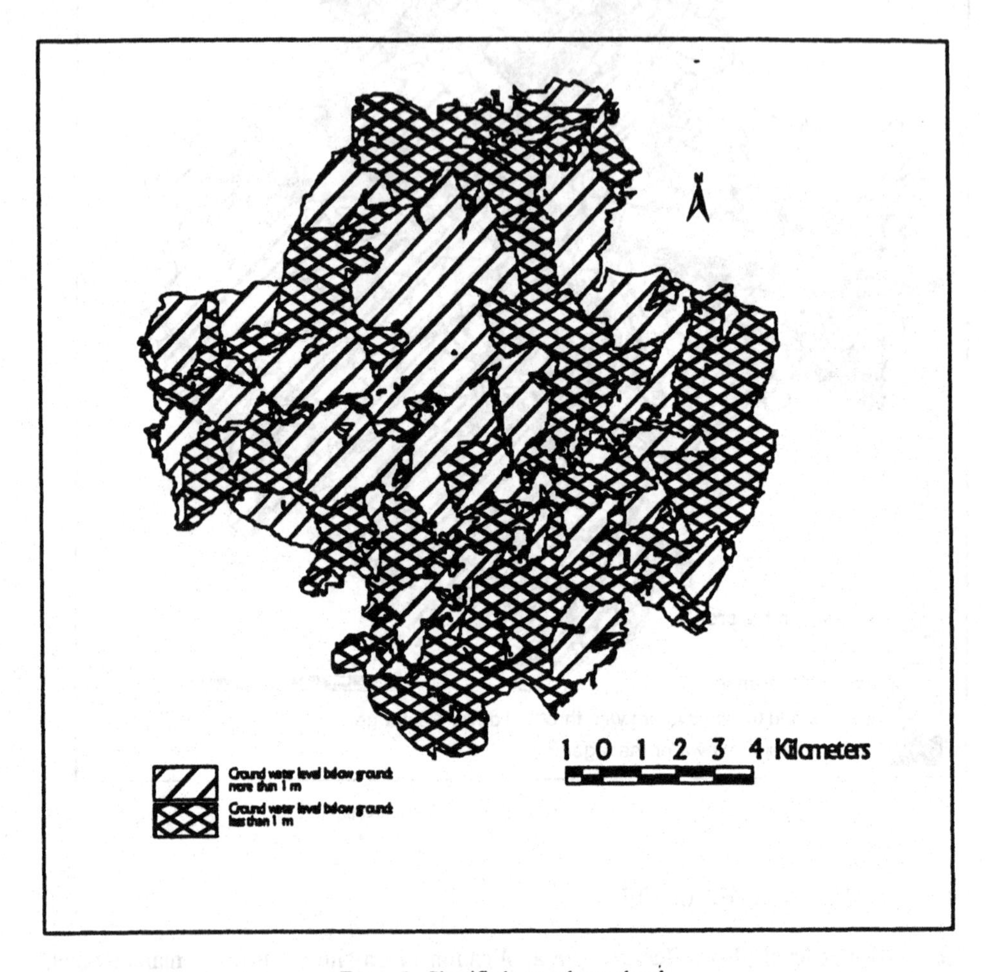

Figure 2. Classified groundwater level.

5. Land Use Factors

Land use factors give information about the potential of disconnection of impervious areas from the sewer system. At first, the impervious area must be divided into different degrees of pollution. The lower the pollution is (i.e. roofs), the better infiltration of stormwater runoff is possible. This information is difficult to obtain. Only few maps with this information are available (interpretation of aerial photos).

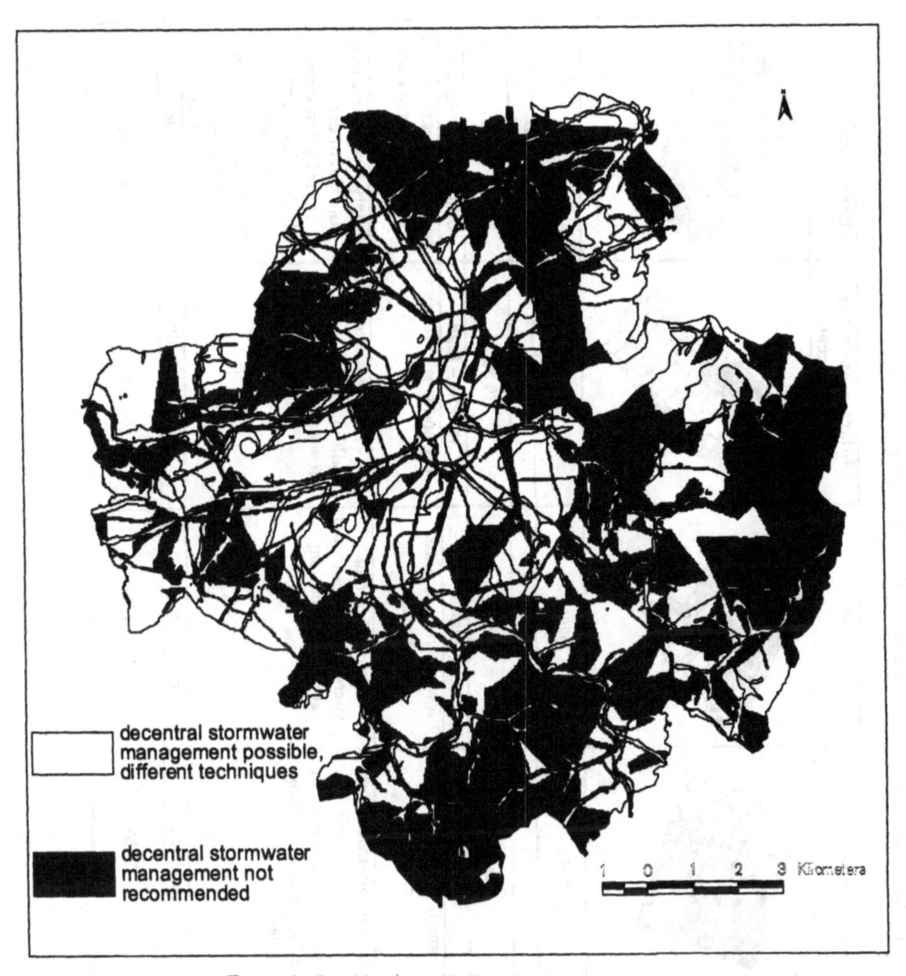

Figure 3. Combination of infiltration capacity and groundwater level.

5.1. TYPE OF BUILDING

The factor 'type of building' gives information about the area, which is available for on-site stormwater management (potential of disconnection). On average, at least 10% of the impervious connected area is demanded for infiltration. Table 2 shows typical urban areas and their potential of disconnection.

The relation between building and garden (or playground) is very important. For example, terraced houses have only limited potential of disconnection, because the small garden is often used for different things (big trees, terraces, etc.). In combination with a zoning (land development) plan, Table 2 allows the interpretation of aerial photographs to determine the potential of disconnection.

TABLE 2: Decentral stormwater management in urban settlements, potential of disconnection

type of building	detached house, big estate	detached house, little estate	flats	Mixed areas in the historical center	Mixed areas outside of center
sketch of piece of land	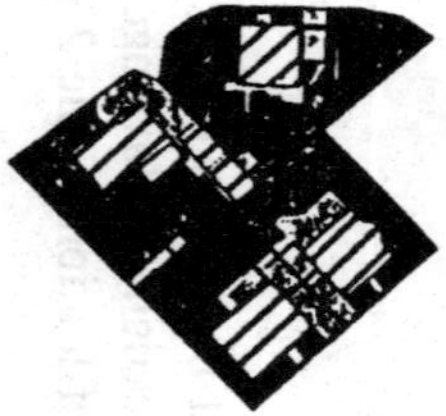		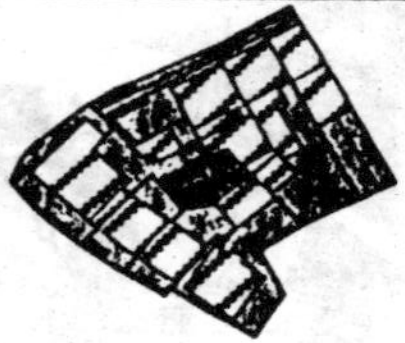	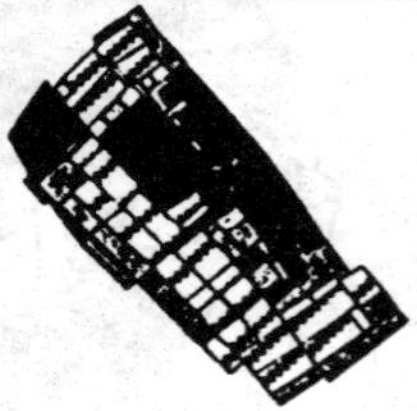	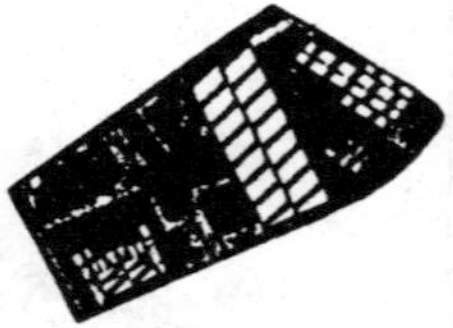
Impervious area, Avers	45 %	49 %	92%	64 %	64 %
Pervious area,	55 %	51 %	8 %	36 %	36 %
Degree of connection	65 %	63 %	99 %	89 %	97 %
Building density	low	middle - high	high	middle – high	different, low to high
Structure of garden and playgrounds	big garden, big estate	small gardens; little estate;	few green areas; little estates	big and small estates	big estates, much area of impervious empty site
Possibilities of decentral stormwater management + suitable - unsuitable	+ building density low; many areas for infiltration; monetary motivation - intensive use of green area	+ monetary motivation - many different ownerships; intensive use of green area; building density often high	- many different ownerships; few potential area for infiltration;	- high density of buildings; towncenter; few potential area for infiltration	+ only few ownerships; potential of unsealing, unused green areas - high building density, polluted imperviousvareareas, few potential area for infiltration
estimation	very suitable	suitable	partly suitable	partly suitable	suitable
potential of disconnection (estimated)	**50 – 70 %**	**30 – 50 %**	**0 – 10 %**	**5 – 20 %**	**20 – 30 %**

5.2. SECURITY DISTANCE

Not only in large scale projects GIS is a useful tool. As illustrated in Figure 4, buildings with basements have to be protected from wetness. Each piece of land and the relation of potential area for distributed stormwater management measures and impervious area can be derived from the map and a security distance set by GIS. The recommended security distance is about 6 m, but can be less (house without basement).

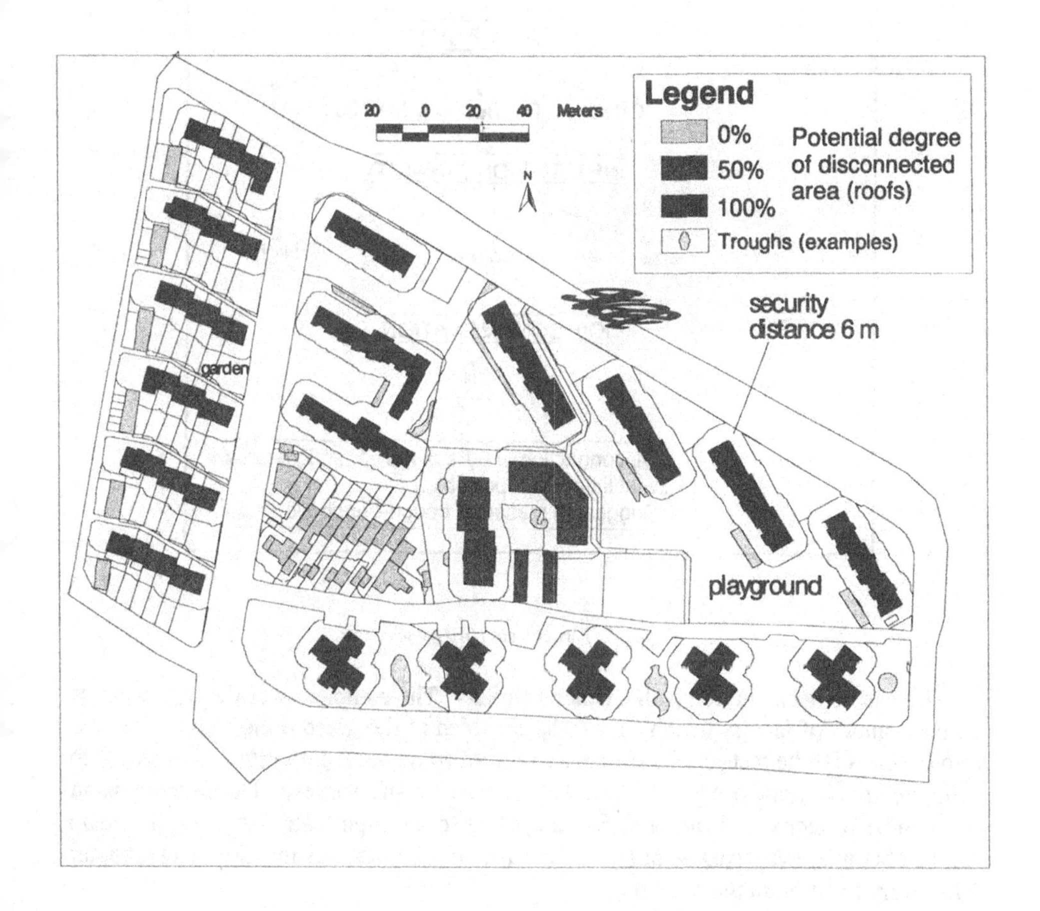

Figure 4. Result map for a project area in Hameln, Germany.

6. Decision Tree

To get a systematic scheme, all these rules used to classify the factors can be integrated in a "decision tree". Figure 5 shows one (simplified) path of a decision tree.

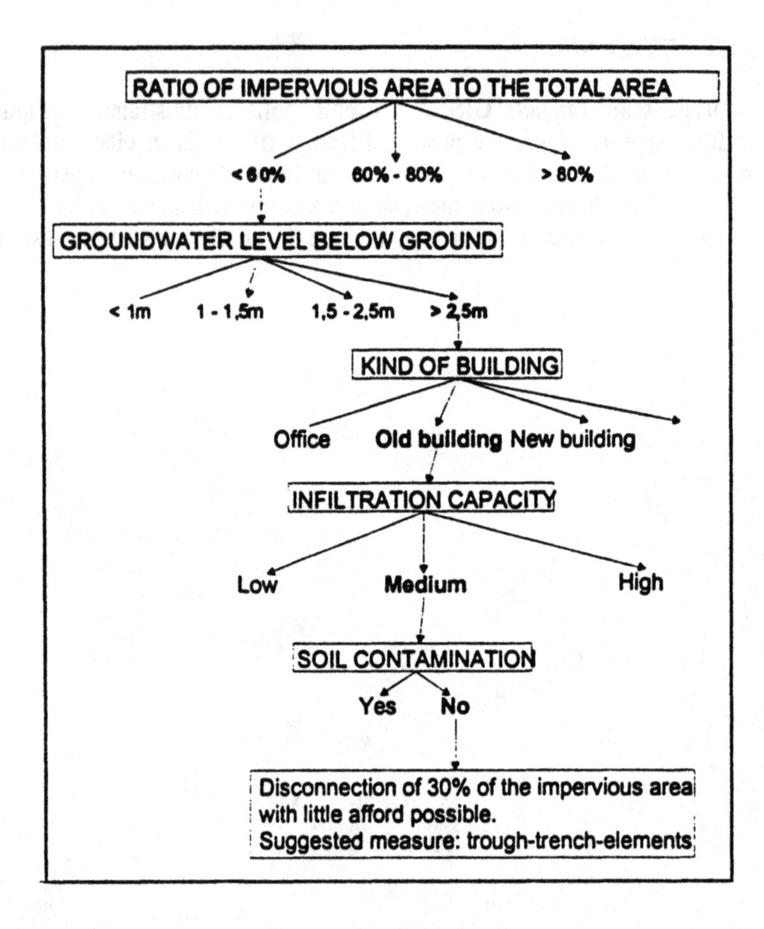

Figure 5. Part of a simplified decision tree.

There is no decision tree with a general validity. The complexity of the tree depends on the number of factors which have to be regarded in the given project area. The tree also depends on the restrictions which are prescribed by local authorities. It is useful to calibrate and validate a decision tree with several on-site surveys. The final decision tree can be implemented into a GIS and applied to the input data. As a result, a map can be obtained, that gives an answer to the four basic questions mentioned in Chapter 3 for every location in the project area.

7. Conclusion

Without a doubt, changes in land use and therefore urbanization has an influence on floods. For smaller catchments, the negative effects of extensive sealing and drainage had been proven. Based on this and other realizations, on-site stormwater management and especially infiltration measures are demanded by law and several regulations. But there are difficulties to translate that demand into practice.

The possibilities of implementing on-site stormwater management measures depend on a large variety of small scale factors. On the other hand, measures are effective only when implemented large scale. This discrepancy makes the large scale planning of on-site stormwater management difficult and asks for new planning tools. It is necessary to add maps that show the requirements for on-site stormwater management especially to River Management Plans.

Several projects have shown that Geographical Information Systems can be a very useful tool for this task. With GIS, it is possible to collect, classify, overlay, analyze and present all the different natural, quality or land use factors even for large project areas in a reasonable time. The examples presented are focused on urban areas only, but it is possible to transfer the knowledge to rural and agricultural areas. For flood protection, the latter is very important.

8. References

1. Sieker, F. (1998) Principles of on-site stormwater management. Handbook of software TRINTSIM Ingemeurgesellschaft Prof. Dr. Sicker mbll (IPS), Berliner Str. 71, 15366 Dahlwitz-Hoppegarten, Germany.
2. Van d. Ploeg, R.R. and Sieker, F. (1999) Bodenwasserrückhalt zum Hochwasserschutz durch Extensivierung der Dranung landwirtschaftlich genutzter Flächen. Wasserwirtschaft (angenommen zur Veroffentlichung).
3. ATV (1997) Sewer statistics publication of the German Association for the Water Environment (ATV), www.atv.de, 1997.
4. LFU (1998) Digital environmental information system of the Saarland, Landesamt fur Umweltschutz, SaarbrUcken, 1998.
5. Field, R. (1993) Integrated Stormwater Management, Lewis Publishers, Boca Raton.
6. Sieker, H. and Klein, M. (1998) Best management practices for stormwater runoff with alternative methods in a large urban catchment in Berlin, Germany, Water, Science & Technology, **38**, (10) 91-97.

SELECTED PROBLEMS OF FLOOD PROTECTION OF TOWNS

PIOTR KOWALCZAK

Institute of Meteorology and Water Management
ul. Dąbrowskiego 174/176
60-594 Poznań, Poland

The objectives of the paper are to indicate basic problems associated with flood control in towns situated in the Odra river basin during the flood of 1997 and to outline possibilities for introduction of improvements on the basis of experiences gained in the course of operations.

The summer flood of 1997 was one of the most extensive recorded floods that had ever occurred in Poland. In the Odra basin, where the flood had the most dramatic course, in the majority of profiles all the recorded states were exceeded.

The flood was caused by extreme meteorological conditions which occurred in the Czech Republic and south of Poland in the Odra river catchment basin. Three periods of extremely heavy rainfalls occurred:

- first – from the 3rd to 10th of July
- second – from the 15th to 23rd of July
- third – from 24th – to 28th of July.

Heavy rainfalls occurred nearly in all parts of the basin but the highest and most intense rains were recorded in the southern parts. Rainfalls of extreme intensity and long duration, which occurred in the first phase of the flood in the upper course of the Odra, followed by rainfalls of varying intensity in the entire catchment basin and rapidly changing hydrological situation (waves of rainfalls, broken embankments) all required undertaking appropriate methods of flood control to protect different towns.

On the segment of the Odra river from the Nysa Łużycka confluence up, the situation was particularly dramatic; towns and cities were flooded both by rains and by rivers and water courses flowing through them. Therefore, despite good hydrologic and meteorological forecasts announced at right time, there were very slim chances for effective flood protection of these towns. Atmospheric precipitation of uncommon intensity and duration led to hydrologic phenomena which were not predicted in any flood scenarios.

Towns in the upper course of the river were not flooded; they were simply crushed with huge waves of water. A good example here may be the town of Kłodzko which was overrun by a 10 m high flood-wave (10 m above medium state).

Other examples of towns and cities flooded in this region can be: Koźle, Opole, Racibórz and the biggest one – Wrocław. Wrocław is situated in the Odra valley; it is supplied in this region by tributaries such as: Oława, Ślęza, Bystrzyca, Piława and Widawa. After the flood of 1903, Wrocław authorities assumed a flow of 2200 m^3/s as

J. Marsalek et al. (eds.), Flood Issues in Contemporary Water Management, 323–332.

the basic one for all design solutions planned for the flood control systems to be developed for this city.

The hydrological system of Wrocław is made up of (Figure 1):

- a flood canal (with a parallel navigable canal) whose main purpose is to discharge the flow of 770 m^3/s (35%),
- the proper Odra river with the discharge of 1300 m^3/s (59%) of which 470 m^3/s should flow through the old Odra, and
- a discharge of 130 m^3/s water to the Widawa polder (6%).

During the 1997 flood the adopted assumptions were verified by the flood wave with the flow of 3650 m^3/s, where most of the discharge flowed through the Old Odra and Widawa. Polders, which made up an essential part of the flood control system, had been utilised earlier (to build housing settlements, drinking water treatment plants for the city of Wrocław) which contributed to increased losses and added to threats but, even more importantly, deprived the flood control system of its essential part.

The main tasks that the rescue workers were faced with were

- to protect the remaining parts of the city against flooding, and
- to minimise losses caused by waters flowing through the streets of the city which formed pools of standing water. The biggest depth of water in the flooded city was 9 m.

It can be said that the mistake made consisted in the excessive concentration (including flood control devices) of the trough of the great water, housing settlements and utilisation of flooding areas for other purposes (discussions with 'the greens' about the need to cut bushes and trees from the trough of the great water). It must, however, be emphasised that despite the exceptional magnitude of the tasks and problems, in many instances emergency services successfully protected a number of towns situated along the Odra river above the Nysa Łużycka confluence.

As a result of bursting embankments and water inundating increasing areas of the river catchment, basin retention increased, causing a considerable reduction in the unit discharge down the river. This type of situation increased chances of towns and cities situated further down the river. Another problem was the fate of towns situated in the neighbourhood of retention reservoirs which had to be emptied in critical situations and frequently towns and villages were flooded suffering heavy losses. A good example of such a town was Nysa (Figure 2). Along the segment of the Odra down from the Nysa Łużycka confluence, there were two other characteristic cases of protection of towns against the flood. The first of them was Słubice (Figure 3) where, as a result of the river rising between dikes, the adjacent areas were below water level (town centre – about 4.5 m).

There were the following threats:

- elevated water level in the river; the town below river level,
- filtration causing water in basements and uncontrolled outflow of water via surface intakes as well as flooding of local depressions,
- no possibilities for draining off surface waters flowing from the area of the town,
- defective operation of the sewerage system, and

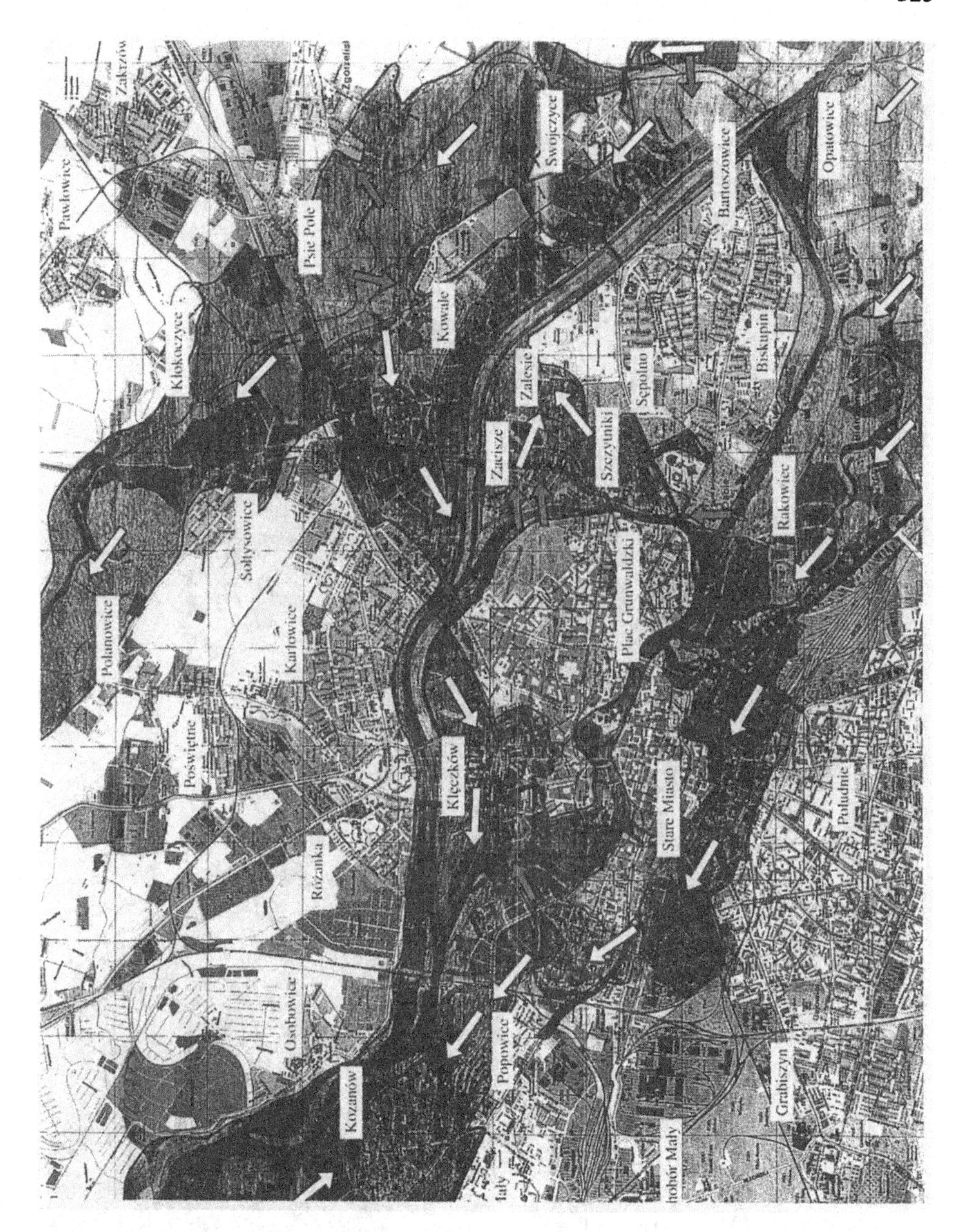

Figure 1. **Flooded area in Wroclaw.**

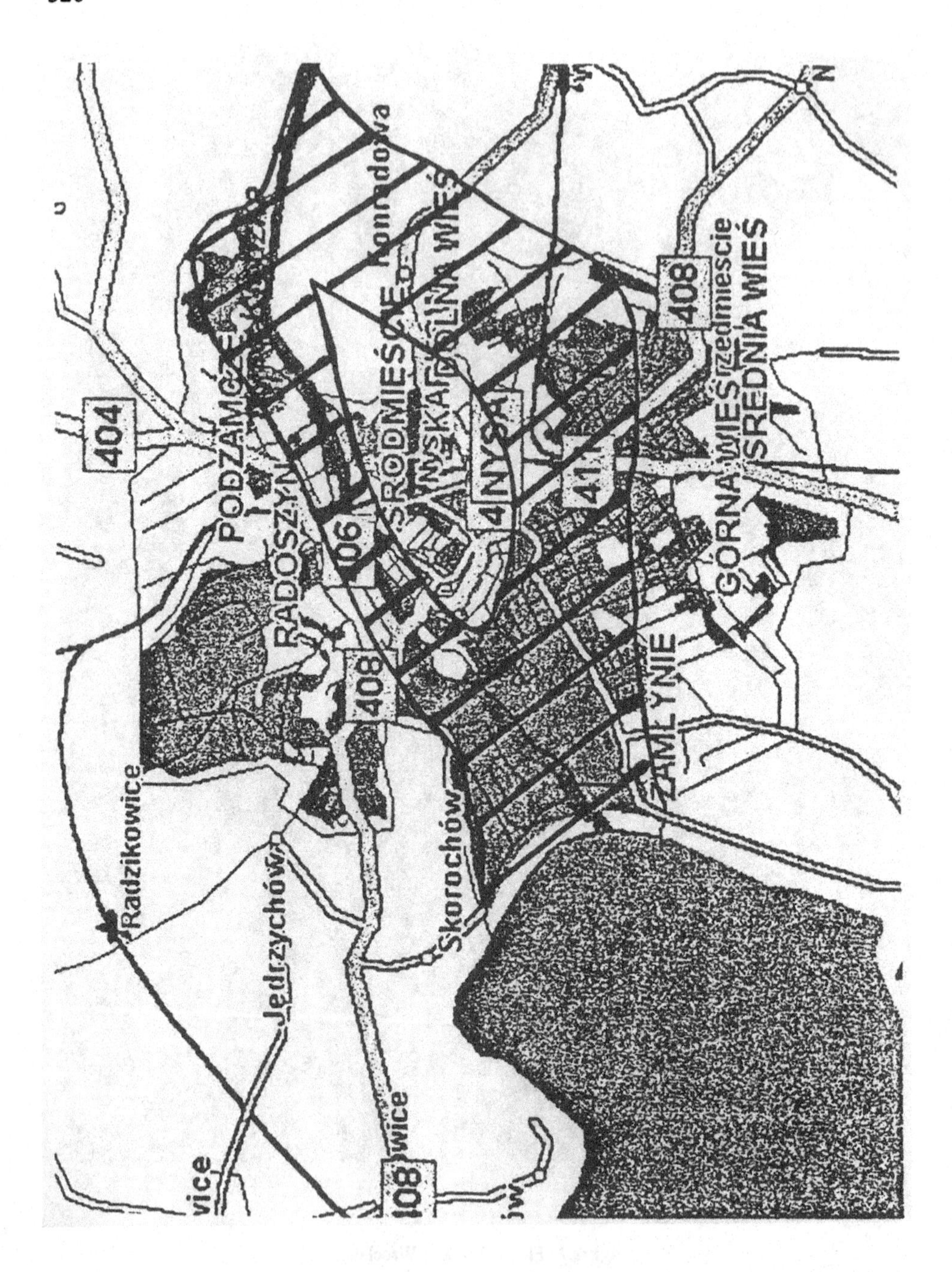

Figure 2. **Flooded area in Nysa.**

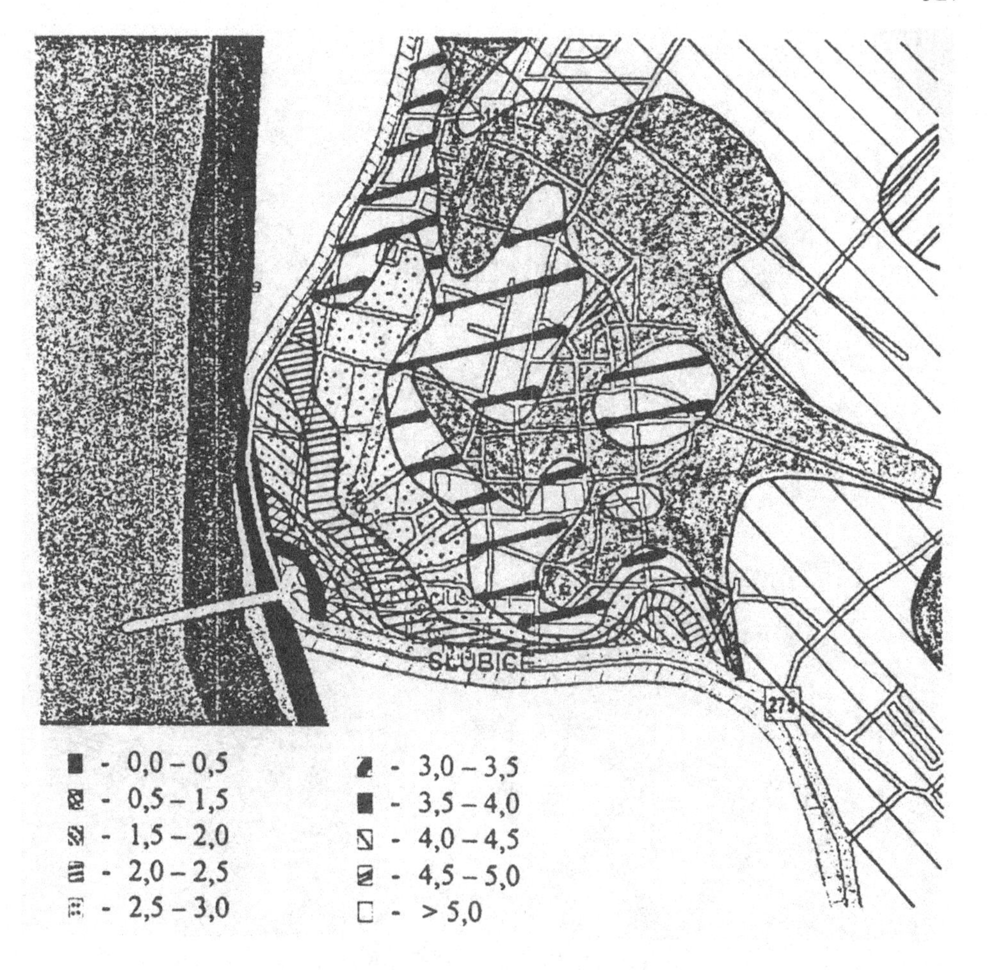

Figure 3. Prognosis of water depth in built-up area of Slubice Town by water level of 700 cm (Source: City Office in Slubice City).

as a result of long-term high water levels (about 6 weeks), earth flood control embankments were threatened; their poor condition was made worse by continuous rains.

The main objective of the flood emergency actions was to protect dikes as even their smallest failure would mean the loss of the town.

After the flood, it was possible to assess both the level of preparation of the town as well as the flood control as exemplary – this refers in particular to such operations as evacuation of citizens, dismantling of fuel stations, operation of services, protection of the property of evacuees,

The second case – Kostrzyn (Figure 4) – is typical for lowland towns. The first line of defence was located on the flood dikes but as water levels continued to rise,

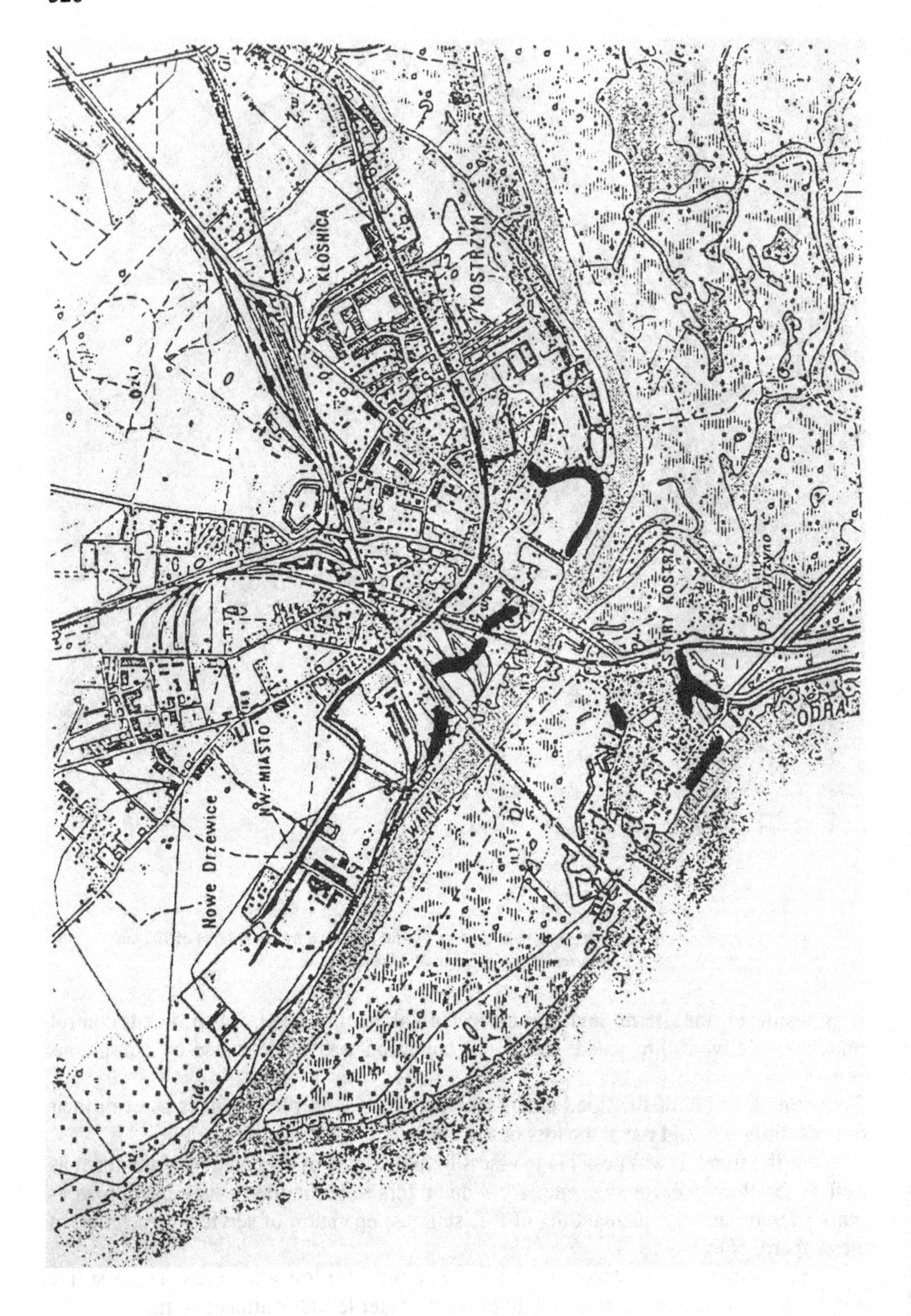

Figure 4. **A diagram of water flood protection for Kostrzyn town (Source: Voivodeship Office in Gorzów city).**

successive lines of defence were built on areas situated higher. In the case of further rises of flood waters, further lines of make-shift dikes were planned within town confines. The main threats were the waters of the Odra and its main tributary – the Warta. The most important task was to maintain successive lines of defence.

There are several remarks concerning towns which were flooded.

- In general, flood control plans assume a positive scenario of the course of the action and do not anticipate disastrous phenomena on a scale that occurred in 1997.
- It is essential to determine the duration of the flooding of individual parts of the town together with estimated velocities of flows in the case of burst dikes.
- It is important to determine the localisation of flood control facilities under construction in order to monitor flooding (enforce the direction of the runoff so that the occurred losses are smaller, local reduction of flow velocity, fewer tasks to be carried out).
- It is essential to determine evacuation routes.
- It is important to indicate the appropriate type of equipment to be used for evacuation, food and drinking water supply, technical and medical assistance for inundated districts (it turned out motorboats were most effective, while army amphibious vehicles, most commonly used in such situations, had limited possibilities of application).
- It is important to determine the possibilities of utilisation of urban structures, or ensuring their protection or shut down.

In the period preceding flooding, it is necessary to solve several basic problems:

- the problem of spatial management,
- ensure that all procedures connected with spatial planning must fulfil strict hydrological conditions of flood control with the general principle – no housing or other elements of important urban infrastructure can be allowed in flooded areas,
- determination of threatened zones in towns and cities,
- work out a hierarchy of flood control investments in major towns,
- introduce appropriate regulations connected with the insurance of properties situated on flooded areas because it appears that only economical factors can guarantee proper utilisation of flooded areas,
- conduct appropriate educational and information campaigns referring to the threats and character of floods and actions and procedures which have to be followed in the case of the threat, occurrence and withdrawal of the flood (One of the most effective methods is the display of signs of great water in towns),
- analysis of the effectiveness of the existing system of warnings, and
- preparation of an inventory of the most threatened elements of urban infrastructure and preparation of proper procedures for waters of definite probabilities of occurrence (protection, evacuation, partial exploitation).

The elements of infrastructure most exposed to danger are:

- system of water supply,
- sewerage and sewage treatment plants,
- waste dumps and pits,
- fuel stores and stations,

- electrical installations,
- telecommunication installations.

Damage or destruction of any of these elements leads to development of new threats.

- It is necessary to prepare new water control plans which would take into consideration the occurrence of a flood on a scale hitherto unpredicted because this will allow to work out operation procedures. Such a program must be verified and upgraded regularly. Lack of this kind of program and documents results in ad hoc decisions in the course of action.
- The above mentioned plans should be prepared on the basis of maps of flood areas as well as the balance of retention possibilities of a given catchment from a water gauge situated upstream of the protected town for different discharge volumes, including a disastrous one.
- All flood control plans in towns should be supplemented with information connected with what should be done in the case of flooding.
- It is also important to prepare a simulation model showing the process of town flooding in different scenarios of dike bursting.
- In the case of urban areas, essential undertakings must be determined,
- Train all urban services in the field of flood control. It is essential to carry out training not only among top officials but also officers and clerks at lower levels, e.g. firemen, since in emergency situations, it is always people of that rank that are in charge of small teams carrying out specific tasks. It important to remember that training should deal with very simple tasks to be performed in the course of emergency actions, e.g. filling sand bags.
- Prepare flood maps and simulation prognoses of a specific probability of occurrence and indicate:
  - threatened areas,
  - areas which can be allowed to be flooded, and
  - areas which must be protected unconditionally.
- The following undertakings of flood control character should be implemented in individual towns:
  - elimination of bridges which could impede the predicted flow of waters,
  - special care to be taken over relief channels,
  - reconstruction of waterways which can have a negative effect on discharge within town boundaries,
  - structural changes in spatial management of urban areas,
  - drainage of districts where housing settlements are situated,
  - construction of polders within town boundaries,
  - taking appropriate measures to ensure good technical condition of water courses and water facilities situated in urban areas, and
  - development of retention facilities within town boundaries.
- Even the smallest hamlet should have its own water mark which would inform emergency services during flood control operations about dangers and which would allow, within existing possibilities, to link it to the system of hydrological predictions.
- On the basis of the collected experience it can be said that a special slot (time) should be reserved in TV and radio programs where official bulletins of the

emergency centre could be broadcast without commentaries because, as it turned out, frequently irresponsible commentaries and inaccurate information resulted in undesirable attitudes developing among population threatened by the flood and constituted a source of false information for the rest of the country. After the flood, the Bureau of National Security, acting on the initiative of the former minister of environmental protection Mr K. Szamałek, organised a seminar with the participation of mass media and participants of the flood services from 1997 where different opinions and arguments were discussed. These problems will be discussed also in individual regions, e.g. in Wielkopolska.

- Work out a data bank of experts in different branches who work not only in a given town but also in the neighbourhood. It appears that the best solution would be to prepare a data bank with specialists from all over the country who could be at the disposal of the emergency centre.

In the case of direct threat and damages, the following actions are desirable.

- First of all, acquire good information and forecasts concerning both hydro-meteorological phenomena and the extent of flooding and damages as well as information about other phenomena which can occur:
  - results of infiltration,
  - gradual inactivation of rural infrastructure,
  - perspectives of necessary evacuation, and
  - other restrictions(supplies, introduction of curfew, deportations, restrictions in public transportation).
- Cooperate with the mass media (TV, papers, radio). As indicated by recent experiences, cooperation with local radio is essential as they broadcast information about events 'as they happen' and this is extremely important both for local citizens and services taking part in the action.
- Locally, it is also important to gather detailed hydro-meteorological information because protective constructions which operate in extreme conditions, apart from high levels of water lifting, are often exposed to the operation of other factors, e.g. atmospheric precipitation, evaporation.

THE DAY AFTER (only the water factor)

- During the flooding, damages and threats occurring in buildings should be neutralised as far as it is possible; construction and building services should provide current information about disasters.
- Permanent technical control of water construction should be assured.
- Measurements of water quality should be carried out and water ponds marked off; data which will allow prognostication of water quality on the flooded area should be prepared.
- It is necessary to determine, on the basis of earlier plans, ways of draining towns and to indicate methods of local neutralisation of contaminated water ponds and pockets.
- It is necessary to prepare equipment and, as the water table drops, carry out drying and disinfecting of building and areas which can pose epidemiological hazards.
- Indicate and make safe places where things from flooded buildings will be temporarily stored.
- Reconstruct internal network of water courses and reservoirs in urban areas.

It should be remembered that despite enormous losses caused by the summer flood of 1997, the flood itself occurred in conditions advantageous for the emergency operation. It seems advisable to carry put a simulation of consequences of a winter flood which would occur in the same hydro-meteorological conditions and parameters and on this basis conduct games of the emergency centres.

URBAN FLOOD CONTROL – TECHNICAL SUPPORT TO THE DIAGNOSIS AND REHABILITATION OF CRITICAL AREAS IN THE CITY OF LISBON

L.M. DAVID, M.A. CARDOSO and R.S. MATOS

Laboratório Nacional de Engenharia Civil
Núcleo de Engenharia Sanitária
Departamento de Hidráulica
Av. do Brasil 101, 1799-066 Lisboa, Portugal

1. Objectives

The rehabilitation of large urban catchments is a complex duty that involves multidisciplinary teams, requires large investments and causes strong social and urban impacts. Therefore, the definition of a rehabilitation strategy plays a very important role and should be based on the integrated analysis of the entire catchment.

For over ten years, LNEC has been working with Lisbon Municipality in order to study some flood-critical areas in the city of Lisbon. More recently (since 1994), a closer partnership has been established in order to develop a methodology and actions for the diagnosis and rehabilitation of the main urban drainage systems. The rather complex and large Alcântara urban catchment was modelled between 1994 and 1996, in the scope of the SPRINT SP98/2 European Project. In 1997, a protocol was established between LNEC and Lisbon Municipality in order to develop detailed studies for the diagnosis and rehabilitation of the critical areas of the city, including some of the Alcântara sub-catchments.

This paper summarises the methodology applied to the Alcântara catchment as well as the experience and results obtained during this period. It focuses on aspects such as data collection, site inspections, models building, experimental surveys and system performance diagnosis. The main flooding causes (structural, hydraulic and operational deficiencies) and solutions to improve the system behaviour, in short and long-term, are also discussed.

2. Description of the Alcântara Catchment

The Alcântara catchment is one of the largest catchments in the city of Lisbon. Its area is about 3200 ha, of which about 2000 ha are within the Lisbon Municipality and the other 1200 ha belong to the Amadora Municipality, upstream from Lisbon.

The main and longest stream in the catchment originally started in the hills at elevation 250 m and flowed along 13 km into the Tagus River at elevation 3 m. After the 1755 earthquake and to the present day, the catchment has suffered a dense urban occupation that has caused significant changes in the natural flow. As a result of

J. Marsalek et al. (eds.), Flood Issues in Contemporary Water Management, 333–342.

increasing development, the stream was progressively channelled (completed 1967), and became known as "Caneiro de Alcântara". In the 1980s, a large wastewater treatment plant was built in the downstream area, and the construction of new roads and highways is permanently inducing new changes in the flow.

The permeable, roof and paved areas in the catchment are about 54, 23 and 23%, respectively. The majority of the drainage system is combined, except for the most recently constructed areas where the system is separate. It serves a population of about 344 000 and the total length of the main domestic and stormwater sewers is approximately 250 km. The sewer system is very complex, with some looped networks and backwater effects caused by tidal effects on the River Tagus downstream. There are about 20 different sewer cross section shapes, with a wide range of dimensions from widths of less than 0.80 m to 8.00 m. Almost all the wastewater is conducted by gravity to the Alcântara treatment plant.

There are serious flooding problems during wet weather in some critical areas. Near the Tagus River there is frequent flooding mainly caused by the lack of capacity of the drainage system and the tidal effects on the river. Another serious problem is the frequency of combined sewer overflow discharges by-passing the treatment plant [1].

3. Methodology

Due to the large size and complexity of the Alcântara catchment, the rehabilitation methodology comprised the following stages:

1. development of an integrated hydraulic analysis to the whole catchment in order to evaluate current system performance, critical areas, main causes for flooding and potential solutions (This activity was carried out in the scope of the SPRINT SP98/2 European Project from 1994 to 1996);
2. the definition of a rehabilitation strategy based on the results of the previous stage [2] (in addition to technical benefits, social, economical and financial aspects were also considered for the definition of priorities in this stage); and
3. development of detailed hydraulic studies in the most critical and complex areas (four studies are currently being developed at this level).

4. Integrated Catchment Analysis

4.1. AIMS AND METHODOLOGY

The Alcântara catchment was selected for the Portuguese prototype case study in the scope of the SPRINT SP98/2 European Project. The main aim of this successful project was to apply and disseminate the Wallingford Procedure for rehabilitation of urban drainage systems. The main concept of the Wallingford Procedure is the evaluation of integrated system performance using mathematical modelling techniques [3]. The Hydroworks model, developed by HR Wallingford, was used to simulate the system.

Two different levels of detail were applied for modelling: a simplified model for the analysis of the whole catchment; and a more detailed model, for the analysis of a looped system in a selected critical area – the BD5 sub-catchment. For each case, the main stages considered were as follows [1]:

i) check of system records,
ii) data collection,
iii) identification of the critical points,
iv) planning of the investigations,
v) site investigation in order to complete the mapping and assess structural and environmental conditions and hydraulic problems,
vi) model building based on collected data and defined objectives,
vii) rainfall and flow measurements in order to verify the model,
viii) model validation,
ix) assessment of hydraulic performance and identification of critical points, and
x) definition of rehabilitation options and simulation of their performance.

4.2. DATA COLLECTION AND MODEL BUILDING

A detailed list of data needed was prepared and data collection was based on mapping and information given by the Municipality. As this information was frequently not up to date, it was necessary to complement it with direct inspection. About 60 working days were spent on site visits during a period of six months, and CCTV was used on several sites. Several structural, hydraulic and operational problems were identified, such as sediment deposition in pipes, pipe breaks, bottlenecks, insufficient or damaged manhole accesses, illegal connections and open channels draining domestic wastewater [1,4]. A list of procedures and a database were developed in order to store, verify and manage all the collected data. Figure 1 shows some database menus.

Simplification criteria had to be used for model building. Only pipes with diameters above 800 mm were considered, unless the flow contribution from smaller pipes was relevant. Nodes were considered in the following sections: i) changes in shape, material or slope of the pipes; ii) special structures; or iii) along long pipes without any change. Figure 2 shows the Alcântara and BD5 models. A summary of the model characteristics is presented in Table 1.

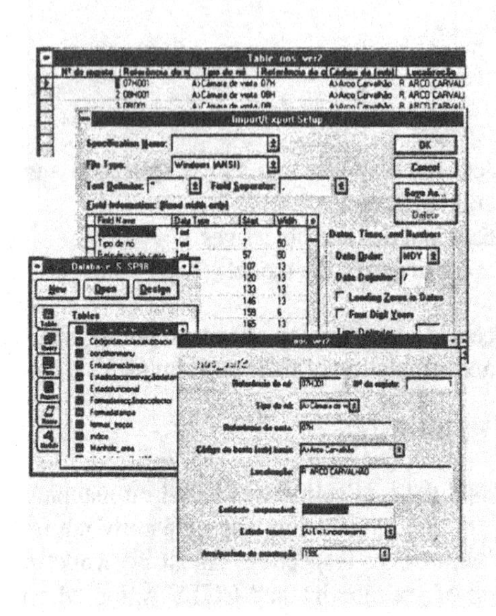

Figure 1. Database menus.

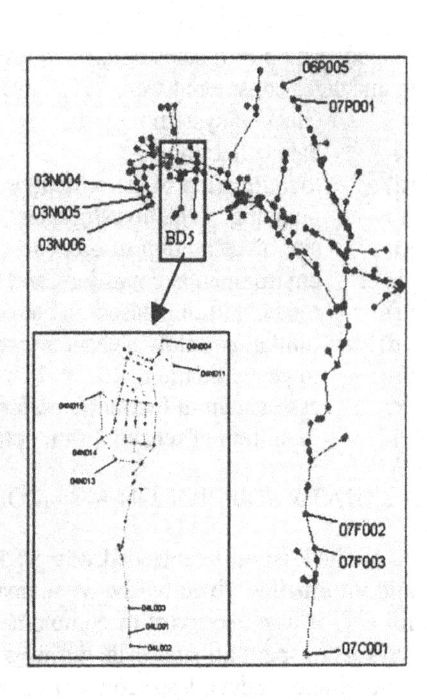

Figure 2. Alcântara and BD5 models.

TABLE 1. Summary of model characteristics.

	Alcântara simplified model	BD5 detailed model
Total area (ha)	2 077	85
Impervious area (ha)	935	33
Number of sub-catchments	132	23
Average slope (%)	8.2	8.1
Number of pipes	219	64
Number of nodes	219	52

4.3. FLOW SURVEY AND MODEL CALIBRATION AND VERIFICATION

The flow and rainfall survey took place from January 20th to March 19th, 1996. A total of 6 raingauges were installed across the catchment in order to give information about the spatial variability; 17 level & velocity gauges were placed in selected sewer sections and one level-gauge in the outlet, which is influenced by tides. About 70 rainfall events with very different characteristics were measured and recorded, which allowed for the collection of a very interesting and useful set of data.

From the measured events, eight were selected for model calibration and two for verification [4,5]. The criteria used to evaluate the calibration, as well as the verification, were as follows: relative error on the hydrograph volume, relative error on the maximum peak, and relative error on other peaks and shape. The levels of accuracy recommended by the Wallingford Procedure [3] are as follows: errors on volume between -10% and +20%; errors on peak between -15% and +25%; and a good reproduction of shape until storm recession occurs.

Figure 3 shows the results of calibration and verification for one event at a sewer section. In most cases, the shape of hydrographs was acceptable and the biggest deviations were observed on peaks, especially in the pervious areas. Volume deviations are not very relevant – the main cause for these is related to the increase of infiltration flow during the event, and the mathematical model does not take this into account (infiltration was considered as a constant input value).

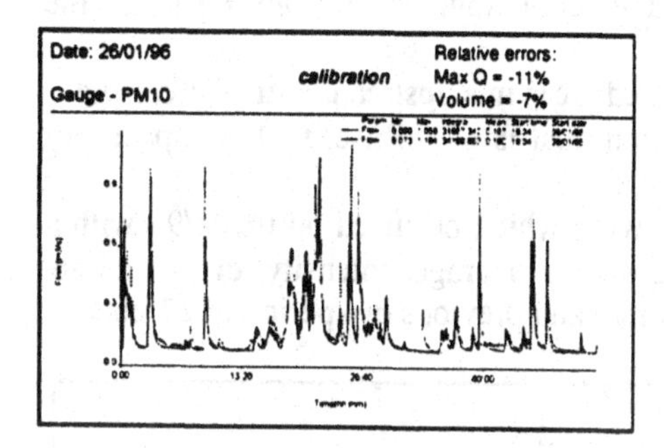

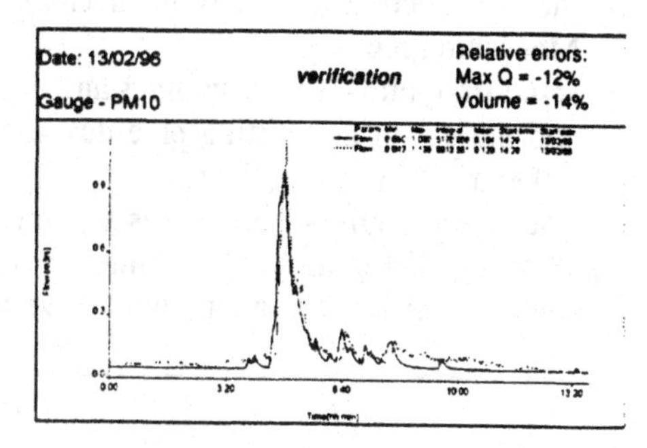

Figure 3. Results of calibration and verification at site PM10.

4.4. SYSTEM DIAGNOSIS AND ANALYSIS OF SOLUTIONS

Results from site inspections, flow survey and mathematical model simulations were all relevant for the identification of critical areas, structural, hydraulic and operational deficiencies and for the system performance diagnosis. The main conclusions concerning the current behaviour of the system and potential solutions were as follows [4,5,and 6].

i) Flooding in several upstream parts of the catchment results mainly from excess of surface runoff due to insufficient number of gutters and inlets; obstructions on gutters and inlets; excess of sediment in sewers; excess of infiltration flows. Improvement on the regular maintenance of inlets and sewers and structural rehabilitation of some areas are required for better local drainage conditions and in order to reduce infiltration flows.
ii) In the upstream and midstream areas of the catchment, the main trunk (Caneiro) has enough hydraulic capacity to transport the flows produced during strong rainfall events. However, some bottlenecks were identified in tributary pipes that require local interventions in order to improve the hydraulic and/or storage capacities of the system. The mathematical model was used to simulate potential solutions in a critical upstream area – some of those were later implemented (Figure 4).
iii) Flooding in the lower parts of the catchment is mainly due to the combination of strong rainfall events with tidal effects, as well as to the excess of deposits in the sewers. Four types of solution were identified for controlling floods in this area:

(1) removal of deposits from sewers and improvement of regular maintenance (short/middle-term solution); (2) disconnection of the downstream sub-catchments from the main trunk (middle-term solution); (3) reduction of peak flows and volumes generated in the upstream sub-catchments (middle/long-term solution); and (4) improvement of storage capacity in the upstream areas, building of storage devices and use of existing pipe storage capacity through real-time control (RTC) management (long-term solution).

Figure 4 compares simulation results for the current system and for the system with the following changes:

i) the construction of a new open channel to drain rain water from a part of the Monsanto green area;
ii) some structural changes in pipes and drained areas in a upstream critical area; and
iii) and the use of an existing pipe downstream, that is out of use, with a capacity of 9 000 m^3, as a storage device.

The selected rainfall event was a strong event which occurred on 08/01/96 with a duration of 754 min, total volume of 63.8 mm, average intensity of 5.8 mm/h, maximum intensity during 2 minutes of 96 mm/h and previous dry period of 27 hours.

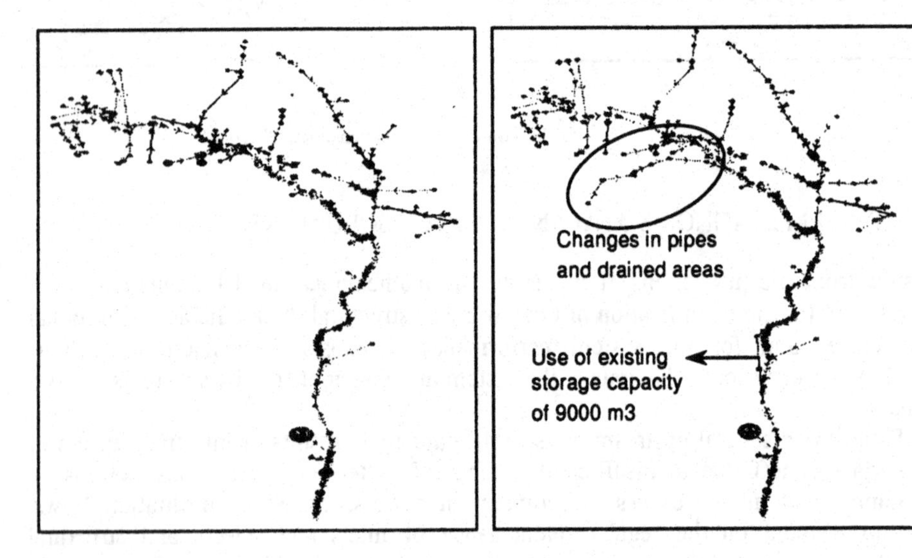

Figure 4. Current model (a), changed model (b).

As can be observed in Figure 4(a), there was flooding and surcharge in the system for the current situation. Floods are marked with circles and were actually also observed in reality. In Figure 4(b) some significant improvements can be observed. Flooding that previously occurred in the changed upstream part of the catchment no longer takes place. In the downstream part, there are some changes in system behaviour but not very significant. In fact, the modelled storage capacity is low compared to the storage capacity needed for stronger events in the whole catchment.

5. Detailed Analysis of Critical Areas

As a result of the previous study, four detailed studies are being developed in the lower parts of the Alcântara catchment, as is schematically shown in Figure 5:

a) a study of the drainage conditions and impact on the existing system of a new urban development to be built – "Bairro PER";
b) a diagnosis and rehabilitation study of a small but very critical downstream area of Alcântara – "Rua Fradesso da Silveira";
c) a diagnosis and rehabilitation study of the west side of the lower Alcântara catchment;
d) a diagnosis and rehabilitation study of the east side of the lower Alcântara catchment.

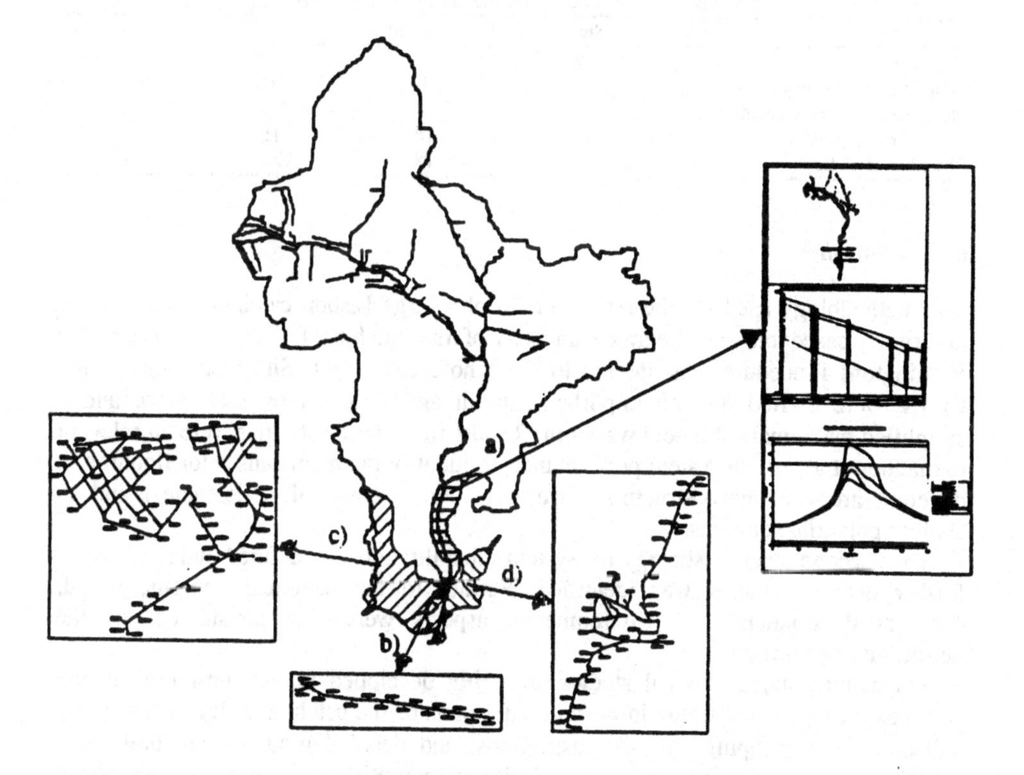

Figure 5. Detailed studies on Alcântara sub-catchments

In study a), the simplified Alcântara model (developed during the SPRINT Project) was used to generate the boundary conditions in the new urban development area and to evaluate its impact on the existing system for several real and design conditions. As a result of the first stage of this study [7], retention ponds are planned to be built and strategies for RTC of their emptying are being developed.

In studies b) and c), detailed models from the selected area were built and run for several real and design conditions. In the connections with Caneiro, water level results from the simplified Alcântara model were used as boundary conditions. Deposition in pipes plays a very important role in these lower, flat areas affected by tides. Solutions such as removal of pipe deposits, improvement of regular maintenance of inlets and pipes, rehabilitation of structural and hydraulic conditions of some pipes, disconnection of some sub-catchments from Caneiro and improvement of the final system discharge capacity into the Tagus river should be implemented [8,9].

The area covered in study d) has already been disconnected from the Alcântara catchment. During site inspections, some severe hydraulic and structural deficiencies were identified, causing deposition and floods. A detailed model was built and solutions to improve the system performance were proposed [10].

A summary of the critical areas and model characteristics is presented in Table 2.

TABLE 2. Summary of critical areas and model characteristics

	Study b)	Study b)	Study c)	Study d)
Total area (ha)	32 ha	3 ha	100 ha	35 ha
Impervious area (ha)	8 ha	3 ha	60 ha	25 ha
Number of sub-catchments	-	9	88	32
Number of pipes	-	18	141	103
Number of nodes	-	18	120	99

6. Conclusions

The methodology used for the rehabilitation of a large Lisbon catchment was briefly described, based on the integrated analysis of the catchment. On a first stage, the Wallingford Procedure was applied to the whole catchment. Site inspections using CCTV were carried out, an important rainfall and flow survey took place and a simplified mathematical model was built. Results from these activities were all relevant to diagnose the current system performance, to identify the main causes for flooding in different areas (namely structural, hydraulic and operational deficiencies) and to analyse potential solutions.

On a second stage, a strategy for system rehabilitation was defined and the need of further detailed studies was identified. Further to the technical benefits, social, economical, financial and environmental aspects were also considered for the definition of priorities.

On a third stage, detailed studies are being developed in the most critical and complex areas, located in the lower and older parts of the catchment. Information was collected from mapping and site inspections, and detailed models were built. The models were run for several real and design conditions, using results from the simplified Alcântara model as boundary conditions.

The main flooding causes are associated with inappropriate surface runoff conditions, deposits in sewers, bottlenecks, pipe breaks and other hydraulic and structural deficiencies, and lack of an integrated management for the whole catchment resulting in severe runoff transfer from upstream to the lower and flat areas affected by tides.

Beyond the short and middle-term structural and hydraulic rehabilitation of sewers and surface drainage conditions, the regular maintenance of sewers and surfaces is fundamental to ensure the expected system behaviour. The implementation of source control solutions and the improvement of storage capacity of the system, building storage structures and/or using the available storage capacity in sewers through RTC management, should play an important role in the future integrated management of the entire catchment.

7. Recommendations

After the implementation of the short and middle term solutions resulting from the described studies, three main activities were identified as of great importance to be developed (long-term solutions):

- monitoring of system performance after the implementation of short and middle term solutions, in order to assess the real benefits achieved and to ensure model validation;
- development of further studies, aiming at the implementation of source control solutions in new or re-urbanised areas, such as retention and/or infiltration ponds, pervious/porous pavements, retention in roofs and terraces and transfer of runoff from roofs, terraces and other impervious areas to green areas; and
- development of studies with the aim of providing additional storage capacity in the upstream and midstream parts of the catchment. A strategy for the construction of storage structures and use of the existing storage capacity in sewers should be established, based on the analysis of the whole catchment and using RTC techniques.

The adopted methodology also highlighted the importance of using objective criteria and appropriate tools for the definition of rehabilitation strategies and decision support. A study is currently being developed in LNEC in order to develop a methodology and tools for this purpose.

8. References

1. Almeida, M.C., Cardoso, A., Pinheiro, I., Matos R. and Rodrigues, C. (1996) Sewerage rehabilitation in Portugal - methodology and practical application, *7th International Conference on Urban Storm Drainage*, Hanover.
2. Matos, R.S., David, L.M., Vale, J., Monteiro, A., Fernandes, F., Teixeira, J. and Simplício, J. (1998) *Controlo de Inundações – Apoio ao Diagnóstico e Reabilitação de Pontos Críticos da Cidade de Lisboa. Relatório 1*, LNEC, Lisboa.
3. IHS (1994) The Wallingford Procedure. Training Course Notes.
4. Cardoso, M.A., Pinheiro, I., David, L.M. and Matos, R.S. (1996a) Application of Mathematical Modelling to Sewerage Rehabilitation in Portugal - A Case Study, *Int. Symposium and Workshop Environment and Interaction*, Porto, *paper* 38, 6 p.
5. Cardoso, A., Pinheiro, I., David, L., Matos, M.R., Berget, S. and Almeida, M.C. (1996b) *Project SPRINT SP98/98: Application of Hydraulic Analysis to Sewerage Rehabilitation in Member States. Report on model building, verification and alternative solutions*, LNEC, Lisboa.
6. Cardoso, A., Pinheiro, I., David, L., Matos, M.R. and Almeida, M.C. (1996c) *Project SPRINT SP98/2: Application of Hydraulic Analysis to Sewerage Rehabilitation in Member States. Final report*, LNEC, Lisboa.
7. David, L.M., Cardoso, M.A. and Matos, R.S. (1996) *Análise e Impacte dos Efeitos de Cheia e de Maré no Bairro PER na Av. de Ceuta - Lisboa. Relatório final*, Lisboa.
8. Cardoso, M.A. and Matos, R.S. (1998) *Controlo de Inundações – Apoio ao Diagnóstico e Reabilitação de Pontos Críticos da Cidade de Lisboa. Modelação Matemática da Zona Baixa de Alcântara – Rua Fradesso da Silveira. Relatório 2*, LNEC, Lisboa.

9. Cardoso, M.A. and Matos, R.S. (1999a) *Controlo de Inundações – Apoio ao Diagnóstico e Reabilitação de Pontos Criticos da Cidade de Lisboa. Modelação Matemática da Zona Baixa de Alcântara. Zona Oeste do Caneiro. Relatório 3*, LNEC, Lisboa.
10. Cardoso, M.A. and Matos, R.S. (1999b) Controlo de Inundações – Apoio ao Diagnóstico e Reabilitação de Pontos Críticos da Cidade de Lisboa. Modelação Matemática da Zona Baixa de Alcântara. Zona Leste do Caneiro. Relatório 4, LNEC, Lisboa.

FLOOD MITIGATION IN URBANISED AREAS

TOMAS METELKA

Hydroinformatics
Hydroprojekt a.s. , Táborská 31, 140 00, Praha 4, Czech Republic

1. Introduction

A rapid industrial development, focussing primarily on the growth of production and consumption, has been seen in European countries over the last few decades. This accelerating cycle resulted in a massive exploration of natural resources, causing large-scale environmental consequences. In turn, this process drives the overall transformation of the global market and the society (including comfort of life, habits, communication, etc...). The effects of the societal transformation and environmental impacts are more and more visible. The growing population concentration in large agglomerations causes a worldwide increase in urbanization and results in difficulties for transport, energy and mass supply, treatment of pollution, managing solid waste, etc.

The impact on the environment is then unavoidable. The change of the agglomeration microclimate (as well as global macroclimate) has an influence among other issues on the fragile equilibrium of water resource balance. Large flood events are seen over the world. These are followed by fresh water shortages. Increasing pollution of underground and surface water prevents water resources from direct utilization. Exponential population growth, especially in the third world countries, demands an increasing amount of fresh water. A global warming will have impacts on the water balances in distinct climate areas.

It seems to be clear that gradual expansion of the mass and energy exchange cycles from the local to global scale necessitates sophisticated management of the natural, economic and social processes aiming at sustainable development from the global point of view. In order to reach this goal, new shared visions have to be invented and followed using adopted processes and modern technology.

2. Effect of Urbanization on Catchment Hydrology

At present, clear tendency is seen in the concentration of the world population in large city agglomerations. A search for better living conditions and employment possibilities constitute the main driving force of this social phenomenon. But, in most cases, the city infrastructure is not ready to accept such increasing population loads. In developed countries, increasing urbanization overloads actual public city services (public transport,

J. Marsalek et al. (eds.), Flood Issues in Contemporary Water Management, 343–349.

security of property, communication, water and energy consumption, wastewater treatment, solid waste disposal, etc.) and results in system failures. The elementary city functions then become more fragile and unstable.

Urbanization activities have adverse impacts on changes of microclimate and local hydrological cycle.

- Local temperature increase changes the precipitation statistics resulting in higher intensity rainfalls.
- Large impervious areas accelerate the runoff process resulting in higher flood peaks occurrence and faster flow celerity.
- Urbanization restricts the natural infiltration process and results in a shortage of groundwater supply.
- Natural river courses are, in many cases, transformed into artificial channels resulting in absence of natural inundation and loss of flood transformation and retardation effects.
- Unsatisfactory groundwater storage results in a decrease of minimal river discharges during dry weather conditions.
- Sewer systems operate as large size drainage and cause a decrease in the groundwater table.

The consequences of these environmental changes are obvious. On the one hand, the increase in frequency and peaks of flood events is seen in urbanized areas including sites outside the river floodplains which were expected to be safe. On the other hand, increased pollution load in rivers causes loss of natural functions during dry weather periods. This situation is not sustainable and demands an integrated solution. The complexity of this task then requires the adoption and use of a modern technology and scientific approaches. Integrated mathematical modeling of natural processes is recognized to be one of fundamental issue for supporting long-term decisions of a sustainable water resource management.

The predictive and integrated roles of mathematical models are key advantages of this approach. The simulation scenarios in terms of "What happens, if" evaluation can be used to predict and optimize possible future system changes, to identify weak points of a particular system and to perform warning while being coupled with sufficient monitoring and decision support tools (in a hydroinformatics framework). The integration of hydrological, hydraulic, chemical and biological processes into the mathematical model makes it easier to understand better the complexity of nature and to apply this knowledge to environmental tasks.

Years of experience worldwide have proven the applicability and advantages of mathematical modeling for solving practical projects. A couple of projects are described in more detail in the following paragraphs. The first one is related to the application of a mathematical model for surface flood protection; the second is focussed on the protection of a sewer system against flooding from a river during periods of high water level.

3. Surface Flood Protection - Botic Case

3.1. SCOPE OF THE PROJECT

The main concern of this study has been focussed on the analysis of the impact of increasing urbanization in the outskirts of Prague on the surface runoff process at the upper part of the river Botic catchment. The catchment has an area of 92 km^2 with a moderate slope; it is partly forested and partly used for agricultural purposes. The catchment is located in south-east part of Prague. The upstream, 10 km long part of the river Botic is a main receiving water in the catchment with average flow 230 l/s. The mid-size Hostivar dam is located on the downstream part of the river. According to the city development plan, a huge urbanization, including large size stores, and shopping centers is planned. Therefore, the impact on the catchment hydrology had to be studied.

The study was divided into two parts. In the first part, the attention was devoted to the analysis of the impact of urbanization on flood routing (volumes, peak flows, travel time); in the second part, emphasis was on the impact on river pollution during dry weather flows.

3.2. APPROACH USED

A pseudo two-dimensional mathematical model, MIKE11, was used as a technological tool in the study including surface runoff, hydrodynamic and water quality modules. A unit hydrograph and SCS method were used while formulating the surface runoff model (See Figure 1). Distinct subcatchments were analyzed from the point of view of land use and impervious areas. Then, a set of rainfalls was used to generate runoff hydrographs from the area as an upstream boundary condition for the hydrodynamic module run. The hydrodynamic model was composed of branches of the main river Botic and its right tributary Pitkovicky creek. All river structures were implemented into the model including the Hostivar dam with its operation procedures (see Figure 2).

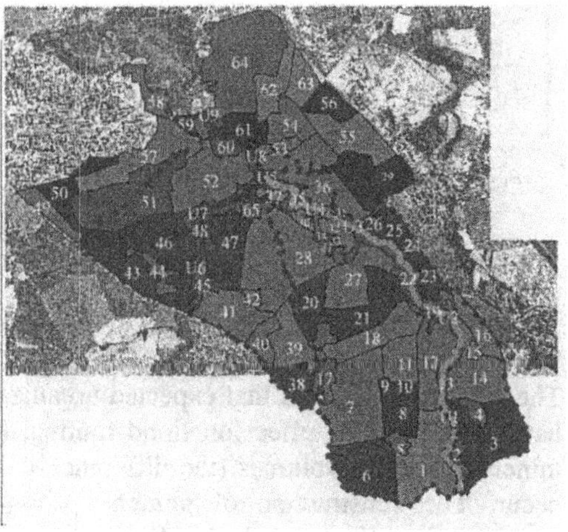

Figure 1. Example of catchment evaluation procedure.

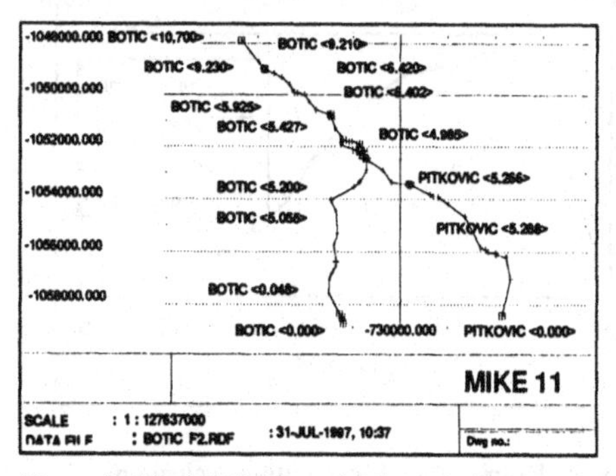

Figure 2. The MIKE11 model formulation setup.

After model formulation and calibration was completed, the current state of the system was evaluated. Then, two basic future scenarios were defined for the river Botic:

- construction of a new retention area (dry polders) (see Figures 3 and 5),
- the retention area was not considered (See Figure 4)

The above mentioned scenarios were then used as a basic setup for the evaluation of the flood routing process as well as for dry weather pollution situation.

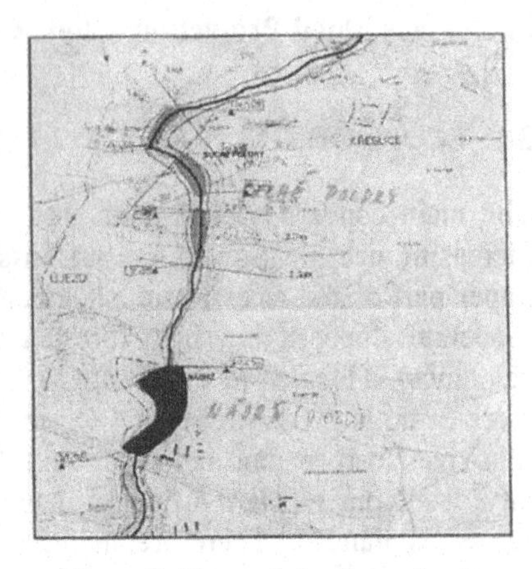

Figure 3. The example of a catchment dry polders location.

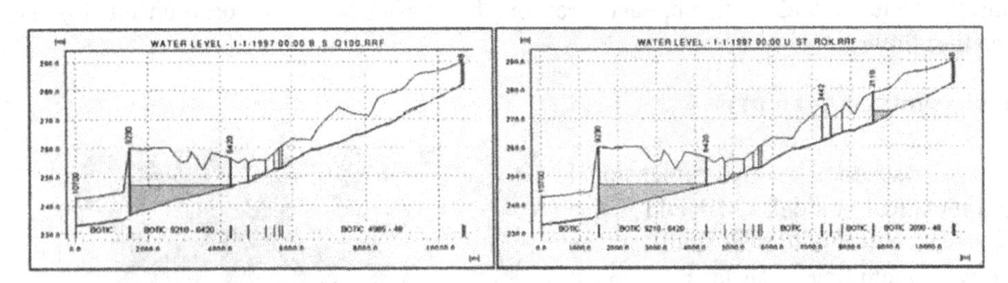

Figure 4. Example of a sideview of the river Botic with Hostivar dam.

Figure 5. Example of a sideview of the river Botic with polders.

3.3. PROJECT RESULTS

The simulation showed that expected urbanization in the upper Botic catchment would have no significant effect on flood routing. However, slight increases in travel time, minor changes in volumes (the difference is less then 30000 m^3) and peak flows will occur. The construction of polders positively affects the flood transformation. The flood peak travel time shows a delay reaching 40 minutes and the peak drops by 2 m^3/s. The overall capacity of polders reaches 170,000 m^3 during the 100 year flood. The peak water level drops 25 cm along the river reach of Botic during the 100 year flood (see Figure 6). Under dry weather flow, a significant impact is seen on water quality in the river.

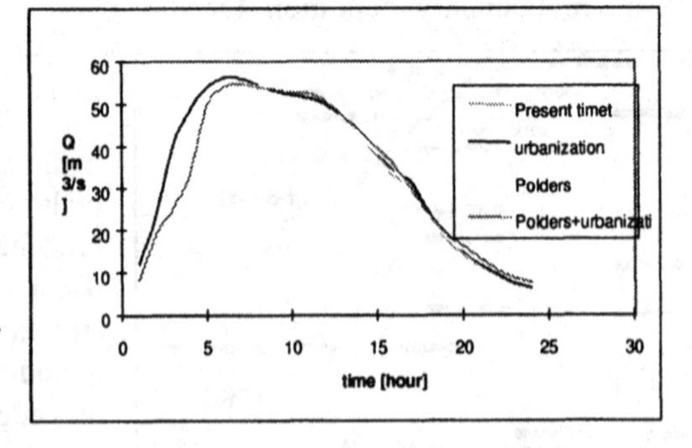

Figure 6. Example of a flood routing evaluation.

In an extreme case (a sudden, local, high intensity storm causing overflows from the sewer system to the river with an average discharge), the disolved oxygen drops down temporarily from 9 to 3-4 mg/l, causing the total flush out of the river. The situation improves after 24 hours.

4. Sewer System Flood Protection - Olomouc Case

4.1. SCOPE OF THE PROJECT

The main concern of this study is focussed on the application of Best Management Practices (BMPs) for the control of combined sewer overflows on urban flooding in the city of Olomouc. The city is an important regional centre with 100,000 inhabitants and is located on both sides of the river Morava and its tributary Bystricka. In addition, the river Morava in the city splits into two connected channels. The city is located on the river floodplains and hence the terrain is flat. The complexity of the river topology together with flatness of the area results in frequent sewer system flooding through combined sewer overflows (CSOs). Olomouc sewer system has about 214 km of sewers and 16 CSOs, all being under operation for more then 30 years. During flood events, many parts of the sewer system become surcharged. Uncontrolled backward discharges through CSOs occur due to the interaction between receiving water and sewer systems. After the flood disaster in 1997, which resulted in substantial loss of life and property damage, protection of the sewer system was identified as one of the city's flood protection priorities.

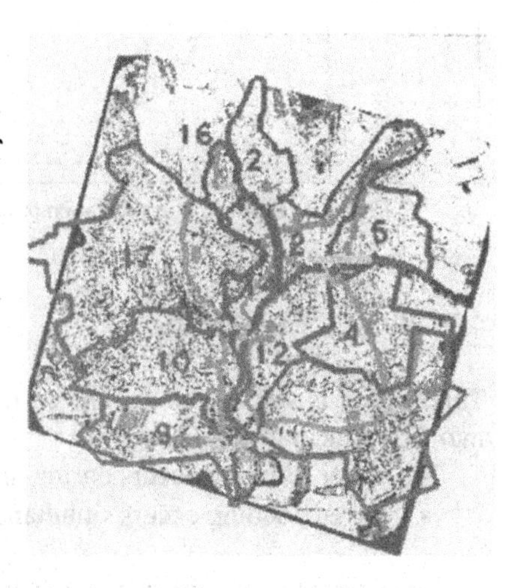

Figure 7. The olomouc catchment plan.

4.2. APPROACH USED

A simplified one dimensional MOUSE model of the sewer system has been used to evaluate possible flooding scenarios and to help develop a sewer flood protection plan.

A simplified model formulation of the trunk sewer system has been performed including principal CSOs important for the system's operation. Estimates of flow in the

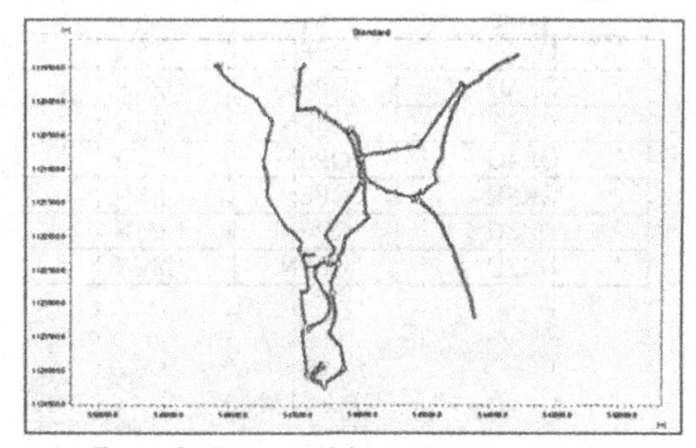

Figure 8. Example of the model schematization.

sewer system and data for hydrodynamic model calibration and verification were obtained from a monitoring campaign carried out during July, August and September, 1998. Precipitation data were obtained from rain gauges placed on selected subcatchments, and local historical rain series from 1978 to 1997. Water levels and/or flow rates were measured continuously and simultaneously with precipitation at the selected CSOs. The rain gauge network consisted of three principal gauges and two gauges that served as reserve. There were 6 H-points and 10 Q-points inside principal CSOs. The measuring period started in mid July and was executed for 3 months.

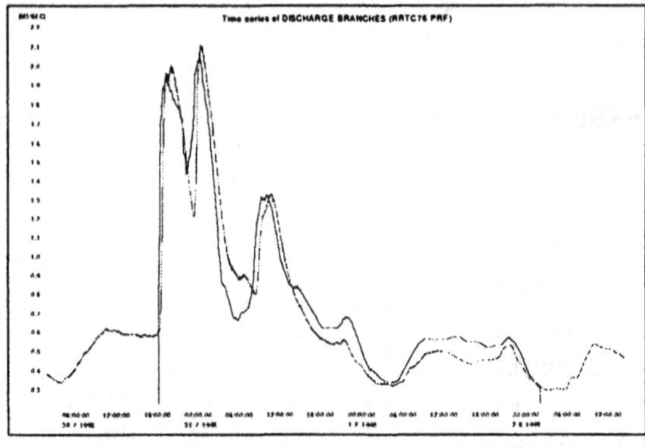

Figure 9. Example of a calibration procedure.

4.3. PROJECT RESULTS

This project is still ongoing. The current set of model scenarios is, in general, divided into two principal situations:

- river flooding occurs during dry weather flow, and
- river flooding occurs simultaneously with a local rainfall event in the city.

River water levels for distinct return period floods were obtained from the Morava Water Board as an output from MIKE11 steady simulation. These data are used as boundary conditions in the sewer system model. Based on results from the simulation, a simple flood protection plan will be scheduled for discharges with different return period (See Table 1 for preliminary schedule.).

TABLE 1. Example of an operation plan

CSO	1.1.1 *Flood discharge with return period*				
name	**Q_1**	**Q_2**	**Q_3**	**Q_5**	**Q_{10}**
OK6C	OPEN	OPEN	OPEN	OPEN	CLOSE
OK6D	OPEN	OPEN	OPEN	OPEN	OPEN
OK4C	OPEN	OPEN	CLOSE	CLOSE	CLOSE
OK4D	OPEN	CLOSE	CLOSE	CLOSE	CLOSE
OK5D	OPEN	OPEN	OPEN	OPEN	CLOSE
OK3C	OPEN	OPEN	CLOSE	CLOSE	CLOSE
OK2D	OPEN	OPEN	CLOSE	CLOSE	CLOSE

This action plan is expected to be improved after new hydrodynamic results from the river Morava simulation have been achieved. After the project completion, the results from this study will enable design and optimization of BMPs with the aim of controlling CSOs. Measurements and modeling will provide the Municipality with information for making decisions on further development and investment policies.

5. Conclusions

The complexity of present environmental tasks demands the use of an integrated approach comprising detailed evaluation of hydrological, hydraulic, chemical and biological processes. The application of modern approaches and up-to-date information technology is then seen as a necessary condition for successful completion of environmental projects and studies. Mathematical modeling of natural processes is argued to be one of the most progressive technique to understand and protect nature for the long term.

6. References

1. Abbott, M.B. (1991) *Hydroinformatics: Information technology and the aquatic environment,* The Avebury technical, Alderschot, UK and Brookfield, USA.
2. Maksimovic, C., Prodanovic, D., Elgy, J. and Fuch, L. (1994) *GIS (or GIM) in water projects - Tools or toys,* Hydroinformatics '94 Delft, 1994, Balkema, Rotterdam, 1994.
3. Gustafsson, A.M. (1994) *Key Numbers for Infiltration/Inflow in Sewer Systems*, 7th European Junior Scientist Workshop on Integrated Urban Storm Runoff, Èernice Castle - Bechynì - Czech Republic.
4. Almeida, M. and Schilling, W. (1994) *Real time control in small urban drainage systems*, 7th European Junior Scientist Workshop on Integrated Urban Storm Runoff, Èernice Castle - Bechynì - Czech Republic.
5. Kuba, P., Metelka, T., and Pliska, Z. (1996) *Impacts of the hydroinformatics technology on the way of master planning*, Proceedings of The 2nd International Conference on Hydroinformatics, Zurich, Switzerland.
6. Metelka, T., Mucha, A., Pryl, K., Kuby, R. and Gustafsson, L.G. (1997) *Urban Drainage Master Plan of the Capital City of Prague - Preliminary Study*, Proceedings of 2nd DHI User Conference, Helsingor, Denmark.

The [illegible] is expected to be improved after new hydrodynamic results from the [illegible] have been achieved. After the project completion, the results [illegible] will enable [illegible] optimization of [illegible] with the aim of [illegible] [illegible]. [illegible] will provide the Municipality with information for [illegible] development and investment policies.

5. Conclusions

The [illegible] approach [illegible] [illegible] [illegible] have [illegible] [illegible] [illegible] [illegible] of natural processes [illegible] more progress [illegible] [illegible].

6. References

1. Abbott, [illegible] [illegible] [illegible] Amsterdam [illegible]
2. [illegible] 1994 [illegible] 1994.
3. [illegible] Czech Republic.
4. [illegible]
5. [illegible] Proceedings [illegible]
6. [illegible] Proceedings [illegible] Denmark.

REGIONAL IMPACTS OF LEVEE CONSTRUCTION AND CHANNELIZATION, MIDDLE MISSISSIPPI RIVER, USA

NICHOLAS PINTER[1], RUSSELL THOMAS[1] and JOSEPH H. WLOSINSKI[2]

[1]*Dept. of Geology, Southern Illinois University*
Carbondale, IL 62901-4324, USA
[2]*U.S. Geological Survey, Upper Midwest Environmental Science Center*

1. Introduction

The Mississippi River travels approximately 3780 km from its headwaters in Minnesota, USA to the Gulf of Mexico, draining roughly 3,210,000 km^2. Since European settlement, humans have significantly impacted both the river basin and the river itself. Primarily under the stewardship of the U.S. Army Corps of Engineers (USACE), river-engineering activities have focused on navigation improvement and flood control and have included snag removal, bank clearing, meander cutoffs, dredging, bank stabilization, channel constriction, levee construction, and dam construction [1].

In this paper, we present the results of a study of the changing relationship between water levels, discharge, and cross-sectional geometry at three gaging stations on the Middle Mississippi River (MMR; Fig. 1) that have been monitored daily for 54-135 years. These three stations are located at Thebes and Chester, Illinois, and at St. Louis, Missouri. The St. Louis gaging station is located 24 km downstream of the confluence

Figure 1. Location map of the Middle Mississippi River, USA, showing gaging stations utilized in this study. Bold lines indicate locations of levees (Modified from Maher, 1964).

J. Marsalek et al. (eds.), Flood Issues in Contemporary Water Management, 351–361.

of the Missouri and the Mississippi Rivers. The Upper Mississippi River Basin comprises approximately 440,000 km^2 (~23%) of the contributing area at St. Louis and the Missouri Basin 1,400,000 km^2 (~77%).

Long-term trends in the stage-discharge relationship of the Mississippi River have been a contentious issue, in both the scientific and political realms, since publication of two papers following record-breaking flooding in 1973. Time-trend analysis of historical hydrologic data by Belt [3] and geomorphic analyses of the reach and its catchment by Stevens et al. [4] were utilized to support the assertion that engineering modifications of the MMR had forced flood stages higher over time. This assertion has been met with a barrage of opposition [e.g., 5,6,7,8,9,10].

The criticism of the assertion of human-induced increases in Mississippi River stages include a range of specific objections and counter-assertions. At the heart of these objections is the tenet, central to hydraulic engineering, that channel constriction increases flow velocities and channel conveyance and thereby should decrease stages for an equal quantity of water. In addition, several authors have asserted that the analyses showing increasing stages at St. Louis are faulty because they utilized older discharge data (pre-1933) based on velocity measurements that are not comparable to later measurements obtained by modern techniques [6,10]. It was also argued that the period of record, when limited to post-1931 measurements, was then too limited for meaningful statistical analyses [7]. Finally and most sweepingly, it was argued that time-trend analyses such as used by Belt [3] fail to separate the effects of engineering structures on flood stage from other changes in the contributing basin [9] nor from other parameters such as water temperature that can introduce significant variability to stages [10].

Severe flooding on the Upper and Middle Mississippi River in 1993 brought the issue of human-induced shifts in flood stage back into the political and public spotlight. During the flood, discharge records were broken at 41 stations and exceeded the 100-year recurrence interval at 45 stations, including St. Louis. The peak discharge at St. Louis in 1993 was 30,586 cms, and the peak stage was 15.11 m, the highest stage to date [11]. Both the 1973 and the 1993 floods have been compared with the flood of 1844, one of the most severe events in the historical record. Discharge in 1993 was roughly 6,200 cms less than that of 1844 (although that discharge estimate has been disputed [6], but the stage in 1993 was over 2.4 m higher.

2. Methods

Daily water-level measurements and discharge estimates date back to 1861 for St. Louis, 1942 for Chester, and 1941 for Thebes. Water-level elevations for all three stations and discharge data from 1861 through 1933 were obtained from the USACE, St. Louis District. The other discharge data were obtained from the Rolla, Missouri office of the U.S. Geological Survey. Data used in this research was collated through 1996. Yearly maximum stages and discharges were also examined for trends using linear autoregressive error-correction procedures [12]. Significance for all procedures was set at $P<0.05$.

2.1. SPECIFIC-GAGE ANALYSIS

The specific-gage technique [13,14] has a number of advantages over the time-trend method described in the previous section. The specific-gage technique utilizes daily stage and discharge data to construct a rating curve for each year, and then all of the years' rating curves are used to track stage over time for each gage. Each rating curve is constructed by plotting stage against log of discharge using all of the daily data for a given calendar year. A second-order polynomial regression was fit to each year's rating curve. The goodness of fit to these regressions was excellent, and almost all the graphs had r^2 values greater than 0.95.

For each year, the regression equation was used to calculate the stages associated with the chosen discharges. Because extrapolating the regression in a given year beyond the range of measured data creates unacceptable error, we did not calculate any stage for a discharge that was either less than the measured minimum for that year or greater than the measured maximum. The yearly rating curves at Thebes demanded special attention because backwater effects of the Ohio River caused loop-shaped deviations of up to 4.5 m (15 ft) in stage. If the r^2 value for a given year was 0.95 or less or if there was a visible anomaly in the rating curve, data associated with flooding on the Ohio River were taken out manually. For all three stations, the final step in this procedure was to plot the stages associated with a fixed discharge from all the years against time (Fig. 2). Four selected discharges are shown: 1) below-bankfull discharge, 2) the discharge at which the slope trend is closest to zero, 3) bankfull discharge, and 4) above-bankfull discharge.

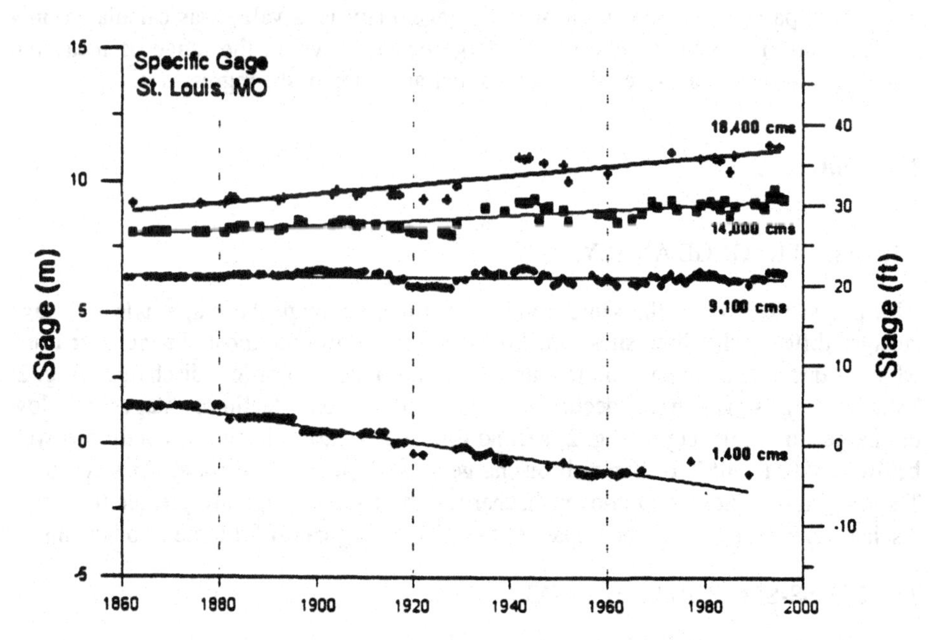

Figure 2. Stage variation over time for four fixed discharges at St. Louis.

2.2. CROSS-SECTIONAL TRENDS

In order to gain additional information into the nature and possible causes of changes in the stage-discharge relationships on the MMR, we have examined cross-sectional data compiled by the USGS for the St. Louis, Chester, and Thebes gages. These data were collected at approximately biweekly intervals since 1933 at St. Louis and for the duration of record at Chester and Thebes, with measurements made as frequently as daily during recent flood events. These data consist of stage, discharge, channel width, cross-sectional area, and average velocity. To identify erroneous data, each parameter at a given station was plotted against discharge, and clear outliers on these graphs were removed.

For each cross-sectional parameter, we tracked changes over time using a pseudo-specific-gage technique. The data was separated by year, and each year's cross-sectional variable was plotted against log of discharge. A best-fit regression was calculated for each yearly relationship. If the regression had an r^2 value less than 0.95, the relationship for that year was not used. For all years with an r^2 value of 0.95 or greater, the equation for the best-fit line was used to generate a value for the variable associated with a given discharge. Three discharges were chosen for each station: below-bankfull, bankfull, and above-bankfull. The below-bankfull discharges were defined for each station as the lowest discharge value (to the nearest 2800 cms [10,000 cfs] increment) for which there was data for at least 20% of the years of record for the majority of parameters. Above-bankfull discharges were defined as the discharge above bankfull stage for which there was data for at least 20% of the years for the majority of parameters. As in the specific-gage analysis, a value was calculated only for the years in which the chosen discharge occurred within the range of measured discharges for that year -- i.e., there was no extrapolation of the regressions.

3. Results

3.1. SPECIFIC-GAGE ANALYSIS

All three stations show the same trends over time, although the magnitudes of stage changes differ at the three sites. St. Louis (Fig. 2) shows the most change over time, which is due at least in part to the longer period of record. For low discharges (Fig. 2; bottom line), stages have declined. The switch from declining stages for low discharges to rising stages (Fig. 2; second line from bottom) begins at conditions well below bankfull: ~65% of bankfull discharge at St. Louis and ~40% at Chester and at Thebes. Above those switch-over discharges, the rises in stage are greater for larger discharges, indicating that the impact of this change is greatest for large flood events.

3.2. CROSS-SECTIONAL TRENDS

Analysis of the cross-sectional parameters reveal strong correlations between the long-term trends in stage with trends in flow velocity and cross-sectional area. For the St. Louis and Thebes gaging stations (Fig. 3; Thebes shown), velocity and area vary systematically with stage for all conditions. When stage is declining over time (for below-bankfull discharges), flow velocity in the channel increased and cross-sectional area decreased; under conditions when stage was increasing (bankfull and above-bankfull discharge), velocity decreased and area increased. The Chester station demonstrated the same pattern as St. Louis and Thebes for all below-bankfull conditions, as well as for bankfull and above-bankfull conditions subsequent to the 1960s, but the opposite relationship with velocity and area is found prior to that time.

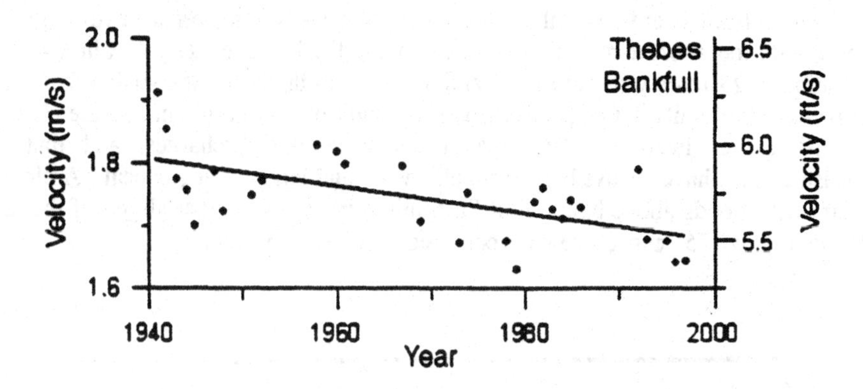

Figure 3. Trends in average flow velocity at bankfull discharge at Thebes.

4. Discussion

Much of the discussion of anthropogenic effects on flood stage has focused on analysis of stage trends over time. We updated the time-trend analyses of Belt [3] utilizing the 23 years of additional data now available. As in 1975, the maximum annual stage data suggest progressive increases in stage, especially for the largest discharges. The additional data also reveal the same limitations that opened the time-trend technique to criticism earlier -- principally 1) the large degree of scatter, and 2) its failure to distinguish among the wide range of possible causal mechanisms.

4.1. SPECIFIC-GAGE ANALYSES OF STAGE-DISCHARGE RELATIONSHIP

Many of the pitfalls of the time-trend technique are avoided using the specific-gage method. The principal advantages of this method are that: 1) it utilizes a far greater quantity of data (daily measurements) than does the time-trend technique (annual values). 2) it reduces scatter by calculating an average annual stage value for a particular discharge, 3) it tracks stage trends for precisely the same discharge rather than simply the largest or smallest value in a given year, and 4) any trends over time thus identified are necessarily linked to the conveyance capacity of the channel at or near the gage. To clarify the last point, the specific-gage method excludes the broad range of upstream factors (e.g., storm intensity, land-use shifts, effects of dams, climate shifts) that may control the magnitude of discharge, focusing instead on what stage results from a given discharge as a result of local conveyance factors in the immediate vicinity of the gage. Specific-gage analysis of historical data from the three stations on the MMR (e.g., Fig. 2) reveal stage trends with very little scatter and free from many of the causal ambiguities in the time-trend results. For example, an upward trend in low-flow stages at St. Louis suggested by the time-trend technique is absent in Figure 2. Although the low-flow data points become scarcer after the 1960s -- because of dam releases and because we chose not to extrapolate our annual regressions -- it can be seen that channel incision has continued unabated to the present.

The original criticism that statistical analysis of historical hydrologic data represents too short a record to be meaningful [7] has been largely blunted by the addition of 23 more years of record and, more importantly, by the switch from annual (time-trend) to daily data (specific-gage). In addition, we point out that the objections to the earlier historical data pertain solely to the discharges, and that stage measurements have always been straightforward and largely unequivocal. A plot of all years with floods above bankfull at St. Louis (Fig. 4) shows that all ten of the largest floods in this 135 year record have occurred in the last 55 years.

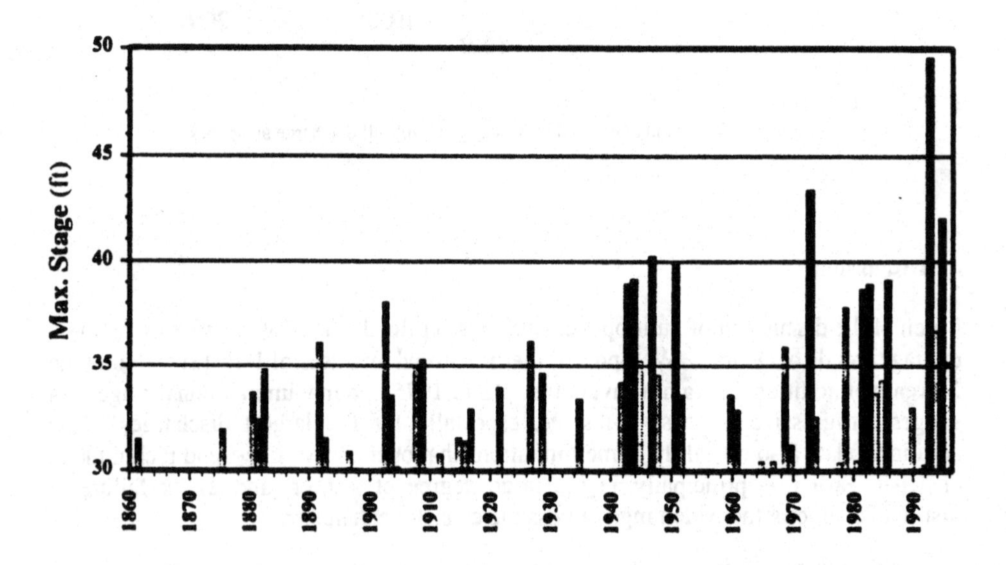

Figure 4. Plot showing maximum stage for all years above bankfull in St. Louis.

4.2. SEASONAL EFFECTS

The original time-trend analysis suggesting rising flood stages was criticized for the large degree of scatter in the data. It was suggested that a major source of scatter is the season in which different floods have occurred [10]. The season in which a flood occurs may affect stage through: 1) the amount of bank and floodplain vegetation, and 2) the water temperature. Colder water is more viscous and carries more sediment, and higher viscosity and sediment concentration tend to decrease channel roughness, thereby decreasing stage [15,16]. In order to quantify the impact of this factor for the stations on the MMR, we examined the effect of seasonality on stage (Fig. 5). The St. Louis and Chester gages, located where the river crosses thick alluvium, exhibited the same pattern illustrated in the figure -- smooth annual curves in which deviations in stage correlate with average water temperature. At Thebes, where the river crosses a bedrock constriction, the substrate may not allow viscosity-dependent smoothing of bed roughness. The water-temperature/viscosity effect does help to explain the extreme stage of the 1993 flood, which stands well above all of the long-term trends lines. The 1993 flood peaked on July 20 and August 1, whereas most floods on the Mississippi are springtime events. This factor helps to explain some of the variability of individual flood stages, but it does not affect the long-term trend.

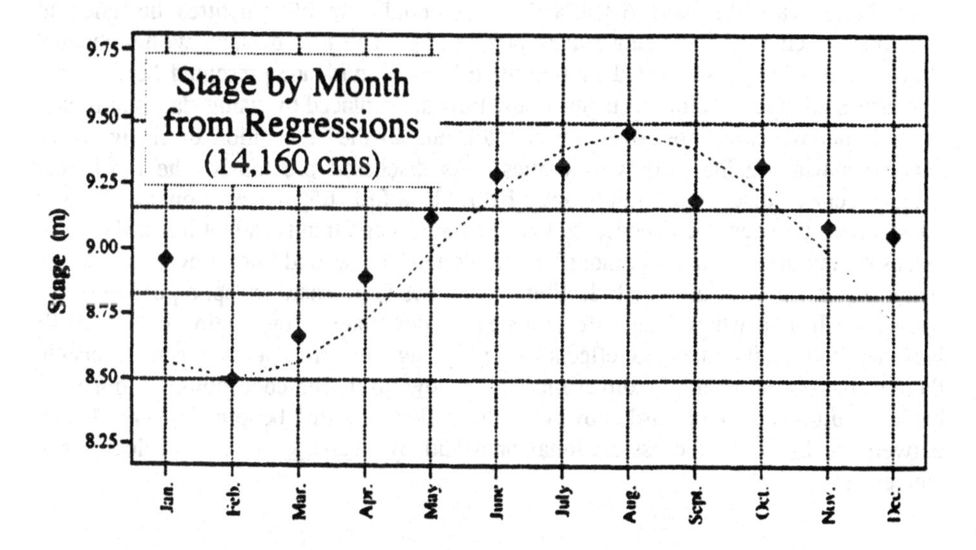

Figure 5. Effect of seasonality on stage.

4.3. POSSIBLE CAUSAL MECHANISMS

Given the long-term trends in the stage-discharge relationship, the challenge is to identify the mechanism(s) responsible. The specific-gage method sheds new light on this question. Because specific-gage analyses track stage changes for a set and fixed discharge, many of the upstream controls such as regional land-use change, dam retention and releases, and even a hypothesized long-term increase in precipitation [17,18] can be eliminated as contributors to the specific-gage trends. The cross-sectional data sheds some additional light on the problem by identifying changes in flow velocity that correlate with shifts in the stage-discharge relationship. However, as with any statistical analysis, correlation does not prove causation. We believe that the results presented here point strongly at certain causal mechanisms, but additional research may be necessary to confirm or refute those links.

Much of the discussion of stage shifts over time has focused on the effects of levees [3,9,19]. Several modeling studies conducted by the USACE and others (e.g., [20,21]) suggest that levees should indeed force flood stages higher, although by amounts less than demonstrated by the research presented here and by previous studies of empirical data. Although such models depend heavily on assumptions about floodplain roughness, their conclusion that there must be additional mechanisms responsible for stage increases is not unreasonable.

Other researchers have discussed the potential role of structures designed to constrict the channel in raising stages [4,2,22,23]. The role of such within-channel structures is strongly supported by two main lines of evidence presented here: 1) the discharges at which declining stages at low flow are replaced by rising stages at higher flows (what we call "cross-over" discharge), and 2) the association of rising stages over time with declining flow velocities. As discussed previously, the cross-over values occur at discharges significantly below bankfull (65% at St. Louis and about 40% at both Chester and Thebes), so that the rising stage trends cannot be attributed to levees or any other floodplain modification alone. The second line of evidence is that rising stages are associated with declining velocities in all cases except a portion of the record at Chester, where local effects appear to dominate. Constriction of the MMR has clearly had its intended effects -- increasing velocities and causing incision, thereby increasing channel conveyance -- for low-discharge conditions. From mid-bank conditions and upward, however, these flood-control benefits apparently are outweighed by loss of cross-sectional area and by decreases in flow velocity and conveyance.

5. Conclusions

Engineering modifications of the MMR are linked to changes in the cross-sectional geometry and flow regime. These changes have been manifested in declining stages for low discharges and rising stages for water levels starting at 40-65% of bankfull and above (Fig. 6). The decline in stages is associated with increasing flow velocities over the same period of time, and likely result from progressive channel incision as previous workers have suggested [3,4]. The increase in stage at higher discharges is clear at all three stations and supports the conclusions of other researchers who have found that flood stages on the MMR have increased over time [3,4,18,24,21]. These increases in stage are associated with proportional decreases in flow velocity, consistent with the hypothesis that dike construction has played a major part in forcing stages higher. The fact that the rising stage trends continue to overbank conditions at all the stations suggests that levees also contribute to this change. This conclusion is consistent with modeling results suggesting that part, but not all, of the observed increases in stage are caused by the exclusion of flood discharges from the floodplain.

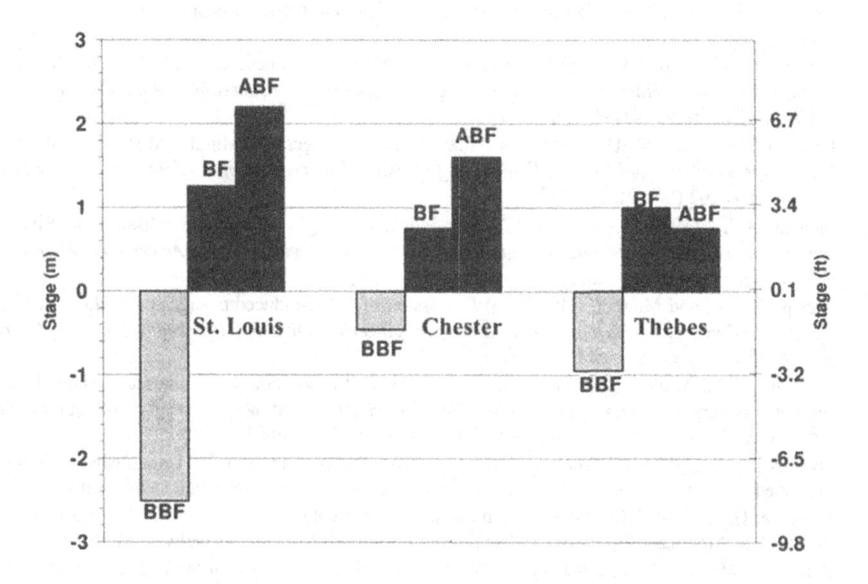

Figure 6. Trends in below bankfull (BBF), bankfull (BF) and above bankfull (ABF) stages (as defined in the text) at St. Louis, Chester and Thebes.

The specific-gage technique eliminates the range of upstream factors that do affect the magnitude of discharge delivered to the study reach but which have little or no impact on the stage-discharge relationship at the gaging stations. Ephemeral factors, such as water temperature and rising versus falling limb of flood hydrographs, do affect the stage associated with a given discharge on a specific date, but these effects rapidly average out and disappear when the specific-gage and related techniques are utilized. The stage trends over time for each station (e.g., Fig. 2) are remarkably free from scatter, and the long-term patterns thus revealed are unambiguous.

Anthropogenic increases in flood stage have broad scientific, technical, political, and legal ramifications. Design heights for levees and other structures are based on past hydrologic records, and systematic changes should be factored into such designs. Most importantly, stage is the parameter that directly controls flood damage when such events occur. Floodplain managers should recognize that rising stage trends translate into expanded hazard zones, increases in probable damage, and heightened flood risk. Engineers, planners, and politicians should seek to reach an equitable balance between 1) the local benefits of flood control and navigation improvements and 2) the elevated flood risk that these structures can cause.

6. References

1. Dobney, F.J. (1978) *River Engineers on the Middle Mississippi: A history of the St. Louis District, U.S. Army Corp of Engineers,* U.S. Government Printing Office, Washington.
2. Maher, T.F. (1964) Degradation study of the Middle Mississippi River, vicinity of St. Louis. *U.S. Dept. of Agriculture* **970**, 424-430.
3. Belt, C.B., Jr. (1975) The 1973 flood and man's constriction of the Mississippi River, *Science* **189**, 681-684.
4. Stevens, M.A., Simons, D.B., and Schumm, S.A. (1975) Man-induced changes of Middle Mississippi River, *J.Waterways, Harbors, and Coastal Engineering Division, Proceedings of the American Society of Civil Engineerings* **101**(WW2), 119-133.
5. Dyhouse, G.R. (1976) Discussion of "Man-induced changes of Middle Mississippi River", *J. Waterways, Harbors, and Coastal Engineering Division, Proceedings of the American Society of Civil Engineerings* **102** (WW2), 277-279.
6. Stevens, G.T. (1976) Discussion of "Man-induced changes of Middle Mississippi River", *J. Waterways, Harbors, and Coastal Engineering Division, Proceedings of the American Society of Civil Engineerings* **102** (WW2), 280.
7. Westphal, J.A., and Munger, P.R. (1976) Discussion of "Man-induced changes of Middle Mississippi River", *J. Waterways, Harbors, and Coastal Engineering Division, Proceedings of the American Society of Civil Engineerings* **102** (WW2), 283-4.
8. Westphal, J.A., Munger, P.R., and Muir, C.D. (1976) Mississippi River dikes and stage-discharge relations, *Rivers '76: Proceedings of the Symposium on Inland Waterways for Navigation, Flood Control and Water Diversions,* American Society of Civil Engineers, New York.
9. Dyhouse, G.R. (1985a) Levees at St. Louis--More harm than good? Unpublished manuscript, presented at American Society of Civil Engineers Hydraulics Division Specialty Conference.
10. Dyhouse, G.R. (1985b) Comparing flood stage-discharge data--Be careful! Unpublished manuscript, presented at American Society of Civil Engineers Hydraulics Division Specialty Conference.
11. Parrett, C., Melcher, N.B., and James, R.W. Jr. (1993) Flood discharges in the Upper Mississippi River basin, 1993. *U.S. Geological Survey Circular* **1120-A**.
12. SAS Institute, Inc. (1993) SAS/ETS User's Guide, Version 6, Second edition, SAS Institute, Cary, North Carolina.
13. Blench, T. (1969) *Mobile-Bed Fluviology: A Regime Theory Treatment of Canals and Rivers for Engineers and Hydrologists,* The University of Alberta Press, Edmonton.
14. Biedenharn, D.S., and Watson, C.C. (1997) Stage adjustment in the Lower Mississippi River, USA. *Regulated Rivers: Research & Management* **13**, 517-536.
15. Fenwick, G.B. (1969) Water-temperature effects on stage-discharge relations in large alluvial rivers, U.S. Army Corps of Engineers, Committee on Channel Stabilization, Technical Report no. 6.
16. Burke, P.P. (1965) Effect of water temperature on discharge and bed configuration, Mississippi River at Red River Landing, Louisiana, Technical Report No. 3, Committee on Channel Stabilization, U.S. Army Corps of Engineers.
17. Guttman, N.B., Hosking, J.R.M., and Wallis, J.R. (1994) The 1993 Midwest extreme precipitation in historical and probabilistic perspective, *Bulletin of the American Meteorological Society* **75**, 1785-1792.
18. Knapp, H.V. (1994) Hydrologic trends in the Upper Mississippi River Basin, *Water International* **19** 199-206.
19. Mount, J., 1998. Levees harm more than help. *ENR (Engineering News Record)* **2/2/98**, p. 59.
20. Hall, B.R. (1991) Impact of agricultural levees on flood hazards. *U.S. Army Engineer Waterways Experiment Station Technical Report* HL-91-21.
21. Koenig, R.L., and Tipton, V. (1993) The flood that wasn't. *St. Louis Post Dispatch,* 12/26/93.

22. Babinski, Z.(1992) Hydromorphological consequences of regulating the Lower Vistula, Poland, *Regulated Rivers: Research and Management* **7**, 337-348.
23. Bonacci, O., Tadic, Z., and Trninic, D. (1992) Effects of dams and reservoirs on the hydrological characterisitics of the Lower Drava River, *Regulated Rivers: Research and Management* **7**, 349-357.
24. Chen, Y.H. and Simons, D.B. (1986) Hydrology, hydraulics, and geomorphology of the Upper Mississippi River System, *Hydrobiologia* **136**, 5-20.
25. Walker, J.H., and Grabeau, W.E. (1995) History of flood and diversion control on the Mississippi River, *Geological Society of America Abstracts with Programs* **27** (6), 85.

RISK OF FLOOD LEVEE OVERFLOWING

M. SOWINSKI

Institute of Environmental Engineering
Poznan University of Technology
Poland, 60-965 Poznan, Piotrowo 5

1. Introduction

The uncertainty for hydraulic structures design can be divided into four basic categories: hydrologic, hydraulic, structural and social-economical. Hydraulic engineers are particularly interested in the first two categories. Traditional approach ignored the existence of hydraulic uncertainty assuming conveyance of the hydraulic structure as deterministic. Risk associated only with the natural randomness of hydrologic event was considered in terms of its return period. In order to integrate hydrologic and hydraulic uncertainties, two types of rateability models have been applied : static models and dynamic models [1]. Static models do not consider the repeated nature of hydrologic events e.g. floods. To this type belong models based on direct integration method [2,3,4,5], first-order second moments method (mean value and advanced) [6,7] and Monte Carlo simulation. Dynamic or time-dependent models can consider repeated application of loading. The objective of these models is to determine the reliability of a structure over a specified time interval (usually its service life) in which the number of occurrences of loading is a random variable. Almost all of types of models specified above were applied in the past to flood levee reliability analysis.

Sources of uncertainty for flood levee systems resulting in failures of these systems can be classified [8,9] as follows:

- overflowing caused by flood water level exceeding the top of the levee ,
- overtopping of the levee by waves induced by wind,
- piping through the body of levee or subsoil,
- instability of inner or outer slope,
- subsidence of the crest due to settlement of the sub-soil or levee,
- erosion of the foreshore by wind or ship induced waves,
- damage caused by drifting ice or ships,
- damage caused by people, animals or plant roots.

The first type of failure mechanism and its first position on a ranking prepared according to frequency is of particular interest to hydraulic engineers; it will be analysed in the paper applying the advanced first-order second moments method (AFOSM).

J. Marsalek et al. (eds.), Flood Issues in Contemporary Water Management, 363–373.

2. Objectives

The main objective of the paper was to present an application of a reliability model for computation of the risk of a flood levee overflowing based on a relatively new AFOSM method which takes into account two types of uncertainty: hydrologic and hydraulic. The other objective was to investigate the influence of the probability distribution of basic variables on this risk.

3. Basic Definitions and Equations

Generally, failure of an engineering system can be defined as the loading L on the system exceeding its resistance R [1,10]. Hence, risk is defined as the probability of the failure

$$RY = p(L > R) \tag{1}$$

Conversely the probability of complementary event, i.e. probability of the failureless performance of the system is called its reliability

$$R = p(L \leq R) \tag{2}$$

The interaction of load and resistance can be described by the performance function of the form

$$Z = L - R \tag{3}$$

This function relates L and R in such way that if R > L, then Z > 0, which is interpreted as "safe state"; if R < L (i.e. < 0) the "failure state" is observed. The boundary separating both states is called "limit state" or "failure surface" and is described by the equation Z = 0. Thus the reliability can be expressed in terms of Z as

$$R = p(Z > 0) \tag{4}$$

If R is a function of some basic variables, denoted generally by $x_1, x_2 .. x_{n-1}$, the performance function depends on those variables and L which is treated as a basic variable and denoted by x_n

$$Z = f(x_1, x_2 ... x_n) \tag{5}$$

In the first-order reliability method, the performance function is expanded in a Tylor series and truncated after the first term

$$Z = f(x^p) + \sum_{i=1}^{n} (x_i - x_i^p)\left(\frac{\partial f}{\partial x_i}\right)_p \tag{6}$$

where subscript p denotes expansion point.

In the mean first-order second-moments method (MFOSM), it is assumed that the expansion point is at the mean value of the basic variables. In the advanced first-order second-moments method (AFOSM), the expansion point is taken on the failure surface of the performance function (which is denoted by *). According to earlier explanations,

$$f(x_1^*, x_2^* \ldots x_n^*) = f(X^*) = 0 \tag{7}$$

Thus, the expected value of the performance function is

$$E[Z] = \sum_{i=1}^{n} [x_i - x_i^*](\frac{\partial f}{\partial x_i}) \tag{8}$$

For the case of statistically independent basic variables, which was assumed in this study, the variance of the performance function is

$$Var[Z] = \sigma_Z^2 = \sum_{i=1}^{n} [(\frac{\partial f}{\partial x_i})\sigma_{x_i}]^2 \tag{9}$$

where σ_z and σ_{x_i} are the standard deviation of Z and x_i, respectively.

The expected value and standard deviation of the performance function are used for determination of reliability index

$$\beta = \frac{E[Z]}{\sigma_z} \tag{10}$$

which is equivalent to the shortest distance in standard space between the point representing system mean state and the failure surface. The system reliability is a function of reliability index β and for normal distribution of performance function it can be determined from relation

$$R = \Phi(\beta) \tag{11}$$

where Φ = the cumulative distribution function.

Substituting equations (8) and (9) into equation (10) yields

$$\beta = \frac{\sum_{i=1}^{n} \{(\frac{\partial f}{\partial x_i}) [x_i - x_i^*]\}}{\sqrt{\sum_{i=1}^{n} [(\frac{\partial f}{\partial x_i})\sigma_{x_i}]^2}} \tag{12}$$

This equation is valid for a linear performance function. In the case of a non-linear performance function, it cannot be applied because location of the failure point is not known a priori. Its determination requires application of an iteration procedure, e.g. such as proposed by Rackwitz [10]. After assumption of initial values for basic variables (usually mean values $\bar{x}_i$) the direction cosines are computed from equations

$$\alpha_i^* = \frac{(\frac{\partial f}{\partial x_i})\ \sigma_{x_i}}{\sqrt{\sum_{i=1}^{n}[(\frac{\partial f}{\partial x_i})\ \sigma_{x_i}]^2}} \qquad \text{for } i = 1, 2...n \qquad (13)$$

They are used by formulation of the set of equations determining new co-ordinates failure point in terms of unknown β

$$x_i^* = x_i - \alpha_i^* \sigma_{x_i} \beta \qquad \text{for } i = 1, 2...n \qquad (14)$$

These equations are substituted into the performance function calculated on the failure surface.

In the case of a non-linear performance function, equation (7) is also non-linear with respect to β and it must be solved by an iteration method. Knowing the value of β, new co-ordinates of the failure point can be computed from the set of equations (14). This procedure is repeated using computed co-ordinates of the failure point as initial values for new iterations until a minimum value of β is found.

If the probability distributions of the basic variables are not normal it is desirable to transform them into equivalent normally distributed variables. Such a transformation is based on the assumption that the values of the cumulative distribution functions (CDF) and probability distribution functions (PDF)of the non-normal distributions are the same as those of equivalent normal distributions at the failure points. Thus the original non-normal CDF denoted by F_{xi} is equated to normal CDF

$$F_{x_i}(x_i^*) = \Phi(\frac{x_i^* - \bar{x}_i^N}{\sigma_{x_i}^N}) \qquad (15)$$

where the superscript N denotes normalised variables.

Similarly, a non-normal PDF denoted by f_{xi} is equated to a normal PDF

$$f_{x_i}(x_i^*) = \varphi(\frac{x_i^* - \bar{x}_i^N}{\sigma_{x_i}^N}) \qquad (16)$$

Thus the standard deviation and the mean of the equivalent normal distributions become

$$\sigma_{x_i}^N = \frac{\varphi\{\Phi^{-1}[F_{x_i}(x_i^*)]\}}{f_{x_i}(x_i^*)} \tag{17}$$

and

$$\bar{x}_i^N = x_i^* - \Phi^{-1}[F_{x_i}(x_i^*)]\sigma_{x_i}^N \tag{18}$$

4. Application of AFOSM Method for Flood Levee System

For a flood levee system, loading is defined as flood flow Q_f and resistance as conveyance of the system Q_c. Thus the performance function can be expressed as the difference between flood levee conveyance Q_c and flood magnitude Q_d

$$f(x) = Q_c - Q_d \tag{19}$$

The problem of levee conveyance determination as a function of basic variables is discussed in the following section, while determination of flood magnitude based on frequency analysis is presented in the next section.

4.1. APPLICATION OF AFOSM METHOD FOR A FLOOD LEVEE SYSTEM

In this study, a simple hydraulic model was applied i.e. model of uniform flow based on Manning formula [5,11]. Lack of accuracy resulting from application of this model can be compensated by introduction of a model correction factor λ. Assuming idealised river cross-section (Figure 1) consisting of a main channel and two symmetrical floodplains, the conveyance of the levee system can be computed [5,11], applying a rule of superposition, from

$$Q_c = \lambda[\frac{1}{n_c} A_c^{5/3} P_c^{-2/3} + 2\frac{1}{n_f} A_f^{5/3} P_f^{-2/3}]s^{1/2} \tag{20}$$

where : subscript c refers to the main channel, subscript f - to the flood plain

- n - Manning surface roughness coefficient,
- A - flow cross-section area,
- P - wetted perimeter.

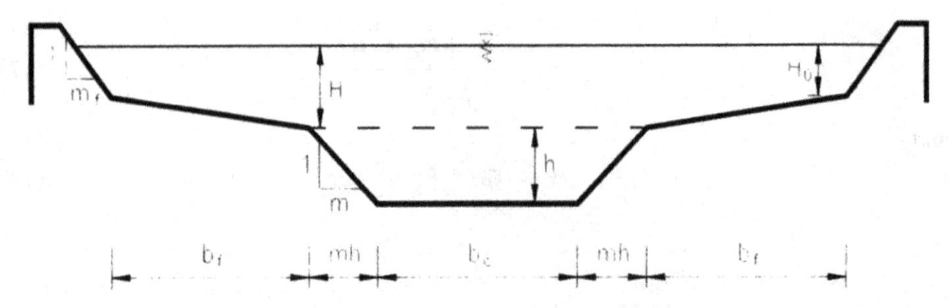

Figure 1. Flood levee system cross section.

The cross-section area and wetted perimeter characterising the geometry of the cross-section are functions of basic dimensional parameters: depth h, width of the bottom of the section b and side slope of banks m. Taking into account the geometric relations of the section (see Figure 1) among those parameters, the following equations were obtained :

- for the main channel

$$A_c = (b_c + mh)h + (b_c + 2mh)H \tag{21}$$

$$P_c = b_c + 2h\sqrt{1+m^2} \tag{22}$$

- for the one floodplain section.

$$A_f = \frac{H+H_0}{2} b_f + m_f H_0^2 \tag{23}$$

$$P_f = b_f + H_0\sqrt{1+m_f^2} \tag{24}$$

Substituting equations (21), (22), (23) and (24) into equation (20) gives expression for Q_c as a function of the basic variables

$$Q_c = \lambda\sqrt{s}\{\frac{1}{n_c}[(b_c + mh)h + H(b_c + 2mh)]^{5/3}(b_c + 2h\sqrt{1+m^2})^{-2/3} +$$
$$+\frac{1}{n_f}[(\frac{H+H_0}{2} b_f + m_f H_0^2)^{5/3}(b_f + H_0\sqrt{1+m_f^2})^{-2/3}]\} \tag{25}$$

4.2. HYDROLOGIC MODEL FOR FLOOD FLOW

Different probability distributions can be applied for description of hydrologic uncertainty associated with the randomness of maximum annual flood flows. If the type I extremal distribution is used for the PDF and CDF of this distribution are expressed by the following equations

$$f(Q_d)=\frac{1}{\alpha_2}\exp[-(\frac{Q_d-\alpha_1}{\alpha_2})-\exp(-\frac{Q_d-\alpha_1}{\alpha_2})] \tag{26}$$

$$F(Q_d)=\exp\{-\exp[-(\frac{Q_d-\alpha_1}{\alpha_2})]\} \tag{27}$$

where α_1 and α_2 are parameters of the distribution computed from formulas given below

$$\alpha_1=\overline{Q}_d-0.577\alpha_2 \tag{28}$$

$$\alpha_2=\frac{\sqrt{6}}{\Pi}\sigma_{Q_d} \tag{29}$$

The flood magnitude for a T-year return period can be determined from

$$Q_d(T)=-\alpha_2\ln[-\ln(1-\frac{1}{T})]+\alpha_1 \tag{30}$$

5. Numerical Example

An example based on data for Warta River upstream from Poznan in Poland was included to trace the stages of calculations by presentation of results of successive iterations and to investigate the influence of type of PDF of basic variables on final solution i.e. on the risk of flood levee system overflowing. Computations were performed for two variants of simple PDF: in the first one, only symmetric triangular distributions were considered; in the second uniform (rectangular distributions were used) with the exception of slope s and model error λ, for which triangular PDFs were assumed due to computational problems.

5.1. INPUT DATA

Characteristic values of the basic variables presented in the Table 1 were estimated on the basis of measurement of river cross sections (geometric characteristics) and simulations of flow by a dynamic model developed in the Institute of Hydraulic Engineering of the Agriculture University of Poznan (Manning roughness coefficient and water surface elevation). They were required for calculation of statistical moments (mean and variance) used in the first iteration by computation of reliability index.

TABLE 1. Characteristic values of basic variables

Section of channel	Symbol	[Units]	Lower value X_L	Value of X for max PDF	Upper value X_u	Coef. of variation
	λ		0.9	1.0	1.1	4.082 E-2
	s		1.2 E-4	1.5 E-4	1.8 E-4	8.160 E-2
Main channel	b_c	[m]	30.0	35.0	40.0	5.832 E-2
	m		2.5	3.0	3.5	6.804 E-2
	h	[m]	2.6	2.8	3.0	2.916 E-2
	n_c		1.9 E-2	2.0 E-2	2.1 E-2	2.041 E-2
Flood plain	b_f	[m]	70.00	75.00	80.0	2.722 E-2
	m_f		1.75	2.00	2.25	5.102 E-2
	H	[m]	5.00	5.20	5.40	1.570 E-2
	H_o	[m]	4.04	4.44	4.84	3.678 E-2
	n_f		2.2 E-2	2.4 E-2	2.6 E-2	3.401 E-2

Parameters of the type I extremal distribution for annual floods were calculated using historical records collected in Poznan for the Warta River in Poznan for the period, 1900 - 1985 (86 years). Their values were estimated as $\alpha_1 = 283.16$ and $\alpha_2 = 208.04$.

The initial (used in the first iteration) value of flood magnitude was determined for a return period of T = 10 years from equation (30) and was found to be $Q_d = 403$ m^3/s.

5.2. RESULTS OF COMPUTATIONS

The failure equation, which was developed according to the procedure described in section 3, was non-linear with respect to β and therefore an iteration procedure was applied for its solution. For both variants, determination of minimum value of reliability index β, what was equivalent to maximum value of risk, required 3 iterations. Results for only the first variant (triangular PDF) are presented in the paper (Tables 2, 3 and 4). All symbols used in tables were explained earlier.

TABLE 2. Results of the first iteration

Variable	x_i	$\sigma^N_{x_i}$	$(\frac{\partial f}{\partial x_i}).\sigma^N_{x_i}$	$\alpha^*_{x_i}$	x^*_i
λ	1.00	4.08 E-2	0.00	2.90 E-3	0.94
s	1.50 E-4	1.20 E-5	84.34	2.89 E-3	1.31 E-4
b_c	35.00	2.04	40.13	1.37 E-3	33.51
m	3.00	2.04 E-1	16.89	5.78 E-2	2.94
h	2.80	8.16 E-2	15.14	2.18 E-2	2.78
n_c	2.00 E-2	2.04 E-1	-18.69	-6.40 E-2	2.00 E-2
b_f	75.00	2.04	28.78	9.85 E-2	73.93
m_f	2.00	1.02 E-1	6.00	2.05 E-4	1.99
H	5.20	8.16 E-2	31.30	1.07 E-3	5.15
H_0	4.44	1.63 E-1	40.17	1.38 E-3	4.32
n_f	2.40 E-2	8.00 E-4	- 0.20	- 8.00 E-5	2.40 E-2
Q_d	403.30	255.1458	-255.1458	-0.8736	451.40

TABLE 3. Results of the second iteration

Variable	x_i	$\sigma^N_{x_i}$	$(\frac{\partial f}{\partial x_i}).\sigma^N_{x_i}$	$\alpha^*_{x_i}$	x^*_i
λ	0.94	3.59 E-4	0.00	0.1014	0.98
s	1.31 E-4	1.10 E-5	71.08	0.1093	1.44 E-4
b_c	33.51	22.522	38.02	0.0584	34.69
m	2.94	2.19 E-1	15.44	0.0237	2.99
h	2.78	8.69 E-2	13.39	0.0206	2.80
n_c	2.00 E-2	2.19 E-1	-16.50	-0.0254	0.02
b_f	73.93	2.25	26.99	0.0415	74.79
m_f	1.99	1.04 E-1	4.99	0.00767	2.00
H	5.15	8.16 E-2	26.36	0.0405	5.19
H_0	4.32	1.80 E-1	37.62	0.0578	4.42
n_f	2.40 E-2	8.00 E-4	-0.02	-0.00003	2.40 E-2
Q_d	1542.82	268.51	-639.3502	-0.9828	1468.47

It can be seen that changes of basic variables determining the hydraulic conveyance (and the conveyance itself) were not significant in successive iterations. But the second component of the performance function - flood flow - increased in the failure point by several times in order to minimize the performance function. It was caused by excessive high conveyance compared to the initially assumed flood flow. Standard deviations and mean values of variables as input data to analysis of uncertainty. It was not included to objectives of this paper but domination of hydrologic uncertainty is evident.

TABLE 4. Results of the third iteration

Variable	x_i	$\sigma_{x_i}^N$	$(\frac{\partial f}{\partial x_i}).\sigma_{x_i}^N$	$\alpha_{x_i}^*$	x_i^*
λ	0.98	4.48 E-2	0.00	0.1412	0.981
s	1.44 E-4	1.30 E-5	92.12	0.1444	1.44 E-4
b_c	34.69	2.11	39.67	0.0622	34.60
m	2.99	0.20	16.17	0.0254	2.98
h	2.80	8.19 E-2	14.43	0.0226	2.79
n_c	2.00 E-2	0.20	-17.78	-0.0279	0.02
b_f	74.79	2.08	27.96	0.0438	74.72
m_f	2.00	0.10	5.61	0.00879	2.00
H	5.19	8.16 E-2	29.82	0.0468	5.19
H_0	4.42	0.17	39.56	0.0620	4.41
n_f	2.40 E-2	8.0 E-4	-0.02	-0.00003	0.02
Q_d	1468.47	620.1091	-620.1061	-0.9723	1646.04

The input data in the second variant (uniform PDF) were the same as in the first one. Due to lack of space, results of iterations are not included in the paper. Only values of reliability index and risk characterizing successive stages of computations are tabulated together with their counterparts from the first variant (Table 5). Comparison of obtained values indicated that PDF of basic variables influence in significant way the risk of flood levee overflowing. As it could be expected, for greater distribution of variables (uniform PDF) the risk was higher than for smaller distribution (triangular PDF).

TABLE 5. Comparison of final results obtained for variants I and II

	Variant I (Triangular PDF)		Variant II (Uniform PDF)	
Iteration	β	RY	β	RY
1	5.314	0	5.22	0
2	2.760	**2.85 E-3**	2.66	**3.92 E-3**
3	3.090	1.09 E-3	3.00	1.34 E-3

6. Conclusions and Recommendations

The problem of risk evaluation for floods overflowing levees was formulated taking into consideration hydrologic and hydraulic uncertainties. It was solved using the AFOSM method. The advantage of this approach over direct integration method is that it does not require an assumption for the PDF of conveyance. However, in this case such an assumption is necessary for basic variables. Example calculations were performed for the simplest probability distributions of basic variables: triangular and uniform. Analysis of results indicates sensitivity of risk to these distributions and the importance of proper choice of them. Finally, the performance of the method was estimated taking into account the number of iterations. In both variants, three iterations were sufficient to find the location of expansion point on the failure surface for which reliability index took minimum value. Further studies should concern collection of information on the actual PDFs of basic variables and examination of their influence on the risk of flood levee overflowing.

7. References

1. Yen, B. C. and Tung, Y. K. (1993) Some recent progress in reliability analysis for hydraulic design, in B.C. Yen and Y.K. Tung. *Reliability and Uncertainty Analysis in Hydraulic Design,* ASCE Publisher, pp. 35-80.
2. Lee, H. and Tung, Y. (1983) Improved risk and reliability model for hydraulic structures, *Water Resources Research* **6**, 1415-1422.
3. Sowinski, M. (1994) Reliability and risk of overtopping of flood levee system, Proceedings of International UNESCO Symposium on Water Resources Planning in a Changing Word, 28-30 June, Karlsruhe, pp. II 113-122.
4. Tung, Y. and Mays, L. (1981a) Risk models for flood levee design, *Water Resources Research* **4**, 833-841.
5. Tung, Y. and Mays, L. (1981b) Optimal risk-based design of flood levee systems, *Water resources Research* **4**, 843-852.
6. Melching, C. S.(1992) An improved first-order reliability approach for assessing uncertainties in hydrologic modeling, *Journal of Hydrology* **132**, 157-177.
7. Sowinski, M. (1996) Application of AFOSM method for risk assessment of Warta River flood levee overtopping, Proceedings of the 1st International Conference on New/Emerging Concepts for Rivers - Rivertech 96, 22-26 September, Chicago, pp. 324-331.
8. Vrijling J. K. (1993) Development of probability design of flood defences in the Netherlands, in B.C. Yen and Y.K. Tung. *Reliability and uncertainty analysis in hydraulic Design,* ASCE Publisher, pp.133-178.
9. Zhou, H.M. (1995) *Towards an operational risk assessment in flood alleviation,* Ph.D. Thesis, Delft University Press, Delft.
10. Ang, A. and Tang, W. (1984) *Probability Concepts in Engineering Planning and Design* , *Vol.2 Decision, Risk and Reliability*, Wiley and Sons, New York.
11. Tung, Y. (1985) Models for evaluating flow conveyance reliability of hydraulic structures, *Water Resources Research,* **10**, 1463-1468.

EMERGENCY FLOOD RESERVOIRS IN THE TISZA BASIN

DR. L. SZLÁVIK

Water Resources Research Centre Plc. (VITUKI)
H-1095 Budapest, Hungary

1. The Flood Hazard Situation in Hungary

Under the particular physio-geographic conditions of Hungary, important and steadily growing concerns have been paid to flood control for centuries. The fundamental cause of the grave flood hazard is that the overwhelmingly flat country is situated in the deepest part of the Carpathian Basin, where the flood waves rushing down from the surrounding Carpathian and Alpine headwater catchments are slowed down, combine with each other, and often result in high river stages of extended duration. Owing to the climate and the physio-geographic situation, floods may occur virtually on any Hungarian river in any season of the year.

Flood plains make up 22.8% (21,248 km^2) of the total country area of 93,000 km^2. In total, 2.5 million people living in about 700 communities in the protected flood plains and are exposed to flood hazard. These plains comprise 1.8 million hectares or one-third of the arable land in the country, over 2,000 industrial plants, 32% of the railway lines, and 15% of the road network. Some 25% of the gross domestic product is generated in this area.

2. The Tisza River Basin and its Floods

2.1. PHYSICAL, GEOGRAPHICAL AND HYDROLOGICAL CHARACTERISTICS OF THE TISZA BASIN

The Tisza river is the largest tributary of the Danube, with respect to both river length (977 km) and the size of the drainage basin. The Tisza collects the waters of the eastern part of the Carpathian Basin. Its round shaped drainage basin covers 157,200 km^2, which represents 20% of the total Danube basin. The width of the catchment in the East-West direction is 520 km, and the North-South length covers 460 km. The highest point in the basin reaches 2506 m above mean sea level, and the mean annual flow at the mouth of Tisza is estimated at 830 m^3/s.

River Tisza has a drainage basin of 9,707 km^2 at the cross-section where it enters Hungary, and 139,078 km^2 at the southern border of the country where the river leaves the Hungarian territory. The catchment increment is 129,371 km^2; 86% of that is

J. Marsalek et al. (eds.), Flood Issues in Contemporary Water Management, 375–384.

contributed by main tributaries (Túr/Tur, Szamos/Somes, Kraszna/Crasna, Bodrog, Sajó/Slana-Hernád/Hornád, Zagyva-Tarna, Körös/Cris, Maros/Mures) which originate mostly outside of Hungary and transfer floods from the surrounding mountainous regions to the main river. The Hungarian part of the catchment is predominantly flatland (46,200 km^2). Floods originate mostly outside of the country and robust flood control measures, such as the construction of flood storage reservoirs, are practically out of question.

The annual mean precipitation over the Tisza Basin varies from 500 to 1600 mm. The peak flow of the Tisza reaches 4,000 m^3/s at the point of entry into Hungary, and the peaks of tributaries exceed the range of 1,000 – 1,500 m^3/s.

The hydrological regimes of the Tisza and its tributaries are extreme. Torrential floods of the Upper-Tisza and tributaries originating in the mountains reach the Hungarian border within 20-30 hours. The decline of these floods requires only several days, while in the middle and lower river reaches high floods of long duration are typical.

The range of water level fluctuations is rather high in the Tisza Basin, usually more than in similar size European rivers. The range of water level fluctuations in the Tisza in its lowland sections ranges from 8 to 13 m, and only the Upper Tisza shows a smaller (5 m) range. The same high ranges are characteristic for most of the tributaries (3 - 10 m).

2.2. GEOGRAPHICAL SITUATION AND POPULATION OF THE TISZA BASIN

Five countries share the Tisza catchment: Ukraine (8.1%), Romania (46.2%), Slovakia (9.7%), Hungary (29.4%), and Yugoslavia (6.5%). Co-operation of these countries in the water field is regulated by bi-lateral transboundary water agreements.

The Tisza Basin has a characteristic distribution of settlements. No major city is situated in the catchment, and 25-30 large towns with 60,000 - 400,000 inhabitants house a population of 2.5 – 3.0 million people. These have been the natural regional centres of the Basin since centuries ago. Out of the 15 million population of the Basin, 80% live in small towns and villages.

2.3. CONSTRUCTION OF THE FLOOD PROTECTION SYSTEM IN THE TISZA BASIN

The present day shape of the Tisza Basin is the result of two centuries of planned water management in the region; 20,000 km^2 of the Basin (30 % of the flatland) used to be permanently covered by water or periodically inundated by floods. One hundred and fifty years of construction resulted in a 4,500 km long embankment system and protection of 27,000 km^2 from inundation (the Hungarian share of this system is 2,900 km of embankments and 18,000 km^2 of land.) The scale of this major flood defence system can be illustrated by its characteristics showing that it is the largest system of this kind in Europe (compare, Po Valley - 2,400 km embankments and the protected area of 12,000 km^2; the Netherlands, 15,000 km^2 of protected area).

Flood risk in the Hungarian section of the Tisza Basin can be described by the fact that this territory is protected by 2,900 km of primary flood embankments representing 70 % of the whole system of flood protection in the country.

Regardless of national borders, the Tisza basin presents a united system of streams, in which land use and other changes in the upstream parts of the catchment also impact on flood conditions in the Hungarian section of the Basin. Construction of flood embankments in the last two decades outside of Hungary was followed by an evident rise of flood water levels on the Tisza and its tributaries. The most obvious example is given by the Körös river where the maximum flood water levels have risen 111 - 172 cm during the last 30 years.

2.4. MAJOR FLOODS IN THE TISZA BASIN

Each major flood has revived the demand and arguments for further development of flood control. This was the case also in the past century, when the floods of 1816, 1830 and 1845 provided the final impetus triggering the comprehensive reclamation project in the Tisza Valley, while the floods of 1855, 1867-68, 1879, 1881 and 1888, each of disastrous proportions, redirected public attention to the importance of continuing and improving the flood defences.

In the present century, such periods of intensive development were prompted in response to the major floods in the Tisza valley: 1919, 1925, 1932, 1939, 1940-41, 1947-48, and 1970. Among the recent events, a major flood travelled down the Körös river in the winter of 1995-96, and the most recent floods occurred in November 1998 on the Upper-Tisza and Bodrog rivers, and were followed by extreme historical floods on the Bodrog and the middle reach of the Tisza in March-April 1999. The last flood to claim lives was the ice-jam flood in the winter of 1955-56 on the Danube. And this success is not attributable to good luck, but rather to carefully planned, methodical development and organisation of flood protection efforts.

In response to major floods in the Körös basin, the method of emergency storage has been applied repeatedly on the Körös rivers during the past decades. Notwithstanding the continuous flood defence improvements and enormous flood fighting efforts, the major floods from 1925 to 1955 could not be contained between the embankments (Table 1). Twenty-one embankment failures and emergency storage cases occurred between 1925 and 1980 (10 in Hungary, 11 in Romania). Between 1981 and 1995, further nine failures or emergency storage events occurred (three emergency reservoirs were opened in Hungary, four in Romania, where also two failures occurred). The experiences gained from a total of 30 flood events are thus available to assess the merits and drawbacks of emergency storage [1]. The "inland delta" between the Fehér- and Fekete Körös rivers was inundated totally or partly on nine occasions during the past 70 years owing to embankment failures, or emergency storage (1925, 1932, 1939, 1966, 1970, 1974, 1980, 1981, and 1995). The area inundated ranged from 10 to 161 km^2, while the inundation water volume ranged from 8 to 200 million m^3.

TABLE 1. Emergency reservoirs built or designated and their uses

Reservoir	River	Distr. Water Authority	Area (km^2)	Volume (10^6 m^3)	State of reservoir	Time inundated
Mályvád*	Fekete-Körös	Körös Region	34.70	75.0	Completed	Aug. 1, 1980 March 13, 1981 Dec. 29, 1995
Kisdelta*	Fehér-Körös	Körös Region	5.50	26.0	Under construction	-
Mérges	Kettős- and Sebes-Körös	Körös Region	18.20	87.2	Completed	July 28, 1980 Dec. 30, 1995
Kutas	Berettyó	Trans-Tisza	38.96	36.5	Designated	Feb. 9, 1966 June 15, 1970
Halaspusztai	Sebes-Körös and Berettyó	Trans-Tisza	21.75	35.0	Designated	July 26, 1980
Ér-menti	Ér and Berettyó	Trans-Tisza	13.52	12.2	Designated	-
Jászteleki	Zagyva	Middle-Tisza	20.00	24.0	Completed	-
Viszneki	Gyöngyös and Tarna streams	North-Hungary	5.56	4.51	Designated	Oct. 21, 1974

Note: All or part inundated because of embankment failures: 1925, 1932, 1939, 1966, 1970, 1974

3. Emergency Flood Storage

3.1. DEFINITION AND ROLE OF EMERGENCY STORAGE

An emergency reservoir is understood to be an area adapted by engineering measures for temporary flood storage. These reservoirs are flooded only in extraordinary situations, to avert an impending failure of the main embankment line and thereby prevent major losses and flood disasters. Flooding and emptying are accomplished by cutting the embankment (usually with a dragline, by blasting or by gated structures). Under normal conditions these areas serve agricultural, or forestry purposes.

The purpose of emergency storage is to retain temporarily part of the flood volume and thus reduce the peak flood level.

The term temporary, rather than normal flood reservoir, is used mainly for economic reasons. The land in the reservoir is not acquired, but designated for such use by the competent authority. This designation entails limitations on land use (e.g. ban on building). Any decision on flooding the emergency reservoir is invariably taken after considering the actual situation. Besides the emergency situation, the decision maker must take into account the potential economic consequences as well.

In the long-term flood control development plan of Hungary completed in March, 1995 and in the Government Resolution of 1995 on the flood control situation and measures based thereon, eight temporary, emergency flood reservoirs of 300 million m^3 in total volume were developed, designed or are under construction in the Körös and Zagyva-Tarna river valleys. These reservoirs are essential parts of the Hungarian flood control system [2]. In critical flood emergency situations, part of the flood flow can be diverted thereto in the interest of protecting endangered communities. These are the rivers which carry violent floods capable of overtopping the embankments. The emergency reservoirs have been flooded nine times since 1966 (Table 1).

3.2. THE CRITICAL SITUATION CALLING FOR EMERGENCY FLOOD STORAGE

The use of emergency flood storage may be warranted in four substantially different situations for each of which examples can be quoted from the Hungarian practice of flood fighting [3]:

(a) at flood stages surpassing the design level for which the defences were built and could safely withstand, to lower the peak of flood hydrograph,
(b) as an instrument for controlling ice-jam floods and to avert impending embankment failures,
(c) to prevent a flood disaster by loss of embankment stability owing to saturation by floods of extended duration or by other defects, and
(d) to control the consequences of an embankment failure.

A detailed scrutiny of the events recorded in the Hungarian practice has led to the conclusion that the case (a), i.e. flooding in the interest of lowering the flood peak, is the case from which the design criteria of emergency flood reservoirs should be derived. Designing emergency flood reservoirs for the other three situations (b-d) is not justified, although such reservoirs sized and built to retain part of the flood volume may be found effective in critical situations caused by ice jam floods or loss of embankment stability, and would reduce the impacts of an embankment failure along the river section influenced by them. These considerations must be remembered in selecting the site and in formulating the design criteria for emergency flood reservoirs.

3.3. THE WATER VOLUME TO BE STORED

The total water volume (W) to be stored was found to be the sum of four sub-volumes:

$$W = W_1 + W_2 + W_3 + W_4, \tag{1}$$

where W_1 - the water volume to be diverted to lower the water level sufficiently on the critical tributary
W_2 - the additional water volume resulting from lowering the water level on the other tributary
W_3 - the "opening correction"
W_4 - the storage space needed to accommodate an additional flood wave travelling down the river while the emergency reservoir is still open.

The design flood hydrograph can be produced by several methods; the key issue is the use of the flood loop rating curves reflecting the characteristics of the flood wave. In calculations related to the sizing and operation of emergency flood reservoirs, the use of the Q-H curve described by a power function and corrected for the flood loop has been found expedient for determining the width of the latter on the basis of the flood loops actually observed.

During the hydrological analysis of emergency flood storage, the water volume (W_1) to be diverted for lowering the water level sufficiently on the critical tributary should be found using the rating curve corrected by the flood loop. The additional water volume W_2 needed to lower the water level on the other tributary does not need to be taken into account, if the flood waves on the two tributaries do not coincide, or if the emergency reservoir is situated at a considerable distance from the confluence of the tributaries. In this case $W_2 = 0$.

Safe diversion of the water volume of the flood peak surpassing a particular level can only be achieved if the peak of the lowered flood peak on the river remains below the design flood level. The water diverted through the opening forms also a hydrograph, so that diversion must be started at a lower water level. This volume component has been referred to as the "opening correction" (W_3). In order to prevent the lowered peak level on the river from surpassing the design level, diversion must start at a lower water level.

The hydrologic characteristics of a particular river will provide information on the recurrence potential (the frequency of repeated floods waves) of flood waves, and with regard also to the rate and method of reservoir emptying, a sound engineering estimate will be possible for the sub-volume W_4 needed to accommodate an additional flood travelling down the river while the emergency reservoir is still open.

3.4. TIMING OF RESERVOIR FLOODING

The studies performed have indicated the "optimal" instant of reservoir opening exists when the rising water level approximates the design water level of the defence structures and the hydrological analysis indicates that the flood peak will attain, or surpass this level.

The analysis of hydrological and river hydraulics conditions of past embankment failures and emergency storage events has demonstrated the importance of exploiting the phenomenon of local drawdown for maximising the effectiveness of emergency storage.

Along the river section(s) upstream of the diversion, the streamflow rate increases suddenly within a few hours. This phenomenon is attributable to the sudden drawdown created in the river channel in the vicinity of the diversion, which is of paramount importance in the hydrology of emergency storage. Thus, opening emergency storage at the correctly chosen instant will produce the desired effect even when diverting a relatively small water volume. By discharging the first 10-20 million m^3 of water during 5-10 hours, the rate of the recession of stages in both rivers under consideration was 140-220 cm. The expected results of emergency storage were achieved in this way. If it would be possible to stop the discharge through the opening, the process of storage filling could be finished.

Evidently, the importance of the drawdown phenomenon will be more pronounced in cases, where the aim is to reduce the water load on the embankments in the immediate vicinity of the reservoir. The method of diverting the flood volume should therefore be such that the "peak-capping" effect is the highest possible. Subsequently, the opening should function as a simple spillway.

The reconstruction of flood hydrographs was generated on the basis of a series of flow measurements. It is an important fact that increase of discharge caused by local drawdown could be realised, when the outflow is below the stage section. The unsteady state in the river is well characterised by the slope of the water surface around the embankment breach or the opening of the emergency reservoir.

The flood hydrographs have demonstrated the "peak-capping" effect. This could be recognised in the time series of stages and discharges. The whole mechanism of emergency storage can be analysed from these data.

The analysis of the flood loops plays an important role in the evaluation of the processes. If the opening reaches up to the stage, the backward-flood loop could be recognised. The emergency storage in Mályvád produced a specific effect on the river; it flowed backward from the downstream (lower) section to the opening of the emergency reservoir.

TABLE 2. Impacts of the emergency reservoirs in the Körös Basin on lowering the flood peaks

	Parameter	Emergency storage in 1974 at the delta – opening in 3 cross-sections	The discharge through the embankment breach in 1980 (Kettős-Körös)	Emergency storage in 1981 at Mályvád - opening in 2 cross-sections	Emer-gency storage in 1995 at Mályvád	Emer-gency storage in 1995 at Mérges
1.	Maximum discharge to the emergency reservoir [m^3/s]	780	750-850	910	150	200-250
2.	Normal maximum discharge of the river at the time of opening the emergency reservoir [m^3/s]	945	842	755	489	767
3.	$M_1 = (2)/(1)$	0.83	1.01**	1.21***	0.31	0.26-0.33
4.	Volume of storage [10^6 m^3]	118	200	75	7.4	38.8
5.	Flood volume over the emergency level I.	488*	550	168	37.0	198
6.	$M_{2,I.} = (4)/(5)$	0.24	0.36	0.45	0.20	0.20

*Total flow of the Fekete- and Fehér-Körös [$10^6 m^3$]

**This is possible because of the local drawdown and back-flow in the Kettős-Körös

***This is possible because of the local drawdown and back-flow in the Fekete-Körös

The indices M_1 and M_2 have been introduced to express the importance of particular emergency reservoirs and to compare their effectiveness in lowering the flood peaks:

– The ratio of the highest discharges:

$$M_1 = Q_{max.\ diverted} / Q_{max.\ river} \tag{2}$$

– The ratio of the diverted water volume to the total flood volume:

$$M_2 = V_{stored} / V_{flood\ wave} \tag{3}$$

Table 2 indicates the impacts of emergency reservoirs on lowering flood peaks. It could be concluded that the ratio of the highest discharges (M_1) may exceed 1. It is possible because the local drawdown and back-flow of the rivers. The efficiency of the emergency storage is described by the value of M_2, which over the emergency level III could reach the value of 1.

3.5. THE IMPACT OF EMERGENCY STORAGE ON THE RIVER SYSTEM

The magnitude of the impact of emergency storage on various sections of the river system will depend on the relative location of the section and the point of diversion. Downstream of the diversion, the impact of emergency storage propagates along the particular branch of the river system and is felt over a long river section, because the flow diverted into the emergency reservoir lowers the hydrograph of the normal (natural) flood wave. The impact of abstraction increases with the value of M_1. The discharge diverted into the reservoir may even be higher than the flood flow in the river ($M_1 > 1$) owing to the local drawdown which may temporarily reverse the direction of flow in the downstream reach towards the diversion [4].

Upstream of the diversion, the drawdown effect is rather strong, but decreases rapidly with distance and practically disappears beyond some limit.

The diversion also impacts indirectly on the tributaries to the river, from which water is diverted. By lowering the water level in the affected river, the slope and consequently the velocity of flow in the tributaries is increased and this results in conveyance of their flow at a lower depth.

3.6. HYDROLOGIC CONSIDERATIONS IN SITE SELECTION AND DESIGN OF EMERGENCY RESERVOIRS

Starting from the experience gained during the nine cases of emergency storage use in the Tisza Basin, between 1966 and 1995, and using further data obtained during the confinement of inundation caused by an embankment failure (Kettős-Körös, 1980), the detailed requirements on engineering measures for flooding and emptying emergency reservoirs have been formulated as follows.

In the case of flooding:

- Controlled activation of emergency storage should be possible at the most appropriate instant, in order to divert the necessary water volume, achieve the desired drawdown effect and make full use of the available storage capacity.
- The effect attainable by drawdown of water stored between the flood embankments should be exploited; that is, instead of simple, gradual overflow into storage, an intensive diversion should be realised to achieve major drawdown within a brief period of time. This requirement is of paramount importance to reduce the load on flood embankments as fast as possible.
- Diversion should be auto-controlled, meaning that once opened, the diversion conveying capacity should increase without taking any further measures (e.g., such as raising the gate, excavating the cross-sectional area, etc.). This should be accomplished via the hydraulic parameters and properties of the diversion structure. During the falling limb of the flood wave, flow diversion into the reservoir should stop automatically. If for any reason the full intended opening cannot be materialised, the drawdown effect should still be achieved by the increased effect of auto-control. In cases where no opening is possible, overflow beyond a predetermined flood level should also produce the desired result.

- Diversion in the case of ice-jam floods should also be possible, if the backwater caused by the jam extends to the river section along which the emergency reservoir is situated.

For reducing the consequences of an embankment failure in the riverine network, or for averting an imminent embankment failure in a river section, whose flow regime can be influenced decisively by a particular emergency reservoir, it should be possible to flood the reservoir without delay and the method of filling, with respect to both location and the engineering measure, should agree as much as possible with normal operation.

The requirements and considerations related to the flooding of an emergency reservoir are complex and often conflicting. Several potential engineering options are also available, such as a gated structure (flood gate), overflow chute, controlled chute, cutting the embankment by excavation, blasting, etc. A critical review of the potential options has led to the conclusion that the controlling requirements in flooding the emergency reservoir are: (i) no diversion must occur below a pre-determined water level, (ii) flooding and the extent of the opening should depend on the decision of the competent authority, and (iii) there should be no need of temporary closure.

In the case of emptying:

- The flooded reservoir area should be drained to the river within a pre-set period of time.
- Drainage should be operated in such way that it does not interfere with embankment restoration in anticipation of another flood wave (and conversely, restoration work must not interfere with drainage operations).
- If the need arises, it should be possible to close the drainage outlet readily, using simple technical means.
- The design and dimensions of the drainage outlet should be adjusted to the successive reduction of water volume to be returned to the river (a solution involving several stages should be adopted).

A critical review of the potential options has led to the conclusion that for draining the emergency reservoir, a multi-stage solution should be adopted, in which the embankment is cut to return the bulk of the water volume, while the return structure (if one was built) is put into operation, while the remaining water, which cannot be drained by gravity, is removed by mobile pumps.

3.7. OPERATION CONTROL AND FLOODING OPERATION SCHEDULE

The general operation control model of the emergency flood reservoirs within the stream network has been developed. It consists of the following three blocks:

(A) forecasting the natural hydrological situation on the rivers; (B) computation of the emergency storage options and the assessment of flooding effects; and, (C) action plan for implementing emergency storage.

It is essential to maintain continuous interaction between the hydrological forecasts and the analysis of emergency storage options.

Flooding an emergency reservoir must be preceded by a set of special technical preparations and careful, thorough organisation. The time required for this sequence of operations may consume a considerable part of the lead time of flood forecasting, especially on rivers having a flashy regime. The preparatory operations must therefore be started at a time, when some of the information needed for decision making on reservoir flooding is still missing. From the analysis of past emergency storage events it has been concluded that the preparatory operations must be planned, ordered and executed in such a manner as to perform the minimum essential activities up to specific times. These activities must be performed reliably, so that the emergency reservoir is prepared for flooding with the reliability required, but without unnecessary expenditures.

A review of the relevant experiences gained during past emergency storage events has indicated the need for setting up special hydrographic observations, measurements and data communications. These should be laid down in guidelines which enter into force, whenever the parameters of the particular flood wave imply the potential need of emergency storage. The aim is to preserve the data and experiences of any abnormal, extreme event of flood hydrology, and emergency storage (or flood confinement) for future use. For this purpose it is essential to perfect the observations and forecasts with respect to both the equipment and methods.

Concerted efforts of the local and national agencies controlling and implementing flood fighting operations are essential. This is hardly conceivable without an effective information system, like the Water Emergency Control Information System [5] established under the author's guidance. The system makes effective communication possible between the various levels of emergency control, ensuring on-line exchange of information and data bases.

4. Practical Application of Results

The results achieved in connection with emergency flood storage have been corroborated in practice and used successfully by others. The results of the study from 1975 have already been used for the application of emergency flood storage. Results of general validity have been achieved and found repeated use in the Hungarian flood fighting practice during the past 15 years (Table 1).

5. References

1. Szlávik, L., Kiss, A. and Galbáts Z. (1996) Hydrological study and assessment of the December 1995, flood in the Körös Valley and emergency storage (in Hungarian). Vízügyi Közlemények **1**, 69-104.
2. Szlávik, L. and Varga, M. (1996) The use of Emergency Flood Retention Reservoirs. The network newsletter, International Network of Basin Organisations, **4**, 4-5.
3. Szlávik, L. (1983) Design and operation of emergency reservoirs (in Hungarian). *Vízügyi Közlemények* **2**, 188-219.
4. Szlávik, L. (1998) Emergency flood storage (in Hungarian). Vízügyi Közlemények **1**, 21-66.
5. Szlávik, L. (1999) Task of preventing and fighting water hazards, concept of developing the Water Emergency Control Information System, implementation and operation (in Hungarian). Vízügyi Közlemények **1**, 5-48.

ACTION PLAN ON FLOOD DEFENCE AND CONTROL STRATEGIES FOR FLOOD RETENTION ON THE UPPER RHINE

H. ENGEL

German Federal Institute of Hydrology - BfG
Koblenz, Germany

1. General Background

The 1980/90s experienced a series of major floods both on a global scale and in Central Europe. Following a 12-year period nearly without floods, a dual event occurred in 1983, with the two flood peaks about seven weeks apart. In March 1988, the riparian dwellers on the Middle Rhine had to cope with the highest discharge measured so far. This gave the first impetus for reviewing man's attitude towards the elementary event of flooding. Finally, in response to very extreme events in the northern Rhine basin (at Christmas 1993 and in January 1995) concrete steps were initiated both at national and international levels.

2. The Action Plan on Flood Defence of the International Commission for the Protection of the Rhine (*IKSR*)

Following an initiative of the Environmental Ministers of the German Federal States, the "Guidelines for Forward-Looking Flood Protection" [1] were drafted and adopted in May, 1995.

In parallel, the Ministers of the Environment of France, Germany, Belgium, Luxemburg and the Netherlands declared on 4 February 1995 in Arles that they deemed necessary the reduction of flood-related risks as rapidly as possible. It was not acceptable to them that situations as came up at that time put people's lives and property and the environment at such great risk. Prior to its adoption, this declaration had been agreed upon with Switzerland.

In the same month, the existing River Commissions on Rhine, Saar/Moselle and Meuse were ordered to establish action plans on flood defence. In response to this order, the International Commission for the Protection of the Rhine (*IKSR*) commissioned the Project Group "Action Plan on Flood Defence" with this work for the River Rhine under consideration of its basin. Ecological improvements of the Rhine and its floodplains should be integrated and promoted by this effort too.

On the one hand, the Project Group harnessed the findings of other commissions and study groups and, on the other hand, made or commissioned its own surveys and investigations.

J. Marsalek et al. (eds.), Flood Issues in Contemporary Water Management, 385–393.

By March 1997, the following had been completed:

- an inventory on flood defences on the River Rhine and
- a survey on warning systems and proposals for improving flood forecasting in the Rhine basin.

By January 1998, the following were available:

- the "Rhine Atlas - Ecology and Flood Protection" and
- the compilation, "Ecologically valuable areas and first steps toward interconnected biotopes on the River Rhine".

2.1. OBJECTIVES OF THE ACTION PLAN ON FLOOD DEFENCE

The *IKSR* Action Plan on Flood Defence [2] was published in March 1998. The intended improvement of the protection of population and property against flooding and the integration of the aim of ecological improvement of the River Rhine and its floodplains presuppose integrated thinking and action at local, regional, national, and transnational levels. Here, contributions from such policy fields like water-resources management, regional planning, nature conservation, agriculture and forestry are indispensable.

The Working Group "Action Plan on Flood Defence" formulated five guiding principles in matters of preventive flood protection.

1. **Water is part of the whole** - everywhere water is an element of the balance of nature and of land uses and must be given due consideration in all policy fields.
2. **Store water** - water must be retained as long as possible in the whole catchment and along the River Rhine.
3. **Let the river expand** - we must return to the river enough room to expand, so that runoff is delayed without posing dangers.
4. **Be aware of the danger** - despite all efforts taken a certain risk will remain. We must learn again to live with this risk.
5. **Integrated and concerted action** - integrated action in the spirit of solidarity throughout the catchment is a prerequisite for the success of the action plan.

On the basis of these conditions, the action plan postulates four essential action targets, which should be understood as political objectives.

1. **Reduce damage risks** - no increase of damage risks until the year 2000, their reduction by 10% by 2005, and by 25% by 2020.
2. **Reduce flood stages** - reduce extreme flood stages downstream of the impounded river reach by 30 cm until 2005 and by 70 cm until 2020.
3. **Increase awareness of flood risks** - increase the awareness of flood risks by drafting risk maps for 50% of the floodplains and flood-prone areas by the year 2000 and for 100% of these areas by 2005.
4. **Improve the system of flood forecasting** - short-term improvement of flood warning systems by international cooperation. Prolong the forecasting period by 50% by the year 2000 and by 100% by 2005.

The percentages given here refer to the base year 1995. The figures of flood-stage reductions are the result of an effectivity assessment of water retention in the Rhine basin [3].

2.2. IMPLEMENTATION OF THE ACTION PLAN ON FLOOD DEFENCE AND ITS COSTS

Cost estimates for the necessary actions for the scheduled project life (until 2020) amount to ECU 12,300 million. Of these costs, nearly ECU 2,000 million must be raised by the year 2000 and another ECU 2,500 million by 2005.

Table 1 shows a breakdown of the costs in the various categories including their achievable flood-protection effects. It is obvious that water retention in the catchment area and along the River Rhine requires by far the highest financial inputs. If one compares the protection effects to be expected with the necessary costs in detail, one finds that technical flood retention produces two thirds of the effects at less than 10 % of the total costs. Some of these technical retention facilities are already operational, while others are in the contract phase. These two parts together provide a retention volume of about 270 million m^3. Another 95 million m^3 are conceivable, reasonable, and desirable. High flood stages can be reduced through water retention by 60 to 70 cm on the Lower Rhine.

The implementation of the measures listed in the Action Plan calls for an international and interdisciplinary policy understanding which is not oriented at the local success of single activities, but at the overall objective to be achieved for the benefit of the whole Rhine basin. The appeal to implement consistently the activities of the Action Plan in their sphere of responsibility addresses all riparian states.

The Action Plan should not be understood as a tied-up bundle of activities, but as a target framework for which the contents are regularly newly defined by new experiences. A first balance on achievements will be drawn by the Governments in the year 2001. The yardstick will be the flood- and damage-reducing effects for a variety of floods from frequent to rare events. The effectiveness of prepared and implemented measures of flood prevention and defence is verified by a collective of model floods which simulate the flood-flow behaviour in the catchment. The tools for this exercise are currently being developed in the context of the EU projects INTERREG IIc with active participation of the *BfG*. Experiences of more than three decades of mathematical runoff modelling and from the cooperation in the *IKSR* effectivity assessment [3] are thus contributed.

TABLE 1. Action plan on flood defence on the River Rhine - Overview on activities 1998–2020, - effects and costs

Activity categories	Flood protection effects	Other effects	Cost estimate [million Ecu]
Water retention in the catchment	**About 10 cm reduction of flood levels**		8400
- Renaturation of rivers (11,000 km)	- Little close-range effects	- Restoration of aquatic and terrestrial habitats	1,160
- Reactivation of inundation areas (1,000 km²)	- Local effects, little effects in the Rhine	- Groundwater recharge, restoration of aquat. & terr. habitats	2,030
- Extensification of agriculture (3,900 km²)	- Little close-range effects	- Groundwater recharge New habitats	1,705
- Promotion of nature, afforestation (3,500 km²)	- Little close-range effects	- Groundwater recharge New habitats	680
- Desealing (2,500 km²)	- Little close-range effects	- Relieving sewers and treatment plants	1,890
- Techn. flood retention (73 million m³)	- Local effects, little effects in the Rhine	- New habitats	935
Water retention on River Rhine			**2,410**
- Reactivation of inundation areas (160 km²)	- Flood-level reduction 15-25 cm	- Groundwater recharge, restoration of aquat. & terr. habitats	1,450
- Techn. flood retention (364 million m³)	- Flood-level reduction 45–60 cm	- New habitats	960
Techn. flood defence			**1,418**
- Maintenance and strengthening of dykes, adaptation to the required degree of protection (1,115 km)	- Reduction of damage risks	- Increased safety for downstream dwellers	
Precautionary activities in the planning area			**60**
- Flood-adapted land uses - Risk mapping	- No increase of damage risks - For 100 % of inundation areas and flood-prone sites	- Prevention of soil erosion - Increasing flood awareness	
Flood forecasting - Improved forecasting - Improved cooperation	- Prolonged forecasting time: 100 % - Improved warning systems	- Increased safety for riparian dwellers	**12**

Certain types of activities are justified not alone by their flood protecting effects but do also meet important targets in other policy fields.

3. Water Retention on the Upper Rhine

The possibilities for improving the protection of riparian dwellers against floods listed by the *IKSR* include also the technical flood retention facilities on the Upper Rhine. Their construction must be seen as a direct consequence of the impoundment of the Upper Rhine by a series of weirs, which had reduced the safety against floods for the downstream reaches. The International Study Commission of Floods on the River Rhine (*HSK*), that was founded in 1968, was able to derive from detailed studies pertinent conclusions and to quantify the effects. In line with its task, the commission also formulated recommendations for "measures against the increased flood risk due to the training of the river". This study took into consideration only such floods which remain below or are equal to the 200-year discharges, but would today overflow the protective dykes (that could safely hold 200-year discharges until 1955). Flow-accelerating effects were left out of account. Figure 1 gives a schematic presentation of changed flow patterns due to the impoundment of the Upper Rhine and the planned effect of the compensation measures.

For attenuation of flood peaks, the *HSK* recommended three different approaches to water retention [4].

- In the special operation mode of the Rhine hydropower stations from Kembs downstream to Straßburg, the water volumes that flow usually through the lateral canal or through the following canal loops are diverted into the natural bed of the river, what causes a time delay and - through retention in the riverbed - a reduction of discharge.
- Impoundment behind retention weirs in the Rhine riverbed allow nearly any desired reduction of discharge per time unit within the limits of the available discharge-dependent storage volume.
- Inundation of presently non-flooded polders reactivates old floodplain areas of the river. This inundation should be regulated by controllable inlet structures.

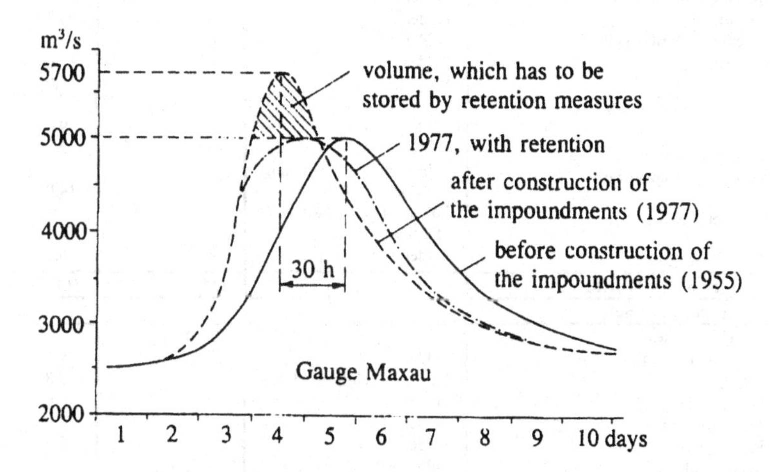

Figure 1. Changing flood hydrographs in the Rhine with focus on flood retention (schematic).

All three forms of retention are only practicable in the impounded reach. On the free-flowing reach of the Rhine this can be achieved only by lateral retention basins along the river.

In 1978, activities to yield a maximum retention volume of about 220 million m^3 were proposed. In 1982, a concept providing for retention of 226 million m^3 was adopted on contract basis (Table 2). Delays in its implementation and new ecological aspects in its assessment have led to changes in some activities, while others had to be abandoned altogether.

Flood protection activities, which have an ecological orientation, were reflected in new proposals. Ecological operation modes of retention capacities, however, require in comparison with the previous concept, wider inundation areas or more retention volumes (Table 2) alone to restore the flood safety of the year 1955 (protection against the 200-year peaks at that time). Additional improvements can be achieved by technical possibilities downstream of the inflow of the River Neckar, namely by better utilisation of existing inundation areas for flood retention or by reactivation of former floodplain areas. Here,

TABLE 2. Retention facilities on the Upper Rhine

Nr.	Retention measure	Type of retention [2]	Concept 1982 [million m³]	Concept May 1996 [million m³]
	France			
1	• *Special operation modus of Rhine hydropower stations* [1]	Operational regulations	45	45.0
2	Erstein	Polder	6	7.8
3	• *Moder*	Polder	5	5.6
	Sum for France		**56**	**58.4**
	Baden-Württemberg:			
4	South of Breisach GW-recharge weir	Weir or lowering of forelands	53	25.0
5	GW-recharge weir Breisach [1]	Weir	10	9.3
6	Breisach/Burkheim	Polder	-	6.5
7	Wyhl/Weisweil	Polder	-	7.7
8	Elz inflow	Polder	-	5.3
9	Ichenheim/Meißenheim	Polder	-	5.8
10	• *Altenheim*	Polder	18	17.6
11	• *GW-recharge weir Kehl/Straßburg* [1]	Weir	37	37.0
12	Freistett	Polder	-	9.0
13	Söllingen/Greffern	Polder	8	12.0
14	Bellenkopf/Rappenwört	Polder or landward shift of dykes	-	14.0
15	Elisabethenwört	Polder or landward shift of dykes	-	11.9
16	Rheinschanzinsel	Polder	-	6.2
	Sum for Baden-Württemberg		**126**	**167.3**
	Rhineland-Palatinate:			
	No defined sites		30 + 14	
17	• *Daxlander Au*	Polder		5.1
18	Wörth/Jockgrim	Landward shift of dykes		8.8
19	Neupotz	Polder		8.6
20	Mechtersheim	Polder		7.4
21	Flotzgrün	Polder		5.0
22	Kollerinsel	Polder		6.1
23	Waldsee/Altrip/Neuhofen	Polder		8.1
24	Petersau/Bannen	Landward shift of dykes		1.4
25	Mittelbusch (Worms II)	Landward shift of dykes		2.3
26	Bodenheim/Laubenheim	Polder		6.0
27	Ingelheim	Polder		3.8
	Sum for Rhineland-Palatinate		**44**	**62.6**
	Sum total		**226**	**about 288**

• *Completed or operational (at the Groundwater-recharge weir Kehl/Straßburg at present only 12 million m³ useable)*

1) Maximum values; the real volumes depend on streamflow in the Rhine

2) The effectivity of all measures in the free-flowing reach depends on streamflow.

controllable polders are practicable as well as landward shifting of dykes and flooding of expanded alluvial areas in the former floodways on the back side of the dykes (Auenzüge).

3.1. CONTROL STRATEGIES

The control strategies for flood retention projects [5] must generally meet the following requirements:

- reduce effectively damage-inflicting floods;
- prevent ecological damage or improve the overall ecological situation on the River Rhine and in the polders;
- regain the full retaining capacity possibly immediately after use (to accomodate possible following floods).

On the impounded river reach, additional demands are as follows:

- minimise the losses in energy generation of hydropower stations during flood retention;
- prevent hindrances to navigation already in the period before the flood.

In order to make best use of the local conditions, for each implementation phase control options have to be defined which promise the optimum effectiveness of the existing retention potential.

At present about 110 million m^3 of technical retention capacity are provided for, of which around 75 million m^3 are operational in the following locations:

- special operation modus of hydropower station in the lateral canal and in the four canal loops downstream to Straßburg;
- the polder of Moder built on French territory;
- the polders of Altenheim and Daxlander Au on the German side;
- the groundwater recharge weir Kehl-Straßburg.

For the settlement of all problems that may occur on the Upper Rhine, Germany and France established a "Standing Commission" which has a "Technical Committee" to deal with all technical issues and integrates, if needed, also pertinent working groups. The rules for the operation of the retention facilities are established by the joint French-German "Working Group Manöver". Moreover, it has the task to update the rules in line with the progress of project implementation.

Control strategies were developed by means of suitable model floods. The operating instruction, that was introduced in 1998 by the "Standing Commission" and remains valid until further notice, comprises also the above-mentioned bundle of activities. The criteria for the various modes of operation are all related to certain thresholds of discharge and their exceedance at gauging stations on the High Rhine and the Upper Rhine.

When a discharge 2,800 m^3/s is observed in the Rhine at Straßburg, the water volume usually stored behind the groundwater-recharge weir Kehl is "(pre-)emptied". In this initial phase at maximum 200 m^3/s are additionally released downstream. As soon as the discharge at Maxau reaches 3,800 m^3/s and shows a rising tendency, the retention of water behind the GW-recharge weir begins with 400 m^3/s at maximum. The retention volume is flow-dependent; the retention gradient is controlled by movable openings in the weir. However, for constructive reasons full-capacity impoundment is not permitted yet (except in explicit emergency cases).

The retention is stopped when the discharge at Maxau drops again below 4,200 m^3/s. If this threshold was not reached, the stoppage becomes effective below Q_{Maxau} = 3,600 m^3/s, provided no new rise is observed at Basel. For a quick recovery of retention volume, the impoundment is emptied immediately after use. However, with a maximum rate of 400 m^3/s this is done in such a manner that the peak discharge prevailing at the beginning of the emptying operation is not exceeded.

The two polders at Altenheim are situated beside the storage of the GW-recharge weir. Simultaneously with the retention behind this weir they are filled at a rate of 150 m^3/s. The stoppage of filling, and their emptying, are controlled by the same rules as those for the operation of the weir.

The special operation mode of the hydropower stations on the Rhine begins when the thresholds of 3,300 m^3/s at the Basel gauge and 4,200 m^3/s at the Maxau gauge are simultaneously reached or exceeded. This action means losses in energy generation, so that it is initiated rather late. It is stopped simultaneously with a possible stoppage of the actions at the GW-recharge weir and in the polders at Altenheim. The re-diversion of flow to the turbines of the lateral Rhine canal is initiated when discharge at Basel is again at least 1000 m^3/s below the previous peak, and when simultaneously discharge at Maxau has dropped below a value of 4,500 m^3/s, but at the earliest 20 hours after the passage of the peak at Basel. In the canal loops, the increase of flow is initiated 20 hours after the beginning of this manipulation in the lateral Rhine canal.

Finally, the polder of Moder is filled when discharge at Maxau exceeds 4,400 m^3/s. Its maximum filling rate is 160 m^3/s; emptying begins when discharge at Maxau drops below 4,000 m^3/s.

This instruction is the result of a great number of simulation computations with a mathematical flow-routing model at the State Agency for the Environment of Baden-Wurttemberg (*LfU*) in Karlsruhe. With the implementation of each additional retention project (the polder of Erstein in France will be completed next) a new, fine-tuned operating instruction will be developed. The aim of the regulation is to prevent overflow of dams as long as possible with the means available (at present a reduction from 5,400 m^3/s to 5,000 m^3/s at Maxau or from 6,500 m^3/s to 6,000 m^3/s at Worms).

Since the conclusion of the contracts on the implementation of the retention capacities on the Upper Rhine, modifications of single measures took place in terms of location, dimension, and planned or permitted operational characteristics, including the new option of ecological flooding. It was necessary to furnish the proof that these modifications will just as well achieve the defined objectives. After completion of all projects, the groundwater-recharge weir and the inlet structures of the polders at Altenheim, as well as the special operation modus of the hydropower stations will be controlled on the basis of a discharge threshold at Maxau. All other facilities become activated according to the discharges measured at the sites or the cumulative discharges in the rivers Neckar and Rhine, unless they are "automatically" activated as in the cases of overflow of dams or dams relocated landwards.

The currently scheduled bundle of measures will make it possible to reduce the 200-year peaks of different flood patterns at Maxau and Worms to the holding capacity of the dams. It ensures moreover that ecological flooding is not interrupted unnecessarily and that water retention is not continued when the flood situation does not urgently require it.

Ecological flooding begins when discharge reaches about 1,500 m^3/s at the site. This means that at discharges in the Upper Rhine of less than 2 years recurrence interval, water retention becomes activated and potentially developing floods of small or medium dimensions further downstream are mitigated.

3.2. ASSESSMENT OF THE RETENTION OPTIONS AND THEIR CONTROL

The measures described here have the exclusive purpose of improving the flood situation of the dam-protected reaches downstream of the last impoundment weir in the Rhine. Regulations for the targeted protection of single objects on the Middle Rhine and Lower Rhine are not part of the contracts. Moreover, this would require forecasting lead times which cannot be achieved at present. Nevertheless, the measures and activities, which are being implemented for the protection of the population on the Upper Rhine, should lead to beneficial effects for the downstream dwellers as well.

The effectivity assessment of flood precautions performed in the context of *IKSR* activities showed that the activation of all retention facilities following extreme situations at site (HQ_{200}) will yield reductions of flood peaks on the southern Upper Rhine in the statistical mean.

The following examples, three flood events during which these measures were activated, demonstrate these effects.

- In 1988, the retention of about 25 million m^3 of water reduced the peak of a flood downstream of the impounded reach by 265 m^3/s (around 23 cm). On the Lower Rhine this effect still amounted to nearly 100 m^3/s.

- In 1990, additional auxiliary measures were activated during a flood event that was relevant only for the Upper Rhine. The effect achieved downstream of the impounded reach was similar to that in 1988.

- In February 1999, 60 - 70 million m^3 of water were retained; this effected a reduction of the peak downstream of the last impoundment by some 50 cm. This reduction shortened the recurrence interval of the peak at Maxau from around 100 to 20 years.

4. References

1. Länderarbeitsgemeinschaft Wasser (LAWA) (1995) *Leitlinien für einen zukunftsweisenden Hochwasserschutz*, Stuttgart.
2. International Commission for the Protection of the Rhine (IKSR) (1995) *Action Plan on Flood Defence*, Koblenz.
3. IKSR (1998) *Wirkungsabschätzung von Wasserrückhalt im Einzugsgebiet des Rheins*, Koblenz.
4. Hochwasserstudienkommission für den Rhein (1978) *Schlußbericht*, Bonn.
5. Engel, H. (1998) Die Rückhaltemaßnahmen am Oberrhein – Steuerstrategien zu ihrem Einsatz, *Wasserwirtschaft 88. Jhg. Nr. 5*, 220 - 224.

DAMAGE CAUSED BY THE 1997 FLOODS, ITS REPAIR AND PROPOSED ALTERNATIVES OF FLOOD CONTROL MEASURES

PAVEL KUTÁLEK[1] and VÁCLAV KOŠACKÝ[2]

[1]Aquatis, JSC
[2]Morava River Board Corporation, JSC
Brno, Czech Republic

1. Introduction

In early July 1997, the weather was characterized by the shift of a large depression from Britain, Scandinavia and southern France to the Czech Republic, and movements of the surrounding cold and the occluded fronts. Much of the subsequent damage and losses were caused by abundant rain in the Morava River spring area on July 4. In the four subsequent days, storm rainfalls occurred not only in the above area, but also in the eastern, central and north-westerly areas of the Morava River drainage area. The most widespread flooding began on 5 July 1997, and the first substantial flush of floodwater came on 7 and 8 July. The situation resulted in the worst flooding in Moravia in all its history. Due to the situation existing in July of that year, a substantial part of the rainwater left that area immediately and the runoff coefficient was very high (0.9 or even higher) during the critical days.

Of the 3,957 km managed by the Morava River Board Corporation(MRBC), almost two thousand km (1,954) were affected by the flooding. Long stretches of water courses were destroyed, with parts of banks being swept away and the growths along the banks and water course branching damaged. Several parallel river beds were formed and protection dikes were damaged. In places, high flow rates caused extensive damage to the valley profiles. The above damage was caused along the rivers Branná, Krupá, Desná, parts of Metra, Morava (from Hanušovice down to Dolní Morava), Roznovská Becva and Bystricka. In addition to river bed damage and destruction of buildings, the floods destroyed parts of roads, bridges, railroad tracts and pipeline and sewerage networks.

In the central reaches of the rivers, and on the Becva and the Morava in particular, dikes were damaged, parts of river banks swept away, and new deposits were made. The biggest changes to the course and shape of the river bed were found in the case of the Becva. Overflowing water devastated towns and villages and road network system.

In the lower sections of the drainage area, the dikes were eroded and damaged. The overflowing water penetrated to areas of uncontrolled inundation where it behaved in an unpredictable manner and stayed there for different periods of time. Its spread was further influenced by local railroads and road beds and dikes. Inundation areas were 1 to 5 km wide (13 km in extreme cases) and up to 2.5 m deep.

J. Marsalek et al. (eds.), Flood Issues in Contemporary Water Management, 395–400.

The damage to the rivers managed by the MRBC and to water management structures along those rivers totalled 1.7 billion Kc, which includes mainly:

- 226 km of scoured banks,
- 136 km of damaged protective dikes,
- 91 km of damaged river bed lining,
- about 280,000 m^3 of deposits in river beds, and
- damage to 132 river stabilization structures.

2. Repair Work

Work to repair the damage and reconstruction work on flood protection systems began as soon as local situation permitted. In the first stage, the focus was on the reconstruction of existing, and on building new, access roads, the removal of barriers (accumulated trees, rubble from destroyed bridges and houses, washings from fields, etc.) and deposits from river beds to allow free passage of water there, and on the repair of the damaged dike systems.

Of the total 251 million Kc invested in repair work carried out in the Morava drainage area in 1997, 130 million Kc came from a government subsidy, 121 million from internal resources of MRBC, of which 105 million Kc was a loan. The total number of reconstruction projects was 105, with the most extensive and technologically demanding ones being the repair of the Morava embankment at Tlumaèov, the Bobrovec embankment at Ostrozká Nová Ves, the Kyjovka's relief channel embankment and the Morava's Nesytské embankment near Hodonín. The embankment and the river bed were repaired upstream of the Hranice weir, the river beds of the Roznovská Becva at Roznov and of the Morava in Hanušovice was repaired (53 million Kc).

The 1998 amount of flood repair work totalled 445 million Kc, and it is assumed that in 1999, projects worth 1,080 million Kc will be carried out. The objective is to complete all the repairs by the end of the year 2000.

In some places, river beds were changed considerably, and environmentalists believe that those changes should be used to benefit the environment. The changes in river beds and floodplains are considered as the basis for the river's natural revitalization. After several studies have been made, a number of reaches of the Becva totalling about 1 to 2.5 km were selected and in these the river will retain its natural course. A more detailed study for the reaches in question is being prepared.

3. Proposed Protection and Prevention Measures

The enormous flooding in July 1997 also triggered discussion on the need for a flood protection system. In co-operation with Aquatis, the MRBC company presented its "Master plan of flood prevention measures in the River Morava drainage area" to the Ministry of Agriculture of the Czech Republic in May 1998. The material draws on experience from the recent flooding and concludes that a significant part of damage was concentrated in zones and parts of the drainage area characterized by inadequate potential capacity of their water works, namely storage reservoirs and other elements of water management systems capable of storing, diverting or carrying necessary amounts of water. The fact that the protective storage capacity of such reservoirs may play an important regulatory role, and that such reservoirs are able to protect downstream areas against floods became particularly obvious during the floods.

The fact was clearly proven during the catastrophic flooding in July 1997. A preliminary comparison of the situation made at the time of the flooding in different parts of the drainage area showed the positive role of existing storage reservoirs and stormwater retention tanks and emphasized the positive role of keeping a part of the discharge in inundation areas. It was also clearly demonstrated how important it is to control the filling and drainage of such inundation areas. On the basis of this experience, the master plan proposed flood control measures for the affected parts of the Morava drainage area. In the preparation of the master plan, the following principles were applied.

- Flood control measures will be designed to protect only settled areas or sites of special public interest; agricultural areas will not be protected.
- Measures will be designed for the biggest flood observed, or for Q_{100}.
- Retention abilities of the countryside (natural inundation areas, man-made stormwater retention facilities) will be made the maximum use of ;
- The new measures will be linked to the flood control systems already in place to complement and extend them.

Measures are being proposed in several groups of equal importance: new storm water retention tanks, flood release basins, river bed modifications to increase their discharge capacity, building new dikes along rivers and dikes along residential areas. The measures proposed thus represent a synthesis of a number of previously prepared local-scale documents and plans adjusted on the basis of experience gained in the 1997 floods.

The upper reach of the Morava and most of its tributaries are characterized by narrow valleys, where better flood control can be achieved by using those natural features to retain a part of stormwater in retention reservoirs. A water reservoir could be built at a site upstream of Hanušovice (a part of the Water Management Plan). To achieve a maximum flood control effect, almost the entire capacity of the reservoir would have to be reserved for retention purposes. With a maximum water depth of 60 or 70 m, its volume would be about 30 or 40 million m^3, respectively. Preliminary calculations have shown that a 40 million cubic metre reservoir in that location could reduce a flood similar to that in July 1997 that culminated at about 200 m^3/s to a harmless discharge of about 30 m^3/s, a reduction of 170 m^3/s. A retention facility of that size would be guaranteed to protect the entire Morava valley at least down to its confluence with the Desná, and it would also influence discharge values further downstream.

Local dikes downstream of Hanušovice could be used to protect Bohdíkov and Raškov. At Ruda and Moravou, a broad inundation area opens up. A dike on the right bank along the residential part of Ruda would protect settled areas but would cause only a minimum reduction of the inundation area. In the same way, Olšany, Bohutín and Chromce could be protected.

Protection for Postrelmov would be provided by a dike around its residential area on the right bank. Leština, lying on the left bank of the Morava, would be locally protected by building new local dikes and increasing the height of the existing ones. An area upstream of the confluence with Moravská, Sázava, is a large natural inundation area. This area might be used as a flood release basin. The levee would be situated upstream of the towns of Dubicko, Bohuslavice and Lukavice, and on the right bank, along the existing railroad track. If the water rose to the elevation of 265 m above sea level (6 m in depth), the minimum volume of the reservoir would be 11 million m^3. The total length of levees would be about 3 km, which would include levees protecting Bohuslavice that lie below the levee.

The effect to the discharge modifications would be combined with the Mohelnice flood release basin. That is one of sites protected by the Water Management Plan and it has been included in a number of plans in the past. The extent corresponds to the VÚV TGM proposal from a period immediately following the 1997 flooding, with a 10 m high levee and a volume of about 30 million m^3. In addition to building dry retention reservoir levees, levees around the town of Treštín and around a railroad tract on the right bank would have to be built. According to preliminary calculations, the Mohelnice dry retention reservoir would reduce the flood discharge by 50 to 100 m^3/s, and the dry retention reservoir upstream of Bohuslavice would help to increase the effect. Although the Mohelnice dry retention reservoir has about the same volume as the Hanušovice reservoir, it is given a higher priority because it may protect a much larger area.

Other possible sites for large volume retention reservoirs in our drainage area are at Hrbety on the Desná, on the Merta upstream of Sobotín and on the Moravská Sázava upstream of Krasíkov.

The flood of the combined water of the Morava, Desná, Moravská Sázava and other smaller tributaries travelling very fast in precipitous reaches spreads over a wide floodplain at Zábreh, and the current gets significantly slower and shallower. In that region, large inundation areas are formed that have an indisputable influence on the transformation of the flood wave. The inundation area upstream of the Kromerí weir has a large retention effect. The gradual reduction in the maximum discharge of the Morava is also due to the existence of inundation areas further downstream. The discharge analysis shows the importance of inundation areas and of the staggering the flood waves from the upper Morava and the Becva. For that reason, any solution that would significantly reduce inundation areas or negatively influence the time sequence of flood waves would be absolutely unacceptable.

It is assumed that levees will be built to protect towns and villages, and that the discharge capacity of the Morava and the Malá voda river beds at Litovel and of the Morava at Olomouc will be increased. An alternative option would be to divert the discharge exceeding the Morava maximum capacity in Litovel to the northern by-pass. Discharge that exceeds the peak capacity of the Morava's existing bed in Olomouc might also be diverted to a similar by-pass.

In all the above alternatives, the inundation levees and river bed reconstruction are calculated for the peak discharge of the July 1997 flooding plus a safety margin of 50 to 100 cm.

The Kromer protection is provided for by increasing the discharge capacity of the river bed to $Q = 700\ m^3/s$ and building levees around the city's perimeter. Excessive discharge is to be diverted to the left bank inundation area. As an alternative, it would also be possible to increase the retention power of the inundation area upstream of Kromer. For that purpose, it would be possible to use the existing gated weir on the Morava and the new gated weir on the Moštenka, which is to be a part of the proposed Kromer system of levees. The two weirs should prevent discharge exceeding the bankful discharge from reaching the city.

Along the Morava between Mohelnice and Olomouc, a number of areas of controlled inundation can be used. A rough calculation estimates their volume at 95 million m^3 compared to existing volume of inundation areas of 68 million m^3.

In the same region, possible flood release basins exist at the Trebuvka upstream of Moravská Trebová, Vranová Lhota and Loštice. Flood release basins on the Bystrice and the Dulní potok were proposed in the 1984 Hydroprojekt Brno projects. Two flood release basins were also proposed upstream of Olomouc at the village of Skrben.

The most expensive and requiring the largest areas would be flood control in the case of Olomouc. Its protection might be provided for either by increasing the discharge capacity of the Morava channel or diverting discharge exceeding the Morava capacity ($Q = 420\ m^3/s$) to a by-pass.

All towns in this section of the Morava are to be protected by local levees.

In the lower reaches of the Morava, flood control measures concentrate on the use of existing inundation areas and proposed levees protecting residential areas from overflowing water. In some areas, the proposal assumes an increase in the channel's discharge capacity.

The situation along the Becva is similar to that in the upper reaches of the Morava. The river passing through relatively narrow valleys with many communication networks, factories and residential areas allows significant changes in flood discharge levels in a limited number of places only.

On the Vsetínská Becva, controlled inundation are possible only at Nový Hrozenkov, at Huslenky, Hovezí and Vsetín. Some towns, for example, Przno or Jarcová, may be protected by levees.

On Ronovská Becva, controlled inundation is possible on the Prostrední Becva, Stríte and Zašová, or at Veselé and Hrachovec. The river also offers some possibilities for rather small flood release basins on the Dolní Becva.

Flood protection of towns along the lower reaches of the Becva is partially provided for by increasing the capacity of river beds in town, and the construction of three new storage reservoirs for the flood wave. They are the storage reservoirs at Teplice nad Becvou, Hranice and Osek. Possible alternatives for Teplice are a reservoir with a 24 m high levee and a maximum capacity of 169 million m^3, or a flood release basin with an 11 m high levee and a maximum capacity of 38 million m^3. After some small modifications, the river at Teplice nad Becvou would be able to carry almost 750 m^3/s. This alternative also assumes that the right bank road will be protected with an embankment wall. Other possible sites for dry retention reservoirs are near Hranice nad Becvou and near Osek nad Becvou, with a total capacity of about 20 to 30 million m^3.

The overall capacity of the entire Prerov section of the Becva river is at present determined by two bottlenecks in its lower part. They are the railroad bridge near the main station and the flood plain, whose size has been reduced by Precheza sludge lagoons to about 450 m^3/s. It is recommended that the capacity of the municipal channel be increased to 600 m^3/s.

In April 1998, the concept of the master plan was discussed with representatives of district authorities, several cities and towns and government authorities. The participants were given general information about possible flood control measures on the Morava and Becva rivers.

The list of measures presented is a survey of options for the protection of the region affected by floods in 1997. The selection of the optimum option must be based on a broadly conceived discussion among both the specialists and the general public. Optimizing the effect of all the projects will be an important part of work on the draft of future protection measures. For that reason, a project to create a model of areas flooded by the Morava and Beèva is under way that will make it possible to simulate the progress of flooding and test the flood control measures proposed. For initial simulations of the flood wave progress and comparisons with the 1997 situation, four basic options of flood control measures have been proposed.

1. "Changes in Land Management and Water Retention in the Basin".
The objective of this scenario is to analyze the effect of changes in the land management in the basin, e.g. changing some arable land to forests and meadows, on discharge parameters. In the model, the changes may be simulated by changing the parameters of precipitations and discharge levels (discharge coefficients, water content levels in the surface layer, time constants for surface discharge, etc.).
2. "Protection of Settled Areas and Places of Special Public Interest by Dikes Only".
The objective is to assess the influence of protecting parts of inundation areas (cities, towns, etc.) against flooding on the peak discharges in rivers.
3. "Protection of Settled Areas and Places of Special Public Interest by Dikes and Retention of Peak Discharges in Teplice, Mohelnice and Hanušovice"
This scenario will analyze the influence of the main retention areas on the transformation of flood discharges. It is also expected that it will show how the management of those man-made retention areas could influence the situation at the confluence of the Morava and the Becva, and how far downstream their effect goes.
4. "The Use of a Part of the Danube - Oder - Elbe Canal in Flood Control"
The last of the scenarios assesses possible effects of a part of the Danube-Oder-Elbe Canal on the progress of flooding.

Preliminary results of simulations will be available in the summer of 1999. They will make it possible to discuss which combination of measures from different scenarios might be the optimum one, and to test the resulting options on the model. The model will make it possible to test the technical part of the solution proposed, i.e. the water levels, discharges, etc. Before the final decision is taken, it will be necessary to take into account a large number of other issues, e.g. a comparison of the extent of damage, impact on the environment, etc.

FLOOD DAMAGES AND INUNDATED LAND RECLAMATION

ANDRZEJ KULIG

Institute of Environmental Engineering Systems
Warsaw University of Technology
Nowowiejska 20, 00-653 Warsaw, Poland

1. Introduction

Floods can be considered on a multitude of dimensions such as social, economic or environmental. The impact on the land surface is extremely diverse and it can involve erosion, changed soil characteristics caused by water stagnating in the flooded area and soil contamination caused by substances found in water or silt. Any reduction of the flood impact on the land surface can be viewed in terms of measures necessary before floods occur (i.e. flood prevention and soil erosion prevention), or measures to contain impacts during flood and restoration measures (i.e. after the flood).

The causes and patterns of the recent floods in Poland (July 1997, April 1998 and July 1998) have been extensively covered in the literature [1] and their impact is well documented. In The Institute of Soil Science and Plant Cultivation and The Institute of Land Melioration and Pastures issued guidelines in 1997 for inundated land reclamation [2,3]. Nevertheless, a number of practical measures both to prevent and reduce the impact of floods (even they have a high priority) still remain to be implemented in Poland. The following sections give an overview of major impacts and reclamation methodologies.

2. Soil Structure Damage Caused by Flood-originated Erosion and Related Reclamation Techniques

The soil profile changes and soil structure degradation consist in washing and denudation of the surface layers, redistribution of fine fractions, rilling, gullying and tunnelling.

The containing of the erosive impact of floods can be viewed from prevention and response perspectives. The erosion is caused by heavy rainfall and it is amplified by the alignment of plots, fields and roads parallel to the slope gradient. The compromised retention capacity of the soil gives rise to increased surface runoff. Consequently, the river water table and flows rise rapidly causing even more destructive soil erosion, relief degradation and agricultural land devastation.

J. Marsalek et al. (eds.), Flood Issues in Contemporary Water Management, 401–404.

Any soil acts as a large water retention reservoir if its profile is well developed and land-use patterns are appropriate. In order to take advantage of this soil capacity, adequate crop cultivation techniques should be taken on diverse relief catchment areas along with a rational distribution of forests, pastures and arable land and fitomelioration and anti-erosion measures.

Mechanical and biological texture formation measures are normally applied on soils eroded by flooding. The choice of specific mitigation techniques will depend on the physiographic conditions, soil types (including the grain-size composition) and the extent of soil degradation and land devastation. Although they are quite sensitive to water erosion (by surface washing and wash-out), loess and loess-like soils are also easy to restore due to the rapid soil formation from soil-less loess deposits. Clayey soils are less sensitive to water erosion but soil formation from soil-less clay deposits poses a much greater challenge. Calcareous soil on slopes and thin mountain soils (thin-soiled land, part of the Carpathian Foothills) are relatively resistant to water erosion but, if it occurs, the damage is usually serious and reclamation very difficult [3].

If the entire soil cover is damaged by flooding, it is critical that the land be covered with deposits of the highest possible quality (top-class soil) and biological land reclamation techniques be applied, i.e. introduction of humus-generating vegetation. It is common practice to cover the devastated land with a layer of humus removed from non-farmed land or with other deposits such as loess, silt, loams, alluvial or brown soil, boulder clay, black-earth etc. Grass and papilionaceous plants should be sown on fields and heavy nitrate, phosphate and potassium fertilisers be applied.

While reclaiming light soils, it is practical to use manure, green manure, ripe straw, turf etc. Light acidic soils will benefit from a restored pH balance as the high soil acidity is likely to mobilise the toxic aluminium, surplus manganese, restrain phosphorous and reduce the bioactivity of micro-organisms.

The restoration of flooded areas also involves adequate management of eroded material washed from slopes and accumulated in gullies, at the bottom of slopes and on valley beds. Adequate restoration measures should be taken to rehabilitate farm roads, which are usually unpaved and often poorly sited. As a result of erosion, these roads are often turned into road gullies up to several metres deep.

3. Waterlogging of Farm and Forest Land and Related Restoration Measures

Stagnant waters alter the physical and chemical properties as well as the bioactivity of farm and forest land. Water flooding and inundation result in oxygen deficiency and the dying of numerous plant species. It has been found that prolonged stagnation of waters in flooded areas causes nutrients to be washed away into the flooding waters.

Land improvement and recultivation techniques are used in order to restore the carrying capacity of waterlogged land. The reclamation of flooded farmland is mainly achieved through agro-melioration measures such as furrowing and deep drainage ploughing [2]. Silt-covered ground has been improved dependent on silt thickness and chemical composition as well as land use. For many cases, as a primary and early restoration measure to the original state, the hydrophilous plants should be sown [4].

4. Chemical Contamination of Soil as a Result of Flooding

The major source of potential soil contamination with chemicals or micro-organisms during floods are flooded industries, especially those which manufacture, handle or store hazardous material (including toxic chemicals), sewage treatment plants and municipal and industrial landfills. Flooded storage tanks containing liquid fuels, pesticides and herbicides, fertilisers and chemicals (solvents, acids etc.) can be a serious concern on a local level. According to the statistics collected by the State Inspectorate for Environmental Protection (PIOŚ), 67 sewage treatment facilities were fully flooded and 101 facilities were partially flooded during the heavy floods in Poland in 1997. Temporary downtime or operating problems occurred in 32 facilities. PIOŚ found that 7 landfills were completely flooded and 36 facilities were flooded partially [5].

The presence of heavy metals (e.g. As, Cd, Cr, Cu, Hg, Pb and Zn), oil derivatives, toxic substances, salinity and biological contamination are the key indicators of the extent of soil contamination.

In case soil contamination is clearly indicated by preliminary studies further research efforts are normally taken to develop a land reclamation or remediation program. Such extended studies aim to identify the following:

- source of contamination;
- type of chemicals and microbiological pollutants involved;
- the surface area and depth of contaminated soil.

The following soil characteristics are critical for developing a remediation (e.g. pollutant removal) programme:

- grain-size distribution;
- thickness of the humus deposit;
- ground water table;
- organic (humus) content;
- nutrient content (N, P, K);
- reaction (pH);
- carbonate content;
- gleying, if observed.

On highly contaminated farmland, the toxic layer is often capped with a cultivable soil layer of 0.4 to 0.6 m or an agro-melioration ploughing is performed down to 0.6 to 0.8 m with parallel deep organic or mineral fertilisation.

Soils and silts have been studied among others under the National Environmental Monitoring programme on farm and forest land and in highly sensitive areas exposed to potentially hazardous sites (sewage treatment facilities and landfills, filling stations, local oil-fired boilers and communities of more than 5,000 inhabitants) [5].

Heavy metal and oil derivative content was selected as the key indicator of long-term soil contamination and microbiological contamination has served as a short-term contamination indicator. Following the July 1997 flood, the National Environmental Monitoring programme sampled soils at 302 sites and silts at 173 sites and a total of 6,400 laboratory analyses have been carried out.

5. Concluding Remarks

The rapid flow of the flood wave in the mountainous and high-lying regions of the former Bielsko-Biala, Jelenia Góra, Katowice, Kraków, Nowy Slcz and Walbrzych Provinces mainly resulted in erosion damage. The low-lying part of the country (former Kielce, Legnica, Leszno Opole, Rzeszów, Tarnobrzeg, Tarnów, Wroclaw and Zielona Góra Provinces within catchment of Oder, Warta and Vistula River and its tributaries) suffered mainly from prolonged stagnation of water and some silt deposition.

In general, PIOŚ has not observed any significant increased presence of heavy metals in the soils and silts. The concentration of lead, chromium, arsenic and mercury was normal. Locally, heavy metal accumulation was observed in the flooding area of the river Ner. Increased concentrations of oil derivatives were observed in the Katowice region.

The studies of the flood damages which occurred in 1997 and 1998 point to the need to include the flood risk in the risk assessment part of the environmental impact assessment performed for newly sited facilities, including sewage treatment plants and landfills. Furthermore, a more accurate assessment of the environmental impact of farming and forest management, in context of flood risk, is critical.

6. References

1. Dziaduszko, Z. and Krzymiński, W. (1998) Meteorological origin of the flood and its hydrological effects in the Odra and Vistula rivers. *HELCOM Sc. Workshop on the Effects of the 1997 Flood of the Odra and Vistula rivers.* Hamburg.
2. *Agrotechnical Guidelines for Flooded Areas* (Polish). Special edition. The Institute of Soil Science and Plant Cultivation, The Institute of Land Melioration and Pastures. Pulawy, Falenty 1997.
3. Józefaciuk, Cz. and Józefaciuk, A. (1997) *Special Guidelines. The 1997 Flood. Mitigating Flood Erosion of Arable Land* (Polish). The Institute of Soil Science and Plant Cultivation, Pulawy, Poland.
4. Siuta, J. (1998) *Land Reclamation* (Polish). The Institute of Environmental Protection, Warsaw.
5. *Assessment of Environmental Impact of Floods* (Polish). Report of the Chief Inspector of Environmental Protection (based on data up to August 25, 1997.). Warsaw 1997.

FORECASTING OF RIVER DISCHARGES IN THE PRESENCE OF CHAOS AND NOISE

V. BABOVIC and M. KEIJZER

Danish Hydraulic Institute
Agern Alle 5, DK-2970 Hørsholm, Denmark

1. Introduction

Constructing models from time series with non-trivial dynamics is a difficult problem. The classical approach is to build a model from first principles and use it to forecast on the basis of the initial conditions. Unfortunately, this is not always possible. For example, in fluid dynamics we have a prefect model in the form of the Navier-Stokes equations, but initial conditions are difficult to obtain. In other cases, we do not have a good model to use. In either case, alternative approaches should be examined. In this contribution, a method inspired by chaos theory for building non-linear models from data — Local Linear Models (LLMs) — is discussed and described. In the second part of the paper, an application to a real-world rainfall-runoff data set is demonstrated.

2. Forecasting

The goal of time series prediction or forecasting can be formulated as follows:

Given:	$y(1), y(2), y(3), \ldots, y(N)$
Find:	$y(N+1), y(N+2), \ldots$

It is supposed that the series is some sampling of a continuous system which may be either stochastic, chaotic or deterministic.

2.1. LINEAR MODELS

One, standard way of generating models capable of forecasting is to use *global* linear models, such as moving average, autoregressive or autoregressive moving average models.

2.2. MOVING AVERAGE (MA) MODELS

Let us assume that we are given an external input series $\{e_t\}$ and want to modify it to produce another (the observed) series $\{y_t\}$. *Assuming linearity of the system* and that the present value of y is influenced by the present and N past values of input series e, the relationship between the input and the output is

J. Marsalek et al. (eds.), Flood Issues in Contemporary Water Management, 405–419.

$$y(k) = \sum_{n=1}^{N} b(n)\, e(k-n) \tag{1}$$

This equation describes a classical convolution filter: the new series y is generated by a linear filter with coefficients $b_0, b_1, \ldots, b_N$. In statistics these models are referred to as N-th order *moving average models,* or MA(N) whereas in the engineering community these models are referred to as *finite impulse response* (FIR) filters because their outputs are guaranteed to go to zero at N time steps after the input becomes zero.

2.3. AUTO REGRESSIVE (AR) MODELS

Yule [1] proposed an autoregressive (AR) class of models defined as follows:

$$y(k) = \sum_{n=1}^{M} a(n)\, y(k-n) + e(k) = \hat{y}(k) + e(k) \tag{2}$$

This is an Mth order autoregressive model [AM(M)] (An M-th order *autoregressive model* AR(M) is referred to as an *infinite impulse response* (IIR) filter, because the output can continue after the input ceases). Depending on the application, $e(k)$ can represent either a controlled input to the system or noise. If e is white noise, the autocorrelation coefficients of the output series can be expressed in terms of model coefficients. The autocorrelation coefficients of the AR model are found by solving a set of *Yule-Walker* equations:

$$\rho_\tau = \sum_{m=1}^{M} a_m \rho_{\tau-m}; \quad \tau > 0 \tag{3}$$

Unlike the situation obtained in the MA case, the autocorrelation coefficients need not become zero after M time steps.

2.4. AUTOREGRESSIVE MOVING-AVERAGE (ARMA) MODELS

The next step in complexity is to introduce both AR and the MA parts into the model; this provides what is commonly called an ARMA(M, N) model:

$$y_t = \sum_{m=1}^{M} a_m y_{t-m} + \sum_{n=1}^{N} b_n e_{t-n} \tag{4}$$

ARMA models have now dominated the field of time series analysis and discrete-time signal processing for more than half a century. If the model is good, it transforms the signal into a small number of coefficients plus residual white noise (of one kind or another).

2.5. THE BREAKDOWN OF THE LINEAR PARADIGM

It has been shown (see [2] for example) that power spectra of globally linear models such as discussed in § 2.1, 2.2, 2.3 and 2.4 and related autocorrelation coefficients contain the same information about a system driven by un-correlated white noise. Thus, *if and only if* the power spectrum fully characterises the relevant features of a time series, will a linear model (AR, MA or ARMA) be the appropriate description. This appealing simplicity of linear models can be entirely misleading even when simple nonlinearities occur. For example, two time series can have very similar broadband spectra but can be generated from systems with very different properties, such as a linear system that is driven stochastically by external noise and a deterministic (noise-free) non-linear system with a small number of degrees of freedom.

3. A Need for Nonlinear Models

Yule's original idea for forecasting was that future predictions can be generated by using the immediately preceding values [1]. An ARMA model — Equation (4) — can then be rewritten as follows:

$$x_t = \boldsymbol{a}\, x_{t-1} + \boldsymbol{b} e_t \tag{5}$$

where $\boldsymbol{x}_t=(x_t, x_{t-1}, x_{t-2}, ..., x_{t-(d-1)})$, $\boldsymbol{e}_t=(e_t, e_{t-1}, e_{t-2}, ..., e_{t-(d-1)})$, $\boldsymbol{b}=(b_1, b_2, b_3, ..., b_d)$ and $\boldsymbol{a}=(a_1, a_2, a_3, ..., a_d)$. Such lag vectors are also called tapped delay lines. There is a deep connection between time-lagged vectors and the underlying dynamics. This connection was first proposed by Packard *et al.* [3] and Takens [4] who published the first formal demonstration of such a connection, and later this was strengthened by Sauer *et al.* [5].

Takens' Time-delay Embedding Theorem [4] can be written as follows: *Given a dynamical system with a d-dimensional solution space and an evolving solution* $\boldsymbol{x}(t)$, *let y be some observation* $y(\boldsymbol{x}(t))$. *Let us also define the lag vector (with dimension d and time delay* τ*)* $\boldsymbol{y}_t=(y_t, y_{t-\tau}, y_{t-2\tau}, y_{t-3\tau}, ..., y_{t-(d-1)\tau})$. *Then, under very general conditions, the space of vectors* $\boldsymbol{y}_t$ *generated by the dynamics contains all of the information of the space of solution vectors* $\boldsymbol{x}_t$. *The mapping between them is smooth and invertible. This property is referred to as diffeomorphism and this kind of mapping is referred to as an embedding. Thus, the study of the time series* y_t *is also the study of the solutions of the underlying dynamical system* $\boldsymbol{x}(t)$ *via a particular coordinate system given by the observable y.*

This can be interpreted less formally as follows. Science is based on the principle of repeatability: each time a system experiences similar conditions — both internal to the system and as exerted externally on the system — we expect the system to exhibit a similar response. Forecasting exploits this principle by using the observed behaviour of a system to predict behaviour when similar conditions recur — and, clearly, a tapped delay line represents a history (evolution) of a system. Even if the equations describing a system are unknown, we can nevertheless use forecasting to learn about the system. Delay vectors of sufficient length are not just representation of the states of a sequences of superimposable linear systems — it turns out that delay vectors can recover the full geometrical structure of a non-linear system. These results address the general problem of inferring the behaviour of the intrinsic degrees of freedom when a function of the state of the system is measured.

Thus, Takens' theorem implies that there exists, in principle, a non-linear autoregression of the form

$$y(t) = g[y(t-1), y(t-2), y(t-3), ..., y(t-T)] \tag{6}$$

which models the series exactly (assuming that no noise is present).

The problem we are addressing in this article is the one of finding this non-linear autoregression. There are at least three possibilities: Genetic Programming — GP — [6,7], Artificial Neural Networks — ANN — [8] and local linear models [9,10]. Here, some results obtained using Local Linear Models are presented.

3.1. LOCAL LINEAR MODELS

Following the approach established by Packard *et al.* [3] firstly a state vector is established $x_t \equiv (x_t, x_{t-\tau}, x_{t-2\tau}, x_{t-3\tau}, ..., x_{t-(d-1)\tau})$. The next step is to create a functional relationship f_T between $x(t)$ and some future state $x(t+T)$.

A very effective way of approximating f_T is by means of local approximation, using only nearby states to make predictions. To predict $x(t+T)$, we firstly find k most similar occurrences of $x(t)$. These are effectively k nearest neighbours of a point $x(t)$ in a d-dimensional embedding space. Then, a local linear interpolation of zeroth or first order is performed, regarding all k neighbours of $x(t)$in the range.

If done in the most straight-forward fashion, finding the nearest neighbour in a sample of size N points in d-dimensional space, requires *O(N)* computational steps. This, however can be reduced to *O(logN)* by partitioning data using a sort of decision tree (see 3.3. The kdtree).

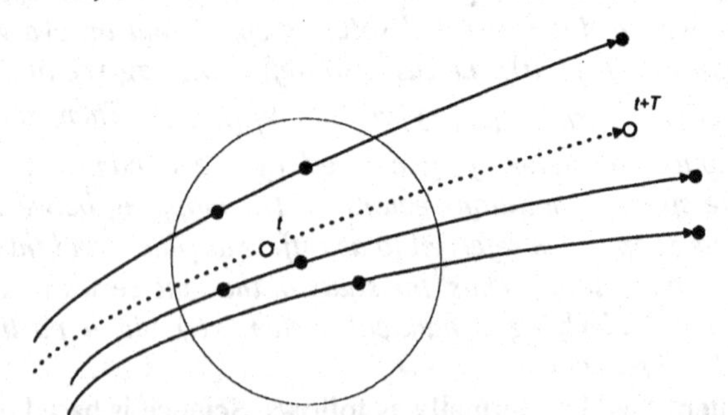

Figure 1. Prediction using nearest neighbours. The black dots are the nearest neighbours of the white dot representing the predictee.

Though linear models are used, the prediction is still nonlinear because for each neighbourhood a different approximation of f_k is made. In this study the following approach has been adopted:

1. The total time series is divided into a training set and a testing set.

2. The training set is used to build the kdtree splitting of phase space (see 3.3 for detailed description). This set functions as the database of past observations.
3. For a new point $y'(t)$, the database is searched for the nearest neighbours of this point.
4. Time delay co-ordinates are used to build local functions that takes each point in the neighbourhood at the time t to the neighbourhood at the next time step (t+T), yr(t) → y(r; t+T); r = 1,2,3,...,k, where k is the number of neighbours and T is the forecast horizon.
5. The prediction y'(t+T) at the time y'(t) is obtained either through averaging:

Zeroth-order interpolation:

$$y(t+T)=\frac{1}{k}\sum_{j=1}^{k} y_j(t+T) \tag{7}$$

or through linear interpolation:

First-order interpolation:

$$y_{t+T} \approx \alpha_0 + \sum_{j=1}^{d} \alpha_n y_{t-(j-1)\tau} \tag{8}$$

3.2. FORECAST

The forecast is performed for a specific *horizon of forecast (T)*. Once this forecast horizon increases to say *2T* there are two possible approaches. One can either iteratively (recursively) apply mapping f_T, using mapping's results for $t+T$ as input at $t+T$ in order to calculate values at $t+2T$. This is a so-called *iterative forecast*. A disadvantage of such an approach is that errors are expected to grow exponentially with each iteration. An other approach may be to perform a so-called *direct forecast* in which new forecast functions are generated for each *T*. The disadvantage of such an approach is that for global approximation techniques to work well, the mapping must be smooth. Higher iterates of chaotic function are by definition not smooth.

The quality of nonlinear forecast also provides an indication of the type of signal that is being considered. For periodic and quasi-periodic signals, the quality of forecast will remain constant or oscillate with the horizon. Forecasts of chaotic signals will show a decrease of the quality because of the divergence of the trajectories as the horizon is increased. It is also seen that for stochastic signals the quality will remain almost constant with the increase of the horizon.

Nonlinear prediction also allows for the identification of optimal neighbourhood size, dimension and time lag as that which will produce the best result. In this study a brute force search algorithm has been employed to produce predictions from a wide range of values of dimension, time lag and neighbourhood size. In the following chapters, results from these exercises will be discussed.

3.3. THE KDTREE ALGORITHM

Most time series consist of large number of data points in a multidimensional space. The traditional indexing techniques are very efficient if the queries are performed using predetermined keys. However, in chaotic systems the queries are of a different kind. The queries are based on the proximity of data points. Therefore, the spatial relations between the points in the multidimensional phase space have a very important role in all these queries.

The kdtree [11,12] is a multidimensional indexing scheme in which *d* dimensional data are indexed in *k* dimensions. According to the kdtree algorithm, the data are organised hierarchically and split into 'buckets' covering the *k*-dimensional space, so that points near to one another are stored together evenly in a physical media. Optimally the 'buckets' should be balanced, so that they contain roughly an equal number of data points. Rather than moving data during the sort process an array of indices is generated and sorted.

Every leaf of the tree consists of a 'bucket' of indices referring to the original vectors. Every node of the tree contains two parameters: one is the dimension along which the split is made, the second the cut-off value for that dimension. Every vector that contains a value lower than or equal to the cut-off value in that particular dimension will be stored in the left branch of the node. Values higher than the cut-off can be retrieved from the right branch.

In building the tree, at each node the variances of all lagged dimensions are calculated and the dimension with highest variance is chosen as the dimension where the split will occur. The dimension with the highest variance is chosen as by splitting on this dimension the variance (spatial spread) in the two branches will be minimised. The corresponding cut-off value will be set to the mean of the values in that dimension such that approximately half of the points will be contained in either branch. This is applied recursively until the number of vectors in a branch falls below the bucket size parameter (often set to 15). Indices to these vectors are then stored in the bucket.

In prediction, one is confronted with new lag-embedded vectors whose evolution needs to be predicted. Searching for *k* nearest neighbours in a kdtree takes place by first finding the bucket that contains the most nearby points. These are readily found by traversing the tree, at each node selecting the branch where the actual value of the vector in the stored dimension exceeds or falls below the stored cut-off value. When this bucket is found the nearest vectors in the buckets are selected by calculating the Euclidean distance.

As these nearest vectors are not necessarily the nearest globally, the algorithm proceeds by going up in the tree again, comparing the largest distance in the current selection with the distance of the vector to the cut-off value stored in the node. If this distance is smaller than the largest distance found, there is a possibility that the other branch of the node contains vectors that are even closer to the target vector than those found so far. In this case, the search will continue along the other branch.

When the distance of the target vector to the cut-off value is smaller than the largest distance found so far, no vector in the un-searched branch will be closer to the target vector, so the search will ignore this branch. In this way, the *k* nearest neighbouring vectors will be found while searching only a small part of the tree. When the tree is balanced and contains all the dimensions in the nodes, the search for the *k* nearest neighbours will necessitate approximately $O(k \log(N))$ comparisons, in contrast with the $O(2N)$ comparisons needed in a naïve search (that is N for calculating the distance of the new vector with all stored vectors and N for finding the *k* lowest values).

4. Data

The present study applies the concepts described above, namely local linear models and embedding of time series, to create a forecast model for a catchment of the river Luznice, in the Czech Republic. The catchment of River Luznice (4226.2 km^2) is a medium-size sub-catchment of the catchment of the river Vltava (24064.4 km^2) which covers roughly 1/3 of the territory of the Czech Republic.

4.1. RIVER LUZNICE

The data used in the process of induction of the models consisted of recorded rainfall intensities and runoffs from the catchment in the period 1991-1995. These data are presented in Figure 2. A summary of statistical information on these data is presented in Table 1.

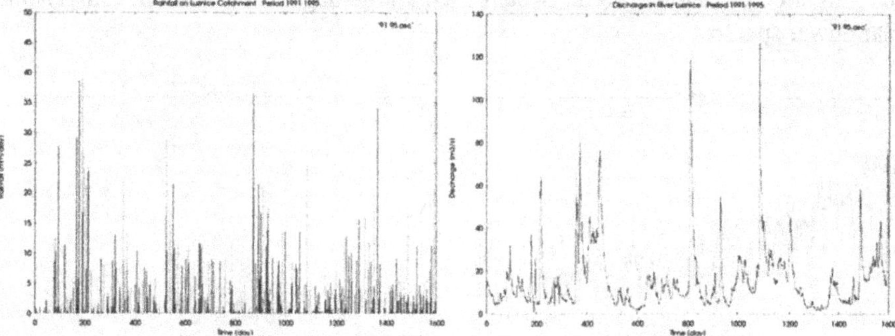

Figure 2. Rainfall and discharges in the river Luznice, 1991-1995.

For the purposes of model induction, the 1991-1995 data were split into two sub-sets. The years 1991, 1992 and 1993 were used in a *training process*, whereas the years 1994 and 1995 were kept as a *validation* set.

TABLE 1. *Statistics for 1991-95 data*

	1991-95 data	
	Rainfall [mm/day]	Q [m^3/s]
Mean	1.59	15.77
Standard Deviation	4.02	14.61
Range	45.40	125.40
Minimum	0.00	1.20
Maximum	45.40	126.60

5. Characterisation and Identification of Time Series

5.1. FOURIER POWER SPECTRUM AND AUTOCORRELATION FUNCTION

The Fourier Power Spectrum provides an easy tool for identifying irregularities in observed signals (see Figure 3). A broadband and continuous power spectrum without any dominant frequency will indicate that the signal has possibly originated from a chaotic system. Thus, any further analysis of the system needs to be carried out taking account of other invariants of motion.

However, it is important to understand that the Fourier power spectrum is not itself a true invariant because new frequencies will be introduced with nonlinear changes to the coordinate system. It is important to understand that multiharmonic outputs do not necessarily indicate that the system is chaotic. They provide only an indication that the system might be of chaotic origin. Systems with large degrees of freedom can generate similar power spectra.

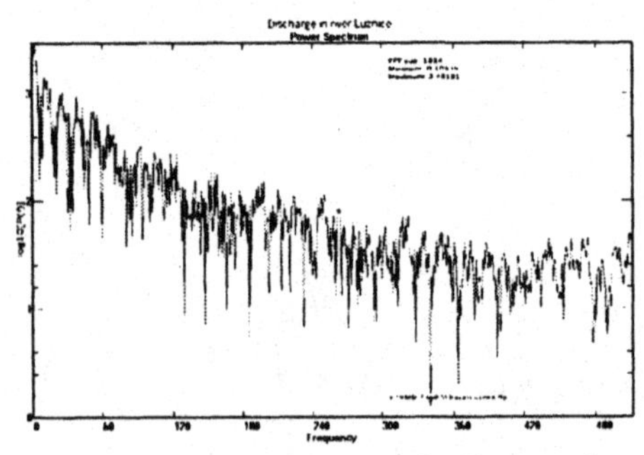

Figure 3. Power spectrum of the discharge in river Luznice. The spectrum is broadband and continuous indicating a nonperiodic signal.

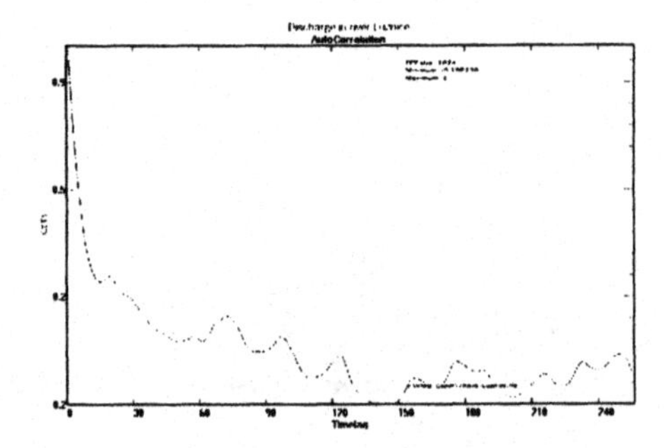

Figure 4. The autocorrelation function plotted against time lags for the discharges in river Luznice. It rapidly falls to zero.

An autocorrelation of a periodic signal produces a periodic function. For a chaotic or a random signal the autocorrelation function will approach zero rapidly (see *Figure 4*). This too is a good indication that the system is either chaotic or random in nature. Both Figure 3 and 4 indicate that discharges in river Luznice may indeed be characterised as chaotic. Therefore, invariants of motion need to be examined in order to confirm a hypothesis of chaotic nature of the flows.

5.2. INVARIANTS OF MOTION—LYAPUNOV EXPONENTS & CORRELATION DIMENSION

The Lyapunov exponents determine the stability of the system. They are indications of the growth of perturbation by the dynamics as a function of the number of steps L advanced along the attractor from the initial perturbation. Lyapunov exponents indicate, on average, how well a prediction could be made of the evolution of the system L steps ahead of the present location. Since chaotic systems are extremely sensitive to initial conditions, Lyapunov exponents provide a good test for chaos.

Global Lyapunov exponents are computed by studying the separation of two points, a_0 and b_0, on two trajectories after some number *n* iterations, that is,

$$\lambda = \lim_{n\to\infty} \frac{1}{n} \lim_{|a_0-b_0|\to 0} \log_2 \left| \frac{a_0 - b_0}{a_n - b_n} \right| \tag{9}$$

Given two initial conditions for a chaotic system, a_0 and b_0, which are close together, the average value obtained in the successive iterations for *a* and *b* will differ by an exponentially increasing amount. If two points are initially separated by a small distance of ε then on average their separation will grow as $\varepsilon e^{\lambda in}$.

Positive Lyapunov exponents indicate that neighbouring points are diverging. This implies the onset of chaos. A zero exponent indicates that the neighbouring points remain at the same distance and the system can be modelled as a set of linear differential equations. Negative Lyapunov exponents indicate that the neighbouring points are converging and the system is dissipative.

The largest Lyapunov exponents can also be used establish a window of predictability of a system in time domain.

$$\mathrm{Pr}\,edictability = \frac{\tau_s}{\lambda}$$

τ_s = time sampling interval

λ = Lyapunov exponent (largest)

Lyapunov exponents are measure of information loss per unit time, expressed in bits. They are average measures and do not address the distribution of information throughout the phase space.

The fractal dimension of a dynamic system is related to the way in which points embedded time series are distributed in *d*-dimensional space, where *d* is the global embedding dimension. The phase space of a chaotic system stretches and folds and remaps into the original space. This remapping leaves gaps in the phase space. This causes the orbits to fill a subspace in the phase space that is not of integral dimensionality. This nonintergral dimension of the subspace points at existence a strange attractor. The dimensions of simple attractors are integers, e.g. for a fixed point the dimension is zero and for limit cycles it is 1. One of the simplest ways of estimating the attractor dimension is by computing the *correlation dimension* [13]. The correlation integral *C(r)* estimates the average number of data points within a radius *r* of the data point *x(t)*. As $r \to 0$, or better, becomes small, the correlation integral is defined as $C(r) \to r^d$. The exponent *d*, defined as the correlation dimension of the attractor, can be obtained from the slope of *logC(r)* versus *log (r)* (Figure *6*).

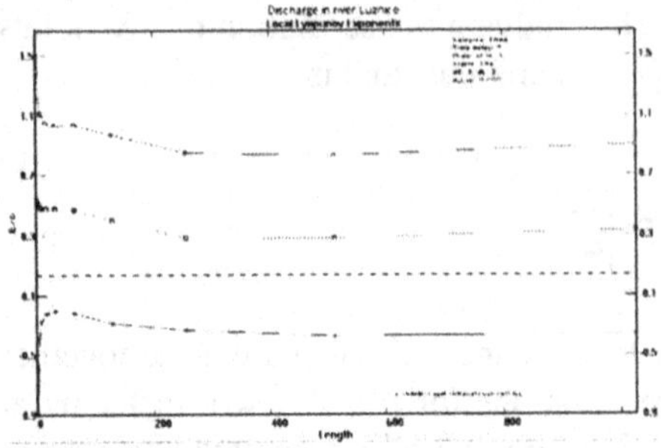

Figure 5. **Average local Lyapunov exponents for discharges in river Luznice.**

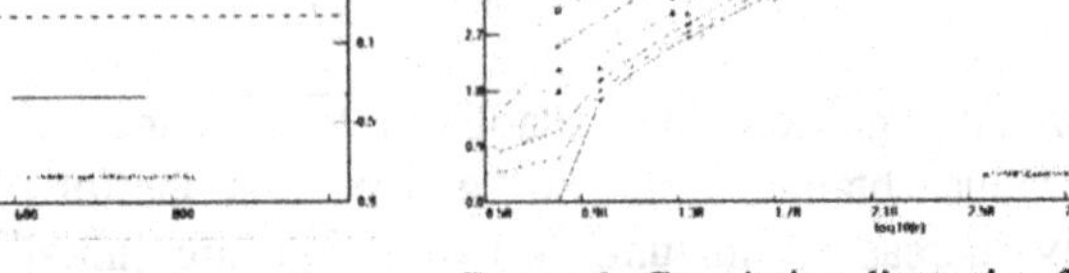

Figure 6. **Correlation dimension for discharges in river Luznice as a function of different neighbourhood sizes and *r* values.**

The stability of the motions on the attractor is governed by the Lyapunov exponents, which then tell us how small changes in the orbit will grow or shrink in time. Figure 5 shows three Lyapunov exponents for the river Luznice. Two positive exponents suggest that the trajectories are diverging exponentially, which clearly indicates that the system is chaotic. The negative exponent suggests that a dissipative mechanism exist within the system. The presence of three exponents also indicates that there are three degrees of freedom, which in turn implies that the behaviour of the system could be described by three differential equations.

The sum of all three Lyapunov exponents is in this case positive. This fact indicates that overall system is diverging and that clear and simple strange attractor can not be identified. One can therefore expect that data may be polluted by a large degree of either dynamic or white noise.

The correlation dimensions different radii r can be estimated from Figure 6. The fractal dimension estimated from the slopes of the curves is in the range 0.9 – 2.3. This indicates non-integer, fractal dimension of attractor and is rather clear indicator of presence of chaotic dynamics.

It leads from discussion above that we can not conclude with a high certainty that discharges in river Luznice exhibit "pure" chaotic behaviour. There are elements indicating chaotic behaviour, but it is not yet clear to which extent noise is present in the data. It is also not clear how this in turn affects forecast skill (this will be further discussed in section 6.2. Local Linear Models) and embedding of time series in phase space (section 6.3. Deterministic versus Stochastic modelling)

5.3. HOW PREDICTABLE IS CHAOS?

From the perspective of time series prediction and modelling, the existence of low-dimensional chaos suggests considering construction of nonlinear deterministic models for time series. If a series is indeed chaotic and low-dimensional, the good news is that it will be *possible to obtain precise short-term predictions* with such deterministic models. The bad news is that it will be *impossible to obtain good long-term predictions* with *any* model, since the uncertainty increases exponentially in time. This divergence of nearby trajectories is a hallmark of chaotic behaviour (see the discussion in section 5.2. Invariants of Motion—Lyapunov exponents & Correlation Dimension).

6. Forecast Skill

6.1. ARTIFICIAL NEURAL NETWORKS AND GENETIC PROGRAMMING

Here, we very briefly summarise results presented in Babovic [14] where both models based on artificial neural networks (ANNs) and Genetic Programming (GP) were presented and discussed.

In the case of GP, analysis demonstrated that a model below with a degree of non-linearity in the rainfall term provides the most accurate results.

$$Q(t) = 1.147 \cdot \left[Q(t-1) + r(t)^{0.20445}\right] + 1.2355 \tag{10}$$

In the case of ANNs, a time-lagged recurrent neural network generated the most accurate results. Statistical measures of accuracy (MSE, denoting mean squared error, NMSE denoting normalised mean squared error, MAE denoting mean absolute error, r_{xy}' denoting linear coefficient of correlation between two sets of data X and Y, and R^2 denoting the square of the Pearson product moment correlation coefficient through data points in X and Y) for the induced models are summarised in Table 2.

Quite obviously, models induced by the means of GP were more predictive. It has to be emphasised here that these are results of modelling rainfall-runoff process: both GP and ANN used rainfall and antecedent runoff as input variables. Thus the induction process resulted in a model of the system. Inaccuracies in the forcing term (rainfall in this case) will inevitably propagate to the resulting runoffs.

TABLE 2. Summary of statistical indicators of accuracy for both GP- and ANN-based models

	Genetic Programming		Artificial Neural Networks	
Performance	Training	Validation	Training	Validation
MAE	1.07	1.08	1.63	1.75
MSE	2.06	4.07	14.51	19.54
NMSE	0.03	0.04	0.064	0.13
MaxAbsErr	16.07	17.30	59.68	56.76
R	0.973	0.964	0.9680	0.9342
R^2	0.947	0.922	0.9371	0.8728

Local linear models, as presented in the next section are purely forecast models. They do not use rainfall as a forcing term, but generate forecasts on the basis of observed runoff only. It may be argued that in this way a valuable source of information is ignored. However, the primary purpose of the present work is to examine forecast skill based on observed data and reconstruct system behaviour from this source only. In application of rainfall-runoff (or for that matter any) models, one requires close-to-ideal forcing to be applied, which is hardly possible. In the case of forecasting, as described here, such a requirement is obsolete, as forecast is based on observed quantities only.

6.2. LOCAL LINEAR MODELS

Local linear modelling requires identification of the optimal neighbourhood size, embedding dimension and time sampling rate (τ). In this study a brute force search algorithm has been employed to produce predictions from a wide range of values of dimension, time lag and neighbourhood size. Results of this investigation are presented in Figures 7 and 8.

Then, a family of local linear models was built for the horizon of $T=1$ using zeroth and first order interpolation (as described in equation (7) and equation (8) respectively). These models were set-up using embedding dimension $d=5$, time sampling rate $\tau = 2$ and size of neighbourhood $k= 50$. Performance on the testing data is presented in Figures 9 and 10.

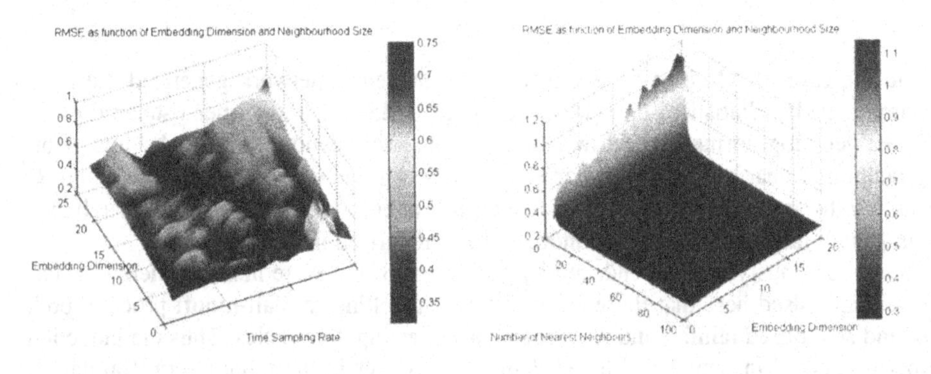

Figure 7. Normalised RMS error as a function of time sampling rate (τ) and embedding dimension *d*.

Figure 8. Normalised RMS error as a function of size of neighbourhood *k* and embedding dimension *d*.

It is clear that averaging of nearest neighbours (Figure 9) causes fairly poor performance on extreme discharges, whereas first-order, linear interpolation provides rather good and stable results.

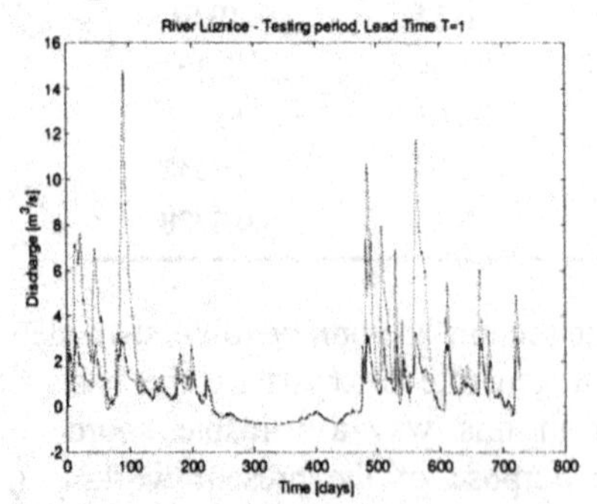

Figure 9. Results of forecast ($T=1$) using embedding dimension $d=1$, neighbourhood size $k=200$ and zeroth order interpolation (averaging).

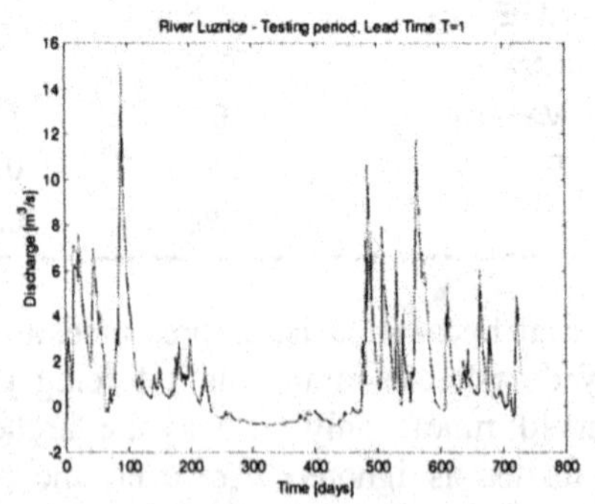

Figure 10. Results of forecast ($T=1$) using embedding dimension $d=1$, neighbourhood size $k=100$ and first order interpolation.

The results for the horizon of forecast of two days $T=2$ are presented in the remainder. Although the largest Lyapunov exponent (Figure 5) indicates that accurate forecast may be issued only for a forecast horizon of $T=1$, the figures below indicate quite reasonable accuracy. Table 3 summarises statistical indicators of accuracy.

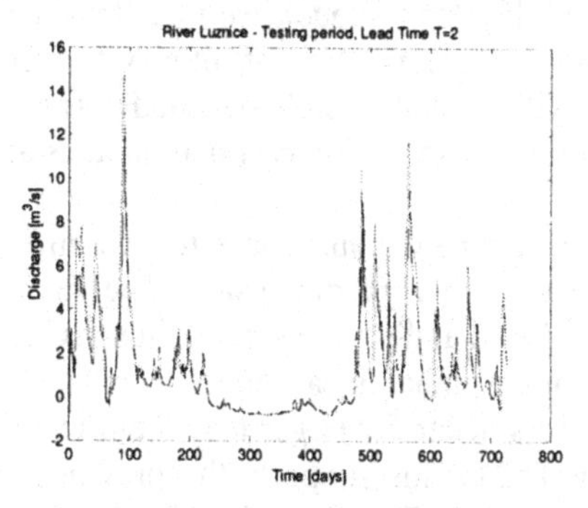

Figure 11. Results of forecast for forecast horizon $T=2$, using embedding dimension $d=1$, neighbourhood size $k=100$ and first order interpolation.

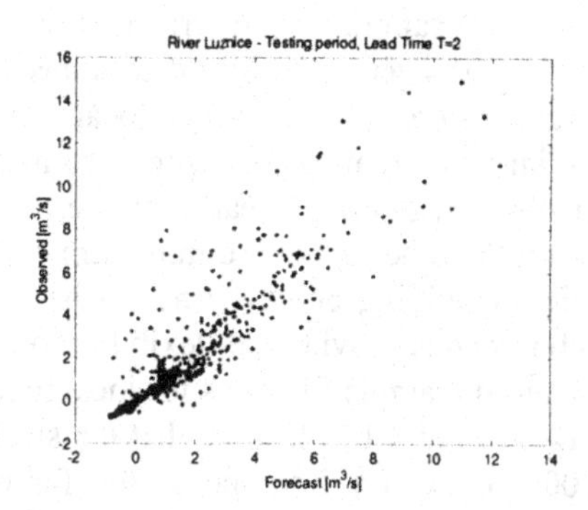

Figure 12. Scatter plot for forecast for forecast horizon $T=2$ (embedding dimension $d=1$, neighbourhood size $k=100$, first order interpolation).

The results are self-explanatory. For the horizon of T=1, this simple forecasting technique outperforms both ANNs and GP (see TABLE for comparison). Even for an horizon of T=2, results are comparable with that of neural networks for forecast horizon of T=1. And all that without any utilisation of the forcing term: rainfall!

TABLE 3. Summary of statistical indicators of accuracy for local linear models for lead times $T=1$ and $T=2$

Performance	Lead Time T=1	Lead Time T=2
MAE	0.278	0.566
MSE	0.682	1.149
NMSE	0.03	0.13
MaxAbsErr	2.24	2.24
R	0.967	0.908
R^2	0.935	0.824

6.3. DETERMINISTIC VERSUS STOCHASTIC MODELLING

In order to understand the properties of the system that generated the time series more fully, the prediction accuracy is analysed as the model class is varied. A "Deterministic versus Stochastic Algorithm" [15] can help distinguish between deterministic and stochastic processes.

The basic idea is the following: a family of local linear models is fitted to data. After each model is fitted, its short-term predictive accuracy is assessed on testing data. The systematic dependence of these out-of-sample errors on model parameters — such as the dimension of the reconstructed state space or the size of the neighbourhood for local linear models — can yield some insights into the dynamics of the underlying system. For example, if the error is lowest for a highly local model, we infer that the time series was generated by a deterministic nonlinear system or by a chaotic system of low dimension. These very local models with small neighbourhoods (near deterministic extreme) should provide more accurate short-term forecast than models at the stochastic (globally linear) extreme.

If, on the other hand, the minimum occurs at larger neighbourhood sizes, we infer that the underlying process was stochastic, or chaotic of high dimension. The small neighbourhoods provide the flexibility to fit the noise in addition to signal, thus giving worse out-of-sample forecasts for local models than for global linear models.

Figure 13 and 14 illustrate that the smallest errors occur for larger neighbourhoods (k=100 for horizon T=1 and k=200 for horizon T=2). This supports the previously postulated hypothesis on the presence of noise in the data. However, the question still remains whether the hypothesis on the presence of chaos in this data should be entirely rejected. In our opinion not. There are some fairly strong indicators of chaotic dynamic behaviour. However, these are not well pronounced due to the presence of noise.

The final explanation may be forwarded as follows. It has been indicated earlier (section 5.2. Invariants of Motion—Lyapunov exponents & Correlation Dimension) that the presence of a strange attractor can not be easily established. The analysis hinted at the presence of a degree of noise in the data. The results of the "Deterministic versus Stochastic" investigation support this conclusion. It seems as if the data does exhibit random behaviour caused either by stochastic influences or high dimensional non-linearities. At the same time the presence of one negative Lyapunov exponent indicates that some degree of convergence exists within the system. This might be the cause for the considerable forecasting accuracy exhibited by the local linear models.

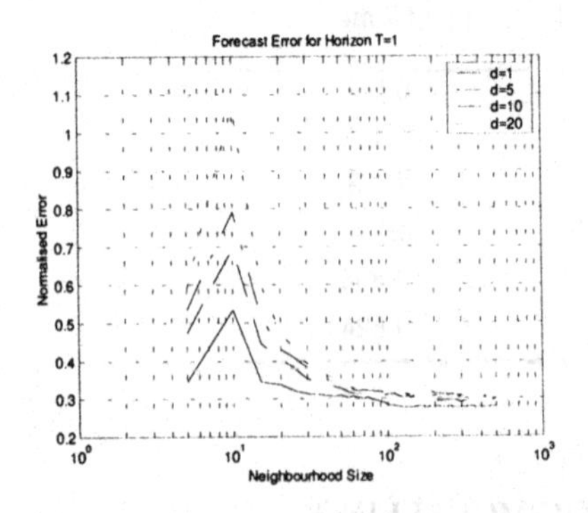

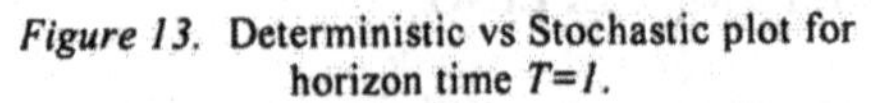
Figure 13. Deterministic vs Stochastic plot for horizon time *T=1*.

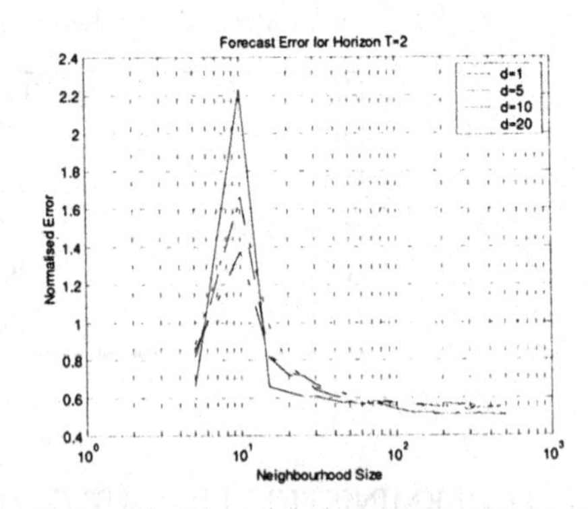

Figure 14. Deterministic vs Stochastic plot for horizon time *T=2*.

Even in the light of all these complications (presence of chaos and noise) the very simple, yet appealing technique of local linear models provides good forecasts of the runoff for the river Luznice— even when it is applied outside the strict boundaries of the theory. To the major surprise of the authors, the technique works better than expected.

7. Acknowledgments

This work was in part funded by the Danish Technical Research Council (STVF) under The Talent Project N° 9800463 "Data to Knowledge - D2K". For more information consult http://projects.dhi.dk/D2K

8. References

1. Yule, G.U. (1927) Philosophical Transactions of Royal Society London, A **227**, 267.
2. Weigend., A.S., and Gerschenfeld, N.A. (1993) The future of time series, in A.S. Weigend, A. S. and N.A. Gerschenfeld, (*ed.*), *Time Series Prediction: Forecasting the Future and Understanding the Past*, SFI Studies in the Sciences of Complexity, Proc. Vol. XV, Addison-Wesley, Reading.
3. Packard, N.H., Crutchfeld, J.P., Farmer, J.D. and Shaw, R.S. (1980) Geometry from a time series, *Phys. Rev. Lett.*, Vol. **45(9)**, pp. 712-716.
4. Takens, F. (1981) Detecting strange attractors in turbulence, in D.A. Rand and L.-S Young, (*eds*), *Dynamical Systems and Turbulence*, Lecture notes in mathematics, Vol. 898, 336-381, Berlin: Springer-Verlag.
5. Sauer, T., Yorke, J.A. and Casdagli, M. (1991) Embedology, *J. Stat. Phys.*, Vol. **65 (3/4)**, 579-616.
6. Koza, J. (1992) Genetic Programming, MIT Press, Cambridge, Mass.
7. Babovic, V. and Abbott, M.B. (1997) Evolution of equation from hydraulic data: Part I - Theory, in *Journal of Hydraulic Research*, Vol. **35**, No 3, pp.1-14.
8. Babovic, V. and Minns, A.W. (1994) Use of Computational Adaptive Methodologies in Hydroinformatics, in Verwey, A., Minns, A.W., Babovic, V., and Maksimovic, C., (eds.), Proceedings of the First International Conference on Hydroinformatics, pp. 201-210, Balkema Publishers, Rotterdam.
9. Farmer, J.D. and Sidorowich, J.J. (1987) Predicting chaotic time series, *Physical Review Letters*, Vol. **59**, No. 8, 62-65.
10. Babovic, V., Keijzer, M. and Rahman, M.S.M. (1999) *Analysis and Prediction of Chaotic Time Series*, D2K Technical Report (http://projects.dhi.dk/D2K)
11. Omohundro, S. M. (1987) *Efficient Algorithms with Neural Network Behaviour*, Complex Systems 1, pp.273-347.
12. Sameth, H. (1984) *The Quadtree and Related Heirarchical Data Structures*, ACM Computing Surveys 16:2, pp. 187-260.
13. Grassberger P. and Proccacia I. (1983) *Measuring the Strangeness of Strange Attractors*, Physica, D9, pp. 189-208.
14. Babovic, V. (1998) A data mining approach to time series modelling and forecasting, in V. Babovic and L.C. Larsen (*eds.*), *Proceedings of Hydroinformatics '98*, pp. 847-856, Balkema, Rotterdam.
15. Casdagli, M.C. and Weigend, A.S. (1994) Exploring the continuum between deterministic and stochastic modelling, in Weigend, A. S., and Gerschenfeld, N.A., (*eds.*), *Time Series Prediction: Forecasting the Future and Understanding the Past, SFI Studies in the Sciences of Complexity, Proc. Vol. XV*, Addison-Wesley, Reading.

ADVANCES IN TRANSFER FUNCTION BASED FLOOD FORECASTING

MATTHEW J. LEES

Department of Civil and Environmental Engineering, Imperial College of Science, Technology and Medicine, London, SW7 2BU, UK.

1. Introduction

Simple transfer function (TF) models were first used for operational real-time flood forecasting in the 1970s after a number of researchers had shown that they could efficiently characterise dominant rainfall-runoff dynamics. A major advantage of the approach at that time was the small requirement for computing power in comparison with more complex conceptual models. Although computing power is no longer a particularly advantageous feature, recent advances in transfer function modelling and forecasting techniques provide the ability to produce state-of-the-art flood forecasts whilst retaining the advantage of simplicity. This paper challenges some popularly held views on transfer function forecasting techniques, such as "a major disadvantage of TF modelling compared to conceptual models is that it is a black-box approach with no physical process explanation", and presents a number of recent advances in non-linear transfer function identification, parameter estimation and real-time implementation, concluding with an illustrative example.

2. Transfer Function Modelling

The general discrete-time transfer function model is a linear dynamic model that can be identified and estimated given input-output time-series data collected from the system to be modelled, which for our current purposes would be sampled rainfall and flow data. The TF takes the form,

$$y(k)=\frac{b_0+b_1z^{-1}+\cdots+b_mz^{-m}}{1+a_1z^{-1}+\cdots+a_nz^{-n}}u(k-\delta)+\xi(k) \quad (1)$$

where $b_0 \cdots b_m$, $a_1 \cdots a_n$ are the model parameters, $y(k)$ is the output at sample *k*, $u(k-\delta)$ is the input at sample $k-\delta$, $\xi(k)$ is a stochastic error term, and z^{-i} is the backward shift operator.

One of the major advantages of the TF over other conceptual models is that the structure is objectively identified as part of the modelling procedure so resulting in

J. Marsalek et al. (eds.), Flood Issues in Contemporary Water Management, 421–428.

parsimonious models. At first sight to many hydrologists, this feature places the TF in the group of 'black-box' models along with Neural Network models since the estimated model parameters do not appear to have a physical interpretation. Indeed, identified TF models can have characteristics such as oscillatory responses that are clearly unacceptable given our current knowledge of hydrological processes. However, as suggested in the data-based mechanistic modelling methodology of Young *et al.* [1], an investigation of any identified models for physical interpretation should also be performed prior to acceptance of the TF as a predictive model of the system. At this stage it is useful to examine the relationship between certain TF model structures, as defined by the triad $[n,m,\delta]$, with more traditional rainfall-runoff models. It is easy to show that discretisation of the well-known linear storage model,

$$\frac{dO}{dt} = \frac{1}{T}[I(t) - O(t)] \tag{2}$$

where $O(t)$ is the outflow at time t, $I(t)$ is the inflow at time t, and T is the residence time, results in the following first order discrete TF,

$$O(k) = \frac{b_0}{1 - a_1 z^{-1}} I(k);\ b_0 = \frac{\Delta t}{T},\ a_1 = 1 - \frac{\Delta t}{T} \tag{3}$$

where Δt is the discretisation step size. Introducing a lag element δ, and rearranging into the difference equation form,

$$O(k+\delta) = b_0 \cdot I(k) + a_1 \cdot O(k+\delta-1) \tag{4}$$

reveals that the model is identical to commonly applied flood forecasting models such as the linear input-storage-output [2].

Having ascertained that the first order TF can be considered to be a representation of common conceptual models one might question the usefulness of the TF approach. However, there are a number of advantages of posing the modelling problem in TF form including the availability of powerful system identification techniques for optimal parameter estimation; the ability to use state estimation methods to improve on-line forecasts; and increased structural flexibility. Most simple conceptual and unit hydrograph (UH) based models do not attempt to include baseflow as an integral model component although it is well known that second order flow dynamics can represent an important component of flood flows. TF modelling objectively reveals the most appropriate model structure for the study catchment as part of the modelling procedure, and the majority of cases studied so far suggest that the most appropriate structure is second order which can be decomposed into two linear stores representing quick and slow flow processes. This parallel storage structure is a common feature of a number of accepted rainfall-runoff models including PDM [3] and IHACRES [4].

The ability to utilise powerful methods of system identification is stated as an advantage of TF models. Obviously, some time-series analysis expertise is required, although much more user-friendly time-series analysis software packages are now available, and identification techniques have been largely automated through the use of model performance statistics. It is recommended that identification and estimation of the TF model is performed using a simplified refined instrumental variable (SRIV, [1]) variant of least squares that has proved to be a robust algorithm for the unbiased estimation of rainfall-runoff TF models.

3. Non-linear Transfer Function Modelling

The rainfall runoff process is inherently non-linear due to antecedent catchment wetness and the effect of dynamic contributing areas on runoff production. Some operational flood forecasting models ignore this non-linearity, instead relying upon real-time model correction, while many incorporate very simple rainfall loss methods of calculating an effective rainfall for input to the routing component of the model (e.g. RORB, [5]). The first non-linear extension of a TF based rainfall runoff model, which is presented in Whitehead *et al.* [6], consists of a rainfall pre-processor that calculates effective rainfall (*er*) as the product of rainfall (*r*) with a catchment wetness index (*CWI*) that is simply filtered rainfall,

$$er(k) = r(k) \times CWI(k); \; CWI(k) = \frac{1}{1 + a_1 z^{-1}} r(k) \quad (5)$$

Here a_1 is related to the time constant (memory) of the filter through $T_c = \Delta t / \ln(-a_1)$. This rainfall loss function is closely related to the antecedent precipitation method of effective rainfall calculation since it can be shown that the output of a first order filter is the summation of exponentially weighted past inputs. Extensions to this loss function have since been made to incorporate the effects of long-term changes in evaporation resulting in the IHACRES model [4], that is one example of a new class of hybrid conceptual-metric (HYCOM) continuous simulation models that combine various conceptual loss functions with a metric TF routing model [7].

While a conceptual loss function is required for simulation purposes, in real-time flow forecasting applications measurements of antecedent flow are likely to provide a better indicator of catchment wetness than the output from a model. Young and Beven [8] propose the following simple loss function,

$$er(k) = r(k) \times f(k - \tau)^{\alpha} \quad (6)$$

where f is flow, τ is a shift parameter that is usually zero, and α is optimised using a non-linear numerical optimisation method such as the simplex method. Excellent results have been reported for this loss function in a number of applications when it is combined with a parallel structure TF model (e.g., [8]).

Obviously the power law form of this loss function is subjective and many other forms could be appropriate according to the specific catchment characteristics. The form of this non-linearity is a current area of research and state dependent modelling [9] techniques that utilise time varying parameter (TVP) estimation and fixed interval smoothing are been used to investigate the form of the non-linearity. The SDM technique for input-output modelling is based on the time varying parameter TF,

$$y(k)=\frac{b_0(k)+b_1(k)z^{-1}+\cdots+b_m(k)z^{-m}}{1+a_1(k)z^{-1}+\cdots+a_n(k)z^{-n}}u(k-\delta)+\xi(k) \quad (7)$$

where the parameters are now estimated over time. Relationships between parameter variations and states provide a non-parametric representation of the non-linearity. For example, it has been found that for the rainfall-runoff case the numerator parameters vary with flow in a manner that can often be approximated as a power law relationship, and since the these parameters act solely on the input variable, the non-linearity can be simplified to an input form. Although SDM modelling provides an objective method of identifying the nature of the rainfall-runoff non-linearity; at present, due to the complexity of the methodology, it is more useful as a research tool to investigate the nature of other suitable non-linear functions that would then be optimised using simpler numerical optimisation techniques. Note that there are clear parallels here between the nature of this non-linearity and the dynamic contributing area concept of models such as PDM [3] and TOPMODEL [10].

4. On-line Updating

It has long been recognised by researchers that on-line forecast correction using the latest available flow data should be an important component of a real-time flood forecasting system. However, with a couple of notable exceptions, many areas of the UK currently rely on forecasting systems that do not have any mechanism for the automatic real-time correction of flood forecasts. Although experienced operators can take account of systematic errors in hindcasts, there is a concern that errors can be made under the pressurised conditions that occur during flooding events.

A number of different correction techniques exist that attempt to improve forecasts by examination of past forecast performance, and as such they can be thought of as an automation of the intuitive correction applied by a human operative. Most techniques can be classified as either state updating, parameter updating or error prediction methods. Error prediction methods can be applied to any type of forecasting model since they are based on the time-series modelling of the forecast error series [11]. State updating and parameter updating can be performed using a Kalman filter (KF) if the model is linear, or by an extended Kalman filter (EKF) if the model is non-linear (see [12] for an introduction to the Kalman filter and a number of real-time forecasting applications). Bell and Moore [11] also present an empirical state updating method for non-linear forecasting models.

The TF model has the advantage that it can be easily converted to the following Gauss-Markov (GM) state space equations,

$$X(k+1)=\boldsymbol{F}\cdot X(k)+\boldsymbol{G}\cdot U(k)+\omega(k);\; y(k)=\mathbf{H}\cdot\mathbf{X}(k)+v(k) \qquad (8)$$

where $\mathbf{X}(k)$ is the state vector, $\mathbf{U}(k)$ is the input vector, $H(k)$ is the observation vector, $\mathbf{F}$ and $\mathbf{G}$ contain the model parameters, and $\omega(k)$, $v(k)$ are independent Gaussian white noise sequences with zero mean and variance $\mathbf{Q}$ and r respectively. Therefore, the KF can be used to estimate an optimal state estimate, which in this case is flow. A major practical problem in the application of the KF is determination of the unknown metaparameters **Q** and r since the relative values of these parameters control the behaviour of the filter: if the variance of the measurement error is specified as too high a value relative to the variance of the model error, then the state estimate will be overly influenced by the noise corrupted output, resulting in poor forecasts. It should be noted here that considerable uncertainty results from the use of rating curve equations particularly at high flows that are often significantly above any rated flow values. Where previously this determination has been a heuristic exercise, recent research has shown how maximum likelihood estimates of these metaparameters can be estimated using a combination of prediction error decomposition and numerical optimisation [9]. This procedure is computationally intensive but since it is an off-line operation it has no impact on the use of the optimised KF on-line.

One major advantage of formulating the TF model as a stochastic GM model is that probabilistic forecasts are also produced by the KF, although for on-line implementation a recursive estimate of the residual variance must also be calculated. This capability could be of great potential use as an automated measure of forecast confidence since the KF generated uncertainty mimics an experienced operative's intuition as it is driven by past forecast performance. A common simplification of the KF state-updating technique is state replacement where the latest available flow measurements replace model output in forecast calculations. For example, the state updated forecasts for a first order model with a 2 sample lead time can be calculated at each step by,

$$\begin{aligned} f(k+1) &= a_1 f(k) + b_0 u(k-1) \\ f(k+2) &= a_1 f(k+1) + b_0 u(k) \end{aligned} \tag{9}$$

with $f(k)$ replaced by the measured flow. This method is of course equivalent to KF state updating with the measurement error variance set to a large value such that the KF estimate closely tracks the output. Although this technique is very simple, it has the disadvantage that probabilistic forecasts are no longer produced.

Parameter updating refers to the on-line estimation of model parameters using a recursive estimator such as the KF with the state vector now containing the parameters. Due to the limited information content contained within individual flood events, it is not possible to robustly estimate a large number of model parameters on-line. In many situations it is sufficient to estimate a scalar gain parameter that does not change the 'shape' of the hydrograph but simply shifts the output up or down. Lees *et al.* [13] report a successful application of adaptive gain updating to a TF based flood forecasting system for the River Nith in SW Scotland.

5. An Illustrative Example: Flood Forecasting for the River Bollyneendorish

To demonstrate the comparative performance of the different TF models and on-line updating techniques, a limited modelling exercise has been performed using data from the River Bollyneendorish in the Galway region of Ireland. This catchment has been selected since the rainfall-runoff response is similar to many fast responding UK catchments, and also because the data is of marginal quality with the rainfall measured using a single tipping bucket gauge, and flow estimated using a transducer situated in a natural section control with a relatively uncertain rating curve. Details of the catchment location, geology and instrumentation can be found in Lees *et al.* [7].

The models have been calibrated on a 16-day period of hourly data during November 1995 and have been validated on a 9-day period from March 1996 that contains the largest magnitude flood on record. Initial TF identification using the SRIV method indicated that a first order model with a 3 hour lag is the most appropriate TF structure. The α parameter of the power law TF is obtained using a simplex optimiser and the SDM modelling technique is used to identify an objective nonlinear relationship that in this case takes the form of a power law relationship with a cut-off at high flows. Figure 1a shows that the SDM time varying parameter is strongly correlated to flow, although in a non-linear form that is captured in finite form using weighted least squares regression that incorporates the parameter uncertainty that is also shown in Figure 1a as 95% confidence intervals. Table 1 summarises the calibration and validation (3 hour ahead forecast) results for the different models and on-line correction techniques.

TABLE 1. Comparative modelling and forecasting results

Model	Calibration		Validation	
	R^2	Peak Error	R^2	Peak Error
Linear TF	0.76	33%	0.56	65%
Power Law TF	0.80	17%	0.64	56%
SDM TF	0.80	4%	0.57	60%
SDM TF + KF	-	-	0.75	22%

It is clear that the non-linear TF models only have slightly better calibration fits in terms of the coefficient of determination (R^2), but have significantly improved performance at peak flows. Figure 1b, which shows the linear and SDM calibration fits, clearly demonstrates this superior fit at high flows and also that the shape of the SDM hydrograph is better due to the incorporation of an effective rainfall calculation. It is particularly important for forecasting applications that the shape of the hydrograph is predicted correctly since real-time correction is most successful when the error term shows persistence.

Validation results show that both the linear and non-linear TF models produce poor forecast results particularly at peak flows. Examination of the rainfall time-series suggests input data error as the main cause of this poor performance. However, the KF state updating method improves the forecasts considerably with the percentage peak error reduced from 60% to 22%. Figure 1c, which shows the results for the SDM model with and without state updating clearly reveals the large improvement in forecast performance provided by the state updating procedure. Finally, Figure 1d shows the SDM forecast with associated 95% confidence intervals which provide a quantitative assessment of forecast uncertainty at any particular point in time.

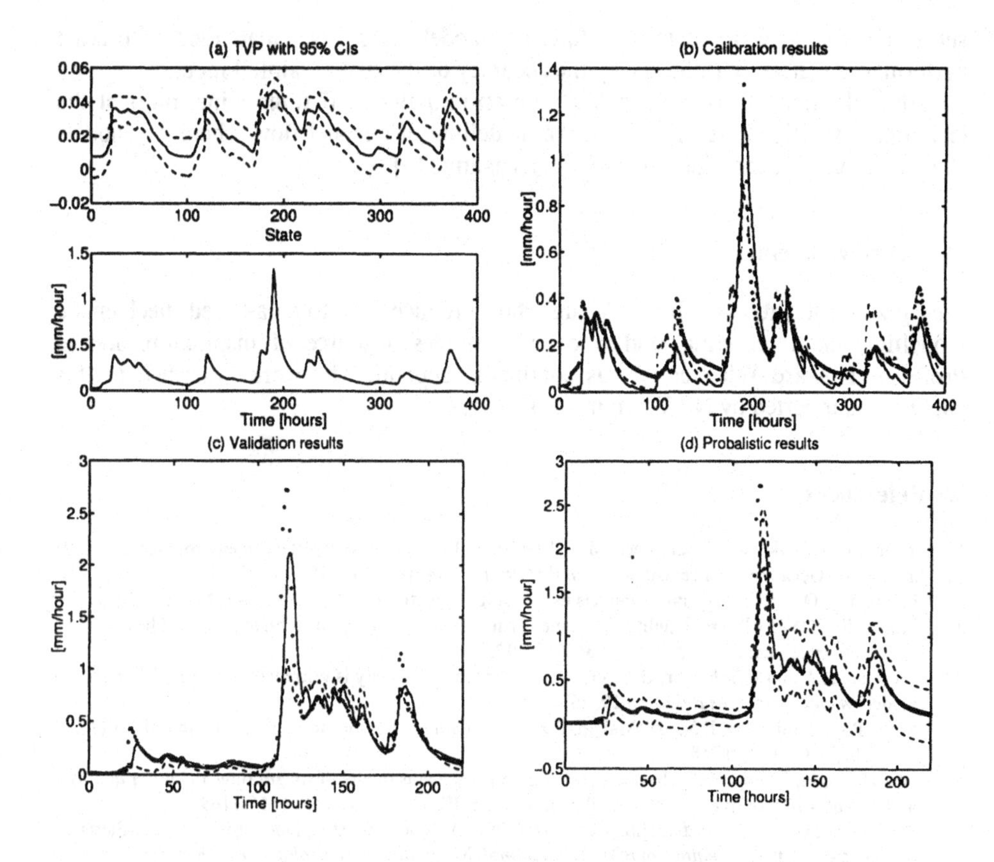

Figure 1. Comparative modelling and forecasting results. (a) SDM results. (b) Calibration results: Actual (dots), SDM (solid), Linear (dashed). (c) Validation results: Actual (dots), SDM + KF (solid), SDM (dashed). (d) Probabilistic results.

6. Conclusions

Physically justifiable structures of the TF model can be considered to be representations of hydrologically acceptable linear storage models that obey the fundamental law of continuity. Formulation of the model in TF form allows powerful methods of system identification and state estimation to be utilised. Non-linear transfer functions improve the performance in catchments where variable antecedent conditions affect flood flows, and KF state estimation provides a powerful method of real-time forecast updating with associated forecast uncertainty. Forecast uncertainty can provide very useful additional information on flooding likelihood, although there is much current debate in the UK as to what form probabilistic forecasts issued to the Public should take. It should be noted here that the use of a sophisticated corrector is no substitute for a well identified forecast model, and also that model forecast performance is heavily influence by the accuracy of the spatial rainfall input.

While the focus of this paper has been on rainfall-runoff forecasting, most of the techniques described are relevant to the modelling of channel flow where TF models can also be used successfully for flood forecasting [13].

7. Acknowledgements

I am grateful to Professor Peter Young who introduced me to data-based mechanistic modelling and forecasting, and who still provides a source of inspiration, and to Professor Howard Wheater for his continuing support. The work described in this paper was supported by NERC grant GR3/11653.

8. References

1. Young, P.C., Parkinson, S. and Lees, M.J. (1996) Simplicity out of complexity in environmental modelling: Occam's razor revisited, *Journal of Applied Statistics* **23**, 165-210.
2. Lambert, A.O. (1972) Catchment models based on ISO-functions, *J. Inst. Wat. Eng.* **26**, 413-422.
3. Moore, R.J. (1985) The probability-distributed principle and runoff production at point and basin scales, *Hydrological Sciences Journal* **30**, 273-297.
4. Jakeman, A.J. and Hornberger, G.M. (1993) How much complexity is warranted in a rainfall-runoff model? *Water Resources Research* **29**, 2637-49.
5. Mein, R.G., Laurenson, E.M., McMahon, R.A. (1974) Simple non-linear model for flood estimation, *Proc. ASCE* **11**, 1507-1518.
6. Whitehead, P., Young, P.C., Hornberger, G. (1979) A systems model of stream flow and water quality in the Bedford-Ouse river – 1. Stream flow modelling, *Water Research* **13**, 1155-1169.
7. Lees, M.J., Price, N., Wheater, H.S., Peach, D. (1998) A flood simulation model for the South Galway region of Ireland, *Proceedings of BHS International Symposium on Hydrology in a Changing Environment (Vol. III)*, John Wiley and Sons, Chichester, pp. 93-104.
8. Young, P.C., and Beven, K.J., (1994) Data-based mechanistic modelling and the rainfall-flow non-linearity, *Environmetrics* **5**, 335-363.
9. Young, P.C. (1993) Time variable and state dependent modelling of nonstationary and non-linear time series, in T. Subba Rao (ed.), *Developments in time series analysis*, Chapman and Hall, London, pp. 374-413.
10. Beven, K.J. and Kirkby, M.J. (1979) A physically based, variable contributing area model of basin hydrology, *Hydrol. Sci. Bull.* **24**, 43-69.
11. Bell, V.A. and Moore, R.J., (1998) A grid-based distributed flood forecasting model for use with weather radar data: Part 2. Case studies, *Hydrology and Earth System Sciences* **2**, 283-298.
12. Wood, F. and Szöllösi-Nagy, A. (1980) Real-time forecasting/control of water resource systems.
13. Lees, M.J., Young, P.C., Ferguson, S., Beven, K.J., Burns, J. (1994) An adaptive flood warning scheme for the River Nith at Dumfries, in W.R. White and J. Watts (eds.), *2nd international conference on river flood hydraulics*, John Wiley and Sons, Chichester, pp. 65-75.

INDEX